Geospatial Intelligence

Geospatial Intelligence

Origins and Evolution

Robert M. Clark

Georgetown University Press / Washington, DC

© 2020 Georgetown University Press. All rights reserved. No part of this book may be reproduced or utilized in any form or by any means, electronic or mechanical, including photocopying and recording, or by any information storage and retrieval system, without permission in writing from the publisher.

The publisher is not responsible for third-party websites or their content. URL links were active at time of publication.

Library of Congress Cataloging-in-Publication Data

Names: Clark, Robert M., author.
Title: Geospatial Intelligence : Origins and Evolution / Robert M. Clark.
Description: Washington, DC : Georgetown University Press, 2020. | Includes bibliographical references and index.
Identifiers: LCCN 2020002221 (print) | LCCN 2020002222 (ebook) | ISBN 9781647120108 (hardcover) | ISBN 9781647120115 (paperback) | ISBN 9781647120122 (ebook)
Subjects: LCSH: Geographic information systems. | Geospatial data. | Digital mapping. | Intelligence service—United States.
Classification: LCC G70.212 .C56 2020 (print) | LCC G70.212 (ebook) | DDC 910.285—dc23
LC record available at https://lccn.loc.gov/2020002221
LC ebook record available at https://lccn.loc.gov/2020002222

∞ This book is printed on acid-free paper meeting the requirements of the American National Standard for Permanence in Paper for Printed Library Materials.

21 20   9 8 7 6 5 4 3 2   First printing
Printed in the United States of America.

Cover design by Jeremy John Parker.
Interior design and composition by click! Publishing Services.

Contents

List of Illustrations   ix
Preface   xi
Prologue   xv
List of Abbreviations   xix

1 Intelligence and Geospatial Intelligence ... 1
The Boundaries of Intelligence   1
Geospatial Terminology   4
The Power of a Single Word   5
Defining Geospatial Intelligence   6

2 A Brief History of Maps and Charts ... 11
Mapping   11
The Silk Road   11
Cartography   12
Photogrammetry   25
Nautical Charts   27
Aeronautical Charts   28
Establishing Claims with Cartography   29
Chapter Summary   32

3 Terrain ... 35
Measuring and Representing Terrain   35
Digital Elevation Models   36
Military Use of Terrain   37
Civil Use of Terrain   41
Oceanographic Terrain   46
Chapter Summary   49

4 Navigation ... 51
Celestial Navigation   52
Radio Navigation   53
Satellite Navigation   56
Chapter Summary   60

5 Geopolitics ... 63
Mahan's Sea Power Theory   64
Mackinder's Heartland Theory   64
German Geopolitik   65
Spykman's Rimland Theory   66
The Continuing Influence of Geopolitical Theories   66

Thematic Cartography 67
Geopolitical Strategy 70
Chapter Summary 73

6 Geographic Information Systems 77
The Cluttered Map 77
Hard Copy Layers 78
Roger Tomlinson, the Father of GIS 79
The Harvard Connection 80
ESRI and Intergraph 80
Interactive Maps and Charts 81
The GIS Choice: Raster or Vector? 83
The Power of GIS 84
The Explosion of GIS Applications 85
Are Paper Maps Obsolete? 89
GIS and GEOINT 90
Chapter Summary 91

7 Geolocation 93
Geolocation Basics 94
Using Imagery 95
Radiofrequency Geolocation 95
Acoustic Geolocation 102
Cyber Geolocation 106
Chapter Summary 107

8 Gaining the High Ground 109
Gettysburg 109
Observation Towers 110
Lighter-Than-Air Craft 111
Exotic Approaches to the High Ground 113
Aircraft 115
Chapter Summary 120

9 The Ultimate High Ground 123
Remote-Sensing Satellites 123
Government Nonmilitary Applications 126
Military Applications 128
Commercial Imaging Satellites 133
Chapter Summary 137

10 Visible Imaging 139
Aerial Film Cameras 139
Satellite Film Cameras 142

Digital Cameras   145
Video Cameras   146
Getting the Image Right   148
Analyzing the Image   153
Chapter Summary   157

11 Spectral Imaging 159
The Infrared Bands   160
The Ultraviolet Spectrum   162
Imaging outside the Visible Band   162
Spectral Imagers   163
Chapter Summary   169

12 Radar Imaging 171
Conventional Radar   172
Side-Looking Airborne Radar   173
Synthetic Aperture Radar   174
Laser Radar   180
Chapter Summary   181

13 The Drivers of Geospatial Intelligence 185
Denial and Deception   186
Fleeting Targets   189
Precision and Accuracy   190
Outside Expertise   191
Characterizing Oceans and Ocean Traffic   194
New Issues   196
A Complete Picture   199
Chapter Summary   201

14 The Tools of Geospatial Intelligence 203
Geomatics   204
Geographic Information System   205
Geovisualization   206
Big Data   209
Data Analytics and Visual Analytics   211
Geospatial Simulation Modeling   213
Chapter Summary   218

15 Sociocultural GEOINT 221
Sociocultural Factors in Conflict Resolution   221
Activity-Based Intelligence   224
Pattern-of-Life Analysis   226
Volunteered Geographic Information   227

Involuntary Geographic Information   231
Chapter Summary   234

16 The Story of the National Geospatial-Intelligence Agency . . . 237
The Defense Mapping Agency   238
The National Photographic Interpretation Center   239
The National Imagery and Mapping Agency's Standup   240
A Tale of Two Cities   241
The Fight to Survive   242
The NGA's Standup   243
Reaching Out   244
Establishing the Boundaries of GEOINT   245
Chapter Summary   251

17 The GEOINT Explosion 253
US Geospatial Intelligence Organizations   254
Five Eyes GEOINT   259
Other National GEOINT Organizations   262
Transnational GEOINT Organizations   265
Chapter Summary   267

18 Non-National Geospatial Intelligence 271
State/Provincial and Local Government   272
Nongovernmental Organizations   276
Chapter Summary   280

19 Commercial GEOINT 283
Geospatial Business Intelligence   284
Strategic GEOINT   288
Operational GEOINT   292
Geospatial Competitive Intelligence   295
Chapter Summary   297

20 The Road Ahead 301
Predicting the Future   302
The Future of Cartography   303
The Tools   305
Applications of GEOINT   313
National-Level GEOINT   319
The Challenge of Ubiquitous GEOINT   321
Chapter Summary   325

Glossary of Terms   329
Selected Bibliography   337
Index   339
About the Author   347

Illustrations

Figures

- 2.1 The Silk Road, overland and maritime   12
- 2.2 The 1154 *Tabula Rogeriana* map of the known world   15
- 2.3 Gnomonic projection centered on the North Pole   17
- 2.4 A Mercator projection of the world   18
- 2.5 Equal area projection of the world   19
- 2.6 A Universal Transverse Mercator grid map of Alaska and eastern Siberia   21
- 2.7 The instruments of cartography   22
- 2.8 Triangulation   24
- 2.9 Map from the Great Trigonometrical Survey of India, 1870   25
- 2.10 Captain Cook's deceptive map of Australia   30
- 2.11 British and US claims in the Northwest Territory   31
- 3.1 Layout for a seismic survey   45
- 3.2 Comparison of hydrographic and bathymetric charts   48
- 4.1 Amelia Earhart's Lockheed Electra   55
- 5.1 Halford Mackinder's 1904 map of Heartland Theory   65
- 5.2 A modern redrawing of Charles Joseph Minard's figurative map of the 1812 French invasion of Russia   68
- 5.3 Richard Edes Harrison's "the not-so-soft underside"   69
- 5.4 Territory controlled by the caliphates, 622–750   70
- 5.5 A depiction of Samuel Huntington's view of cultural blocs and fissures   73
- 6.1 Examples of map layers used in GIS   82
- 6.2 Raster and vector depictions of a scene   83
- 6.3 Raster navigation chart   87
- 6.4 Electronic or vector navigation chart   87
- 7.1 Geolocation by the angle of arrival   96
- 7.2 Geolocation using the time difference of arrival   98
- 7.3 Tracking a submarine with a hydrophone array   104
- 8.1 A tethered aerostat radar system   114
- 8.2 The Abrams P-1 Explorer   117
- 9.1 The first picture of Earth from space   124
- 9.2 Satellite orbital patterns   125
- 9.3 The first TIROS-1 image   127
- 9.4 US commercial satellite imagery companies   135
- 10.1 German World War I aerial film cameras   140
- 10.2 US Army Air Corps camera and aircraft   141

10.3 Model A-2 cameras being loaded into a U-2   142
10.4 A C-119 recovering a CORONA film bucket   144
10.5 Orthographic and perspective views of two structures   151
10.6 Film interpretation light table   153
10.7 Osama bin Laden's Abbottabad compound   155
11.1 The optical spectrum   161
11.2 A false color image of Las Vegas   164
11.3 German Marder 1A5 tank with infrared camouflage   165
11.4 Classes of spectral imagers   167
11.5 Comparison of spectral signatures   169
12.1 Radar Image of the Normandy coast, 1944   172
12.2 A US Army OV-1B Mohawk with a side-looking airborne radar system   173
12.3 US Geological Survey side-looking airborne radar image of San Juan   174
12.4 Laser radar image of Effigy Mounds National Monument   181
13.1 The Caspian Sea Monster   194
13.2 World shipping traffic   195
14.1 Thematic map of the ivory trade in 2017   207
14.2 The likely residence of the Leeds crime perpetrator   209
15.1 The human geography of Yugoslavia, 1995   223
17.1 Office of Strategic Services 1941–42 map of the Russian front   258
20.1 Maslow's hierarchy of needs   303

Tables

4.1 Global Navigation Satellite Position Accuracy   59
5.1 Comparison of Factors of State Power   72
10.1 NIIRS Rating Scale   150

Preface

This book tells the story of how the current age of geospatial knowledge evolved from its ancient origins to become ubiquitous in daily life around the globe. Within this historical frame, the book weaves a rich tapestry of stories—about the people, events, and ideas that affected the trajectory of what has become known as geospatial intelligence, or GEOINT.

The explosive development of geospatial intelligence as an entity followed much the same pattern as that of the internet, with similar transformative effects. The internet originated from a need for scientists and engineers in several countries to share research results, and it relied on a synthesis of technologies (computers and communications). GEOINT has a more complicated history, long predating the name itself; but it gained wide international attention after an initiative of the Department of Defense and the US intelligence community broke through many decades of division and merged mapping and imaging organizations. That merger led to a synthesis of mapping, imaging, and geolocation practices with information technologies to create the governmental and commercial phenomenon that we now know as GEOINT.

The history of geospatial intelligence is the story of three things that have been with us since we organized into tribes: war, commerce, and taxes. These are constant threads throughout the journey from the roots of ancient charts to today, where we carry our geospatial information around with us in our smartphones, automobiles, and, most currently, our wristwatches.

And the far-reaching growth isn't over. Many applications of GEOINT have developed in recent years, and these are explored here. Geospatial intelligence has many chapters yet to come, and they will be at least as exciting as the past ones. There are many new breakthroughs unfolding, and this book addresses a few of them.

GEOINT comprises many disciplines and technologies. Of necessity, most of these subjects receive introductory or overview treatment in this narrative. Entire university courses are devoted to subjects such as cartography, imagery, remote sensing, and geographic information systems that are covered here, sometimes in a single chapter. The book's chapters take a primarily chronological view, explaining the historical origins and subsequent evolution of the various disciplines that have contributed to the current body of geospatial knowledge and now are driving new developments. Starting from their roots in ancient maps and charts, it traces the development of geospatial applications in human activities. It describes how a number of disciplines, enabled by information technology, converged, filled a need, and continue to do so.

My first introduction to geospatial intelligence was as a young United States Air Force (USAF) intelligence officer in the Strategic Air Command, planning for targets to attack and flight routes to use to avoid air defenses for our wing's B-52 heavy bombers in the event of a nuclear war. After joining the Central Intelligence Agency (CIA) as an analyst, I became involved in a different type of geospatial intelligence: working with imagery analysts at the National Photographic Interpretation Center (NPIC) on unidentified facilities and intelligence enigmas that were found in overhead imagery—a problem that continues to bedevil national-level GEOINT analysts. After leaving the CIA, I worked as a private contractor and consultant in planning for development of new imaging systems for the National Reconnaissance Office (NRO).

This book takes a US perspective on geospatial intelligence. After all, this is my background. But that point of view is not intended to downplay the major advances in the field that have come from around the world. A number of the advances across the centuries, in fact, have come from across the Atlantic and the Pacific, and many are highlighted throughout the text.

The book draws on a wide range of sources, including my personal experiences. Extensive *related* materials on geospatial intelligence are covered, from disciplines such as geographic information systems, geography, geographic information science, cartography, remote sensing, photogrammetry, geodesy, and geophysics. While a plethora of general sources (including common historical knowledge) inform the book, those most relevant focus on the path that lead to the burgeoning field of geospatial intelligence. Intelligence practitioners can spend their entire careers in highly specialized disciplines, and many books are devoted to topics covered only briefly here. Instead, this book is a general guide for students, intelligence customers, and analysts alike, with references to lead the reader to more in-depth studies and reports on specific topics or techniques.

This book was made possible by a diverse set of professionals, whom I both respect and admire. My foremost appreciation goes to Donald Jacobs, senior acquisitions editor, for inviting me into the Georgetown University Press family and stewarding me through its rigorous process. I am especially grateful to Lieutenant General (Retired) James Clapper for giving me his valuable time for an interview and the excitement of interacting with him as he attempted to have a simple breakfast in a diner. I thank Stu Shea, chief executive officer of Peraton, for sharing his inexhaustible enthusiasm for the subject matter and also his lengthy interview. Many of the important insights herein have come from two colleagues as we collaborated on developing an introductory course in geospatial intelligence: Todd Bacastow at Pennsylvania State University and Tim Walton at James Madison University. My special thanks to them and to Peter Oleson, who provided invaluable historical context. Many people throughout the US intelligence community, the military, and academia have provided wisdom that I have incorporated; I cannot name them all, but I appreciate their help. On the personal front, my daughter, Allison Clark, contributed from her experience with local government applications of geospatial intelligence. Above all, I'm indebted to my wife and partner in this effort, Abigail, whose conversations and extensive revisions made this a better book.

All statements of fact, opinion, or analysis expressed are those of the author and do not reflect the official positions or view of the Central Intelligence Agency (CIA) or any other US government agency. Nothing in the contents should be construed as asserting or implying US government authentication of information or CIA endorsement of the author's views. This material has been reviewed by the CIA to prevent the disclosure of classified information. This does not constitute an official release of CIA information.

Robert M. Clark
Wilmington, North Carolina

Prologue

In 2002 Stu Shea was the general manager of Northrop Grumman's Space and Intelligence business unit, and he was again confronting a puzzle that had perplexed him for two decades. As a young geographer with an interest in geology, he had studied cartography. He was fascinated by the way that people perceived maps and how their eyes moved when looking at a map. He had started work at Rome Research Corporation in 1982, where his job required scribing maps, drawing on the basic principles of mapmaking but incorporating aerial photography. He had to work with both cartographers and imagery analysts, and he was learning both skills.

But the two groups operated independently. Not only did they not work together, but they seldom even talked to each other, which didn't make sense to Stu. You had to have aerial photographs to produce maps; the days of creating a map just by on-the-ground surveying were long past. And you had to have maps to help interpret what you were seeing in an aerial photograph. Didn't the two groups need each other?

By 1982 Stu had pictured in his mind where things were going—or should be going. An electronic messaging network called ARPANET had come into existence in 1969. The Defense Department's Advanced Research Projects Agency (ARPA) started the program based on a radical new idea: linking computers and researchers in a virtual network. By 1979, ARPANET, now controlled by the Defense Communications Agency, was being widely used to share information among universities and government research institutes. Stu could see that ARPANET, as it became more widely available, would allow people to pull up maps for display—"maps on demand," as he called it. A US company named ESRI was producing maps in electronic form. A British geographer, Roger Tomlinson, had recently built something called the Canadian Geographic Information System, a set of electronic maps of Canada. And commercial imagery from satellites was coming; it was a long way off, but it was on the horizon.

A convergence was happening, Stu reasoned—or it *could* happen. Information technology was enabling what could become a geospatial revolution. But there was just one problem: the two key groups who could make that occur—the cartographers and the imagery analysts—still remained physically and psychologically separated. They all were dealing with spatial data—things, people, and events that were located on Earth. Everything that happens on Earth has a geographical reference. They relied on the same source for their product: imagery from above. Wasn't it natural for the two professions to work together?

Stu moved on to doing cartography work for the US national security community—the Defense Mapping Agency, Central Intelligence Agency,

National Security Agency, and the National Reconnaissance Office. He headed a succession of labs working on cartography and imagery projects and kept encountering the same situation he'd seen in his first job. In contractors' offices and government agencies alike, the pattern was sacrosanct: mapping and imagery professionals lived in separate worlds. Stu was familiar with the term **stovepipes** from his dealings with the intelligence community. It is a term that describes the compartmentation and separation of collection organizations by discipline that unfortunately hinders the sharing of intelligence across organizations. But stovepipes between cartography and imagery made no sense. Stu constantly wondered, How did we get to such an illogical separation in the first place? And what can we do to fix it?[1]

James Clapper was asking himself the same question. As a US Air Force lieutenant general, he had a distinguished career and long experience in intelligence, rising to become director of the Defense Intelligence Agency before retiring in 1995. After six years in industry, duty called again, in a fashion that he could not refuse. He returned from retirement to become director of the National Imagery and Mapping Agency (NIMA) on September 13, 2001—two days after the terrorist attacks on the World Trade Center and the Pentagon.

Clapper quickly realized that he'd inherited an organization that was culturally divided, as symbolized in its name. The product of separate agencies that were still both physically and psychologically separate, they needed to somehow unify under a shared mission. But there was no time to think about that. The US was engaged in a new kind of war, and it would be fought inside the country and around the world. It needed maps, and lots of them. Its forces needed intelligence support in responding to a threat that could appear and disappear within minutes. It needed intelligence from imagery, and it had to arrive in the combat theater quickly. He had experienced firsthand General Norman Schwarzkopf's displeasure regarding the poor imagery support that Schwarzkopf had received in operations in Kuwait and Iraq during Operation Desert Storm in 1991. Imagery had arrived at Schwarzkopf's headquarters too late to be of use. That couldn't happen again. NIMA had to deliver. NIMA's imagery analysts would go where they were needed to support this war.

By 2002, Director Clapper had accomplished the Herculean task of putting the support for deployed troops in place. He had time to think about a new vision for NIMA—one that would close the cultural gap within the organization. It would take a unifying discipline and doctrine that would capture the imagination of his people, one that the imagery advocates and the mapping advocates could rally around. NIMA was doing better in supporting Afghanistan combat than its predecessor had done in Desert Storm, but Clapper knew that he still was managing a divided house.[2]

The other national intelligence agencies had their core disciplines, captured in acronyms: human intelligence (HUMINT), signals intelligence (SIGINT), measurement and signature intelligence (MASINT), and open-source intelligence (OSINT). His agency already had an acronym: imagery intelligence, or IMINT. But that neither described the agency mission nor captured his vision. Clapper's agency was really about *geospatial intelligence*—the synthesis of imagery and mapping—or should be.

Both Stu Shea and James Clapper had a single word that fit the vision: *GEOINT*.

Notes
1. Stu Shea, interview, August 7, 2018.
2. James R. Clapper, *Facts and Fears* (New York: Viking Press, 2018), 80.

Abbreviations

AIS	automatic identification system
AOA	angle of arrival
ASG	allied system for geospatial intelligence
BI	business intelligence
CCD	charge-coupled device
CIA	Central Intelligence Agency
COMINT	communications intelligence
D&D	denial and deception
DARPA	Defense Advanced Research Projects Agency
D/F	direction finding
DIA	Defense Intelligence Agency
DMA	Defense Mapping Agency
DoD	Department of Defense
DWOB	Doctors Without Borders
ELINT	electronic intelligence
EOSAT	Earth Observation Satellite Company
ESRI	Environmental Systems Research Institute
FEMA	Federal Emergency Management Agency
GEO	geosynchronous equatorial orbit
GEOINT	geospatial intelligence
GIS	geographical information system
GPS	Global Positioning System
GRAB	galactic radiation and background (satellite)
HEO	highly elliptical orbit
HSI	hyperspectral imagery
HUMINT	human intelligence
IMINT	imagery intelligence
IP	internet protocol
IR	infrared
ISIS	Islamic State of Iraq and Syria
LEO	low Earth orbit
LIDAR	laser radar
LOP	line of position
LWIR	long-wave infrared
MASINT	measurement and signature intelligence
MEO	medium Earth orbit
MWIR	mid-wave infrared
NASA	National Aeronautics and Space Administration

NATO	North Atlantic Treaty Organization
NGA	National Geospatial-Intelligence Agency
NGS	National Geodetic Survey
NIIRS	National Imagery Intelligence Rating Scale
NIMA	National Imagery and Mapping Agency
NIR	near infrared
NOAA	National Oceanic and Atmospheric Administration
NPIC	National Photographic Interpretation Center
NSA	National Security Agency
NSG	National System for Geospatial Intelligence
NRO	National Reconnaissance Office
OSINT	open-source intelligence
OSS	Office of Strategic Services
PMESII	political, military, economic, social, infrastructure, and information
POL	pattern-of-life
RDF	radio direction finder
RFID	radio frequency identification
SAR	synthetic aperture radar
SIGINT	signals intelligence
SLAR	side-looking airborne radar
SOSUS	sound surveillance system
SPOT	Système Pour l'Observation de la Terre (France)
SWIR	short-wave infrared
TDOA	time difference of arrival
TIROS	television infrared observation satellite
UAV	unmanned aerial vehicle
UGF	underground facility
USGS	US Geological Survey
USGIF	US Geospatial Intelligence Foundation
UV	ultraviolet
VGI	volunteered geographic information

Intelligence and Geospatial Intelligence

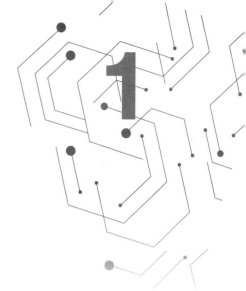

An understanding of geospatial intelligence has to begin with a basic understanding of the field of *intelligence*. Intelligence activities are conducted by national governments, military forces, terrorist organizations, law enforcement, corporations, and even by private individuals. The term can cover a lot of ground, and it's worth taking a moment to explore exactly what ground it covers.

The Boundaries of Intelligence

First, let's understand what intelligence *isn't*. It's not about firing guns or igniting explosives, nor about engaging in hand-to-hand combat. Those action-oriented scenarios play well in the movies, but they are not portraying the practice of intelligence. That sort of media depiction is known in the business as covert action—which is a curious term to use for actions whose effects are anything but covert.

What intelligence actually *is* has long been a subject of academic debate. However, it has a generally accepted meaning in governments and militaries. The common, and narrowest, definition is that intelligence is a process of acquiring and analyzing secrets by secret means.[1]

A broader and more useful definition is that intelligence is about *reducing uncertainty in conflict*.[2] "Conflict" is used here in the most general sense. It can consist of any competitive or opposing action resulting from the divergence of two or more parties' ideas or interests; it does not necessarily include physical warfare. If competition or negotiation exists, then two or more groups are in conflict. There can be many distinct levels, ranging from friendly competition to armed combat. So the practice of intelligence allows a decision-maker (the customer of intelligence) to gain an information advantage over an opponent or potential opponent.

In this context, "reducing uncertainty" requires that we obtain information that the opponent prefers to keep concealed. For example, you may gather information to reduce uncertainty about whether to evacuate for a hurricane, but that is not intelligence: there is no opponent trying to conceal the information in order to lead you in the wrong direction.

Just how do you provide the intelligence to reduce uncertainty for decision-making? In its simplest form, there are two requirements: you need some sources that will provide what is referred to as **raw intelligence**, and you need someone (or some group) to analyze the raw intelligence to produce the conclusions that reduce uncertainty. Let's start by describing the sources.

Intelligence Sources

Intelligence requires analysis of information derived from a number of sources. The US Office of the Director of National Intelligence (DNI) has defined six categories of sources, called collection disciplines, and most countries follow a similar classification:

- **SIGINT. Signals intelligence** is derived from signal intercepts comprising, however transmitted, either individually or in combination all communications intelligence (COMINT), electronic intelligence (ELINT), and foreign instrumentation signals intelligence (FISINT).
- **IMINT. Imagery intelligence** includes representations of objects reproduced electronically or by optical means on film, electronic display devices, or other media. Imagery can be derived from visual photography, radar sensors, and electro-optics.
- **MASINT. Measurement and signature intelligence** is technically derived intelligence data other than imagery and SIGINT. The data results in intelligence that locates, identifies, or describes distinctive characteristics of targets. It employs a broad group of disciplines, including nuclear, optical, radio frequency, acoustics, seismic, and materials sciences.
- **HUMINT. Human intelligence** is derived from human sources. To the public, HUMINT remains synonymous with espionage and clandestine activities; however, most of HUMINT collection is performed by overt collectors such as strategic debriefers and military attachés.
- **OSINT. Open-source intelligence** is publicly available information appearing in print or electronic form, including radio, television, newspapers, journals, the internet, commercial databases, and videos, graphics, and drawings.
- **GEOINT. Geospatial intelligence** is the analysis and visual representation of security-related activities on the Earth. It is produced through an integration of imagery, imagery intelligence, and geospatial information.[3]

There are all sorts of academic arguments within the national intelligence community about which sources belong in each category and whether new collection disciplines should be added, including such things as CYBERINT or the less glamorous TRASHINT (defined as diving through dumpsters for information). For now, let's stick to the generally accepted norm. (Because GEOINT is the topic of this book and a somewhat different creature, a more comprehensive definition appears later in this chapter and throughout the text.) First, however, mere information from these sources does not produce intelligence. You need an intelligence analyst for that.

Intelligence Analysis

Perhaps the best-kept secret about intelligence concerns analysis. It is one of the most challenging, exciting, and rewarding professions—though admittedly, at times, the most frustrating. When well done, it often resembles a Sherlock Holmes adventure.

A good analyst sifts through all the available information on an issue, looking for important (significant) patterns in the mass of raw intelligence that the sources provide, and derives meaning from that material to help a leader make an important decision.

But to do it well, you need to think through an issue critically, which means following steps something like these:

- *Approach the issue*. Be open-minded and work to be well-informed. Examine your own beliefs and values. Avoid looking at the problem, the issue, or the evidence from the viewpoint of *your* beliefs and values.
- *Define the issue*. Analyze problems systematically. Develop a model of the problem.
- *Generate hypotheses*. Formulate ideas succinctly and precisely. Define terms in a way appropriate for the context. Rank the relevance and importance of ideas. Understand the logical connections between ideas.
- *Test the hypotheses*. Evaluate the evidence for and against a hypothesis. Judge the credibility of sources. Ask appropriate clarifying questions. Determine the quality of an argument, including the acceptability of its reasons, assumptions, and evidence. Identify inconsistencies and mistakes in reasoning.
- *Develop and defend conclusions*. Draw conclusions when warranted, but with caution. Identify and support your conclusions and assumptions.[4]

The result of this analysis process can take one of three forms: descriptive, anticipatory, or prescriptive intelligence:[5]

- **Descriptive.** This type of intelligence describes what has happened or is happening now. It's sometimes referred to as situational awareness. It usually requires that an analyst identify threats and opportunities that a decision-maker must deal with. And it typically requires looking at political, economic, and social factors, among others.
- **Anticipatory.** This is sometimes referred to as predictive intelligence, but we really can't predict the future. There are many possible futures, after all. So in intelligence, we try to identify a few alternatives that are most probable. Analysts do that by carefully examining the dominant forces or factors that could shape the future. That requires in-depth research, and it results in several possible future scenarios. We call them probabilistic because they make use of probability distributions and the laws of probability.
- **Prescriptive.** This type of intelligence goes one step farther than either descriptive or anticipatory and recommends a course of action for a decision-maker. In US national and military intelligence, analysts historically have been restricted from advising, mostly due to a cultural separation between intelligence and decision-makers. Few if any national policymakers or military commanders would welcome an intelligence report that concluded with "You should do the following things to be successful." Typically, the customers must consider factors other than intelligence in their decision-making. However, as intelligence teams have become more collaborative, with decision-makers participating more interactively in the intelligence process, the thinking has been changing; prescriptive intelligence at the national and military levels is likely to become more common. The target-centric approach and technology have enabled the beginning of this culture shift.

With that brief introduction to intelligence, let's look specifically at geospatial intelligence, which is both narrower and, in another sense, broader. It helps to begin by introducing terms used in geospatial matters.

Geospatial Terminology

Geospatial intelligence is one of many words that begin with "geo," which means "relating to the Earth." The words define different and sometimes overlapping scientific disciplines, for example:

- **Geography.** The study of Earth's physical features, atmosphere, and human activities as they affect and are affected by the physical features and atmosphere;
- **Geodesy.** Understanding the Earth's shape, orientation, and gravitational and magnetic fields; and

- **Geology.** Dealing with the Earth's physical structure, materials, and processes acting on them.

In addition, basic terms critical to understanding geospatial intelligence appear frequently in the mapping, imaging, and intelligence professions. Before going further into what geospatial intelligence is about, let's establish their meaning:

- **Cartography.** The art and science of graphically representing a geographical area, usually on a flat surface such as a map or chart.
- **Remote sensing.** Acquisition of information about an object or phenomenon from a distance—usually from an aircraft or satellite.
- **Photogrammetry.** The art and science of making measurements from photographs, typically for surveying and mapping.
- **Signatures.** The distinctive features of any entity—including phenomena, equipment, objects, substances, or areas—that allow the entity to be identified and characterized. Signatures appear in many forms: chemical, acoustic, radiofrequency, and optical, among others.

The term "geospatial intelligence" has not been easily or uniformly defined; in fact, there are a number of definitions most often based on the professional context. Since its first use it has taken on a much broader meaning to become a powerful concept.

The Power of a Single Word

It is sometimes possible to capture an abstract concept in a single expression and in the process to *reify* it, that is, to make it tangible and give it power or emotional content. In the 1960s, US Air Force leaders developed such a concept, calling it the *triad*. The term referred to a strategic nuclear arsenal comprising land-based intercontinental ballistic missiles, strategic bombers, and a submarine fleet carrying submarine-launched ballistic missiles. The term *triad* (sometimes *strategic* triad or *nuclear* triad) was used to promote the idea that all three components were needed to synergistically deter Soviet aggression. The word became shorthand for a national strategy, a budget justification, an article of faith in parts of the US military, and, like the "high ground" (discussed in chapter 8), almost a theological concept that continues to be used today.[6]

Geospatial intelligence, or its acronym, GEOINT, has become such a term. In fact, it is more widely known and probably has more impact than *triad* ever did. Unlike triad, GEOINT has been extended well beyond US national security into the educational and commercial worlds. It now has international reach. And like triad, it carries political, economic, and organizational consequences.

Just what is geospatial intelligence? And does GEOINT mean the same thing?

Defining Geospatial Intelligence

The original definition of geospatial intelligence established at about the time that the US National Geospatial-Intelligence Agency (NGA) came into existence was narrowly focused. The NGA defined it as "the exploitation and analysis of imagery and geospatial information to describe, assess, and visually depict physical features and geographically referenced activities on the Earth. Geospatial intelligence consists of imagery, imagery intelligence, and geospatial information."[7]

The Office of the DNI's definition of geospatial intelligence today stresses similar concepts. It is "the analysis and visual representation of security related activities on the Earth. It is produced through an integration of imagery, imagery intelligence, and geospatial information."[8]

In both definitions, geospatial intelligence (like intelligence generally) is both a process (exploitation and analysis) and a product (visual representation); and both definitions indicate specific sources or inputs (imagery, imagery intelligence, and geospatial information). What is most interesting is that the two definitions include the word *analysis*—unlike the five collection disciplines (called "the INTs"), which do not. Geospatial intelligence, it seems, is different; it in fact makes use of all the INTs—a subject to which we'll return in chapter 16.

During the twenty-first century, geospatial intelligence has become a hot topic in government and commercial sector programs as well as in academia. In the process, it has broken free of the constraints of the NGA and DNI definitions to become a part of everyday lives everywhere. The term, as it is now applied, does not necessarily require imagery or imagery products, though they are often used. It has developed a much broader meaning than just information collected and employed by a secret service.

GEOINT, of course, is the acronym that refers to GEOspatial INTelligence, though some distinguish the two as different *terms*—"GEOINT" being considered to have more of a military or government intelligence slant. Depending on the community or profession, the acronym GEOINT may be referred to as an actual term that takes on varying degrees of scope, both narrow and expansive. While academic debates rage on about what should or should not be included, as a term its meaning is arguably still evolving. For convenience, geospatial intelligence and GEOINT are used interchangeably in this book.

This chapter provides a limited introduction to the concept of geospatial intelligence as a means to understanding the historical narrative that follows.

We'll return to the definition in chapter 16 to discuss the political, economic, and organizational consequences of its meaning. For now, an illustration of geospatial intelligence in action should be helpful. One can be found, of all places, in two vignettes from a decades-old Hollywood movie.

In an early scene of the 1997 film *GI Jane*, the staff at a US Navy intelligence center is trying to extract a Navy SEAL team that is carrying wounded members after a firefight. The primary satellite communication is blocked by the terrain, leaving the staff unable to communicate with the SEAL team. The operations officer is about to switch to another satellite, but Lieutenant Jordan O'Neil (Demi Moore), a Navy topographical analyst, argues that the move won't succeed. She views the terrain around the team on her display, and notes that the tide is going out, which means that their escape minisubmarine will soon be stranded. O'Neil picks the single route the team can take to escape—which allows communication only through the *primary* satellite. After some debate, the unit commander allows her 9 minutes before they switch satellites—and of course, just as they are about to switch, the SEAL team comes out where O'Neil has predicted they would and communicates using the primary satellite.

Subsequently, Lieutenant O'Neil is chosen to be the first woman to go through rigorous training to become a member of the SEALs, and the plot chronicles her struggles to survive a number of brutal encounters with her trainer, Command Master Chief John James Urgayle (Viggo Mortensen).

Fast forward to the end: the SEAL team's final training exercise in the Mediterranean is interrupted when an actual crisis develops. A US KH-12 satellite, carrying a quantity of weapons-grade plutonium on board, has crashed in Libya. A Ranger team has landed and seized the plutonium but is encountering a hostile border patrol, and the SEAL team trainees are ordered in to help extract the Rangers. In the ensuing firefight, it is Command Master Chief Urgayle (of course) who is alone and wounded. The team, on the beach, must determine his likely route of escape in order to retrieve him. Pulling out a map and once again drawing on her topographical expertise, O'Neil identifies the route that he will take to escape. And again (of course), she is right—despite the opinions of other team members that a different route is more likely.[9]

Movies, of course, dramatize a situation. What is an illustration of geospatial intelligence in the real world? Let's take a look at one recent example.

In November 2018, a commercially produced satellite image of al-Dawadmi, deep inside Saudi Arabia, revealed a military base that appeared to be intended for testing ballistic missiles. Two features of the site immediately caught the attention of ballistic missile experts reviewing the imagery:

1. It had two launch pads.
2. The missile test stand appeared to be a smaller version of one used by China.

Drawing on their knowledge of nonimagery sources, the experts concluded that

- China is assisting Saudi Arabia in a program to develop long-range ballistic missiles (China had previously sold variants of its Dong Feng ballistic missile to the Saudis).
- The orientation of the two launch pads indicates that likely targets of missiles launched in anger are Israel and Iran—especially Iran, given Saudi concern about possible Iranian nuclear developments.
- Long-range ballistic missile development typically correlates with nuclear weapons development. So Saudi Arabia likely is engaged in, or at least interested in, developing nuclear weaponry for the missile warheads.[10]

The basic idea of what geospatial intelligence is (both process and product) can be found in these fictional and true illustrations: GEOINT requires analyzing a situation and creating a product that

- is anticipatory—it deals with the future;
- involves some type of human activity;
- draws on knowledge of the Earth (its surface, whether on land or sea, and the objects on or beneath that surface); and
- provides an information advantage to someone who must make a decision.

It's important to recognize that in both the movie and the real-world examples, neither the map nor the image alone qualify as GEOINT. In each case the analysis, based on information from experience or from other sources, was required. So, for example, details about a volcanic eruption, alone, do not qualify. But the data, combined with human analysis that identifies likely lava flows and debris fallout from a volcanic eruption, used to warn residents in the area, definitely qualifies.

Though geospatial intelligence is a term of recent origin, its underpinnings have a long and interesting history, the subject of this book.

Notes

1. This is the widely accepted view, though intelligence also is about acquiring and analyzing openly available material, as we'll see in a moment.
2. Robert M. Clark, *Intelligence Analysis: A Target-centric Approach* (Washington, DC: CQ Press/Sage, 2020), 22.
3. Office of the Director of National Intelligence, "What Is Intelligence?" n.d., www.dni.gov/index.php/what-we-do/what-is-intelligence.
4. Clark, *Intelligence Analysis*, 72.
5. David E. Bell, Howard Raffia, and Amos Tversky, *Decision Making: Descriptive, Normative, and Prescriptive Interactions* (Cambridge: Cambridge University Press, 1988), 9–11.

6. US Senate, "Modernizing the Nuclear Triad," July 11, 2019, www.rpc.senate.gov/policy-papers/modernizing-the-nuclear-triad.
7. US Code, Title 10, section 467 (10 USC. §467).
8. Office of the Director of National Intelligence, "What Is Intelligence?"
9. Hollywood Pictures, *GI Jane*, 1997.
10. Jon Gambrell, "Experts, Images Suggest a Saudi Ballistic Missile Program," n.d., www.apnews.com/092c1656a26b484e8e912d1960c65a98.

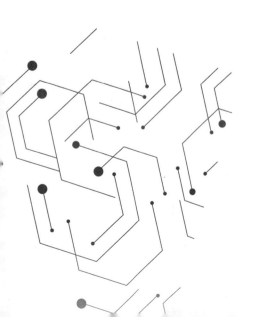

A Brief History of Maps and Charts

The preface cites war, commerce, and taxation—activities of those typically in charge—as the primary drivers of the need for geospatial intelligence. However, it was an internationally diverse field of thinkers—scientists and mathematicians from many disciplines—who developed the art and science of mapmaking, or cartography. They were aided by a wide range of technological developments, again from around the world and over many centuries. Let's look at the beginning of the process: the first maps.

Mapping

The earliest maps looked nothing like what we think of as a map; they looked more like pictures. They showed the immediate vicinity of home and points of interest—locations of food, water, and dangers for their producer. Later on, the maps grew to show small areas such as nearby villages, hunting areas, trade routes, and the locations of hostile tribes—things that were important to the mapmaker's community. As people explored and expanded their reach, the area of maps broadened accordingly. The need to find safe trade routes was the major driver in extending the mapping area. One of the longest such routes of its time was the "Silk Road" between China and Europe.

The Silk Road

By at least 1000 BCE, adventurous traders had developed a network of trade routes through the Eurasian steppes linking China to Europe, India, and Southwest Asia. Persian king Darius in the fifth century BCE built one of the main arteries, the Persian Royal Road. It had postal stations along its route, providing messengers with fresh horses to allow quick mail delivery. The Greek historian Herodotus greatly admired the speed and efficiency of the Persian service,

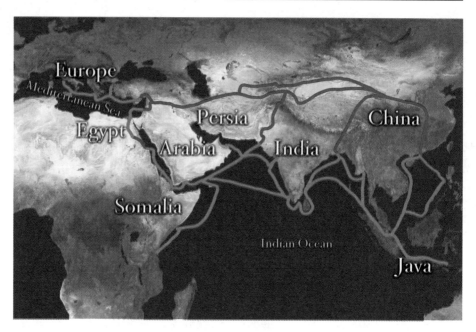

FIGURE 2.1 The Silk Road, overland and maritime. *Source:* US National Aeronautic and Space Administration.

praising it in these words: "Neither snow nor rain nor heat nor gloom of night stays these couriers from the swift completion of their appointed rounds."[1] If that sounds familiar, it should; with a few word changes, it's the unofficial motto of the US Postal Service, chiseled into the granite above the entrance to the New York City post office when it opened in 1914.

Around 100 BCE much of the road network had been mapped (fig. 2.1). From 130 BCE to 1453 CE, it was used to carry a mixture of goods—including silk, spices, and horses—westward. Eastward to China flowed jewels, metals, carpets, amber, and glass. Unfortunately, diseases also traveled the road along with the merchants, the bubonic plague that swept Europe in 542 CE being perhaps the worst.

In the 1800s, the trade routes became known as the Silk Road, a reference to China's major export of silk material. Goods also traveled by sea between China, India, and Europe. The maritime route also shown in figure 2.1 came along later than the overland route, and we'll revisit both routes in chapter 5.

Cartography

Cartography—the science and art of graphically representing a geographical area—would become a primary underpinning of geospatial intelligence.

It developed as a science in ancient Greece; in about 330 BCE, Aristotle had already made the first logical arguments that the Earth is a sphere, based on observations of events around him. His arguments may be summarized as follows:

- Certain stars can be seen only from certain parts of the Earth. As you move northward or southward, new stars appear in that direction.
- A lunar eclipse shows us a shadow of the Earth, and it is circular.
- All parts of the Earth are pulled toward the center and compressed; a globe is the logical result.[2]

Eratosthenes (276–194 BCE), a Greek mathematician and astronomer, went one step further; he estimated the circumference of the Earth.

At the age of thirty years, Eratosthenes began working as a librarian, and later became chief librarian, of the Library of Alexandria—probably the most important center of learning in the ancient world. In the third century BCE, the library contained a report about a remarkable yearly occurrence in the town of Syene (now Aswan), in southern Egypt. On each June 21st, when the sun was at its highest point at noon, towers cast no shadows and the sun shone into the deepest wells; it was directly overhead. But on the same date and time in Alexandria, the sun cast a shadow.

The description led Eratosthenes to correctly conclude that the Earth was spherical. He measured the altitude of the noontime sun at Alexandria at its yearly maximum, and found it to be about 7 degrees from overhead. Using geometry and the distance from Alexandria to Syene (5,000 stadia), he made the first calculation of the circumference of the Earth. It was fairly accurate; estimates are that his calculation was 0.5 to 15 percent high. The uncertainty results from a question about the length of a stadion. A Greek measurement of distance, 1 stadion was known to be 600 feet; but the length of a foot differed in different parts of the Greek world. And though Eratosthenes was able to measure the sun angle accurately, his measurement of the distance from Alexandria to Syene was off.[3]

Eratosthenes also developed a technique for mapping the Earth. He created a set of longitude and latitude lines, and defined the equator as running through Rhodes, a prime meridian as going through Alexandria, and additional lines at irregular intervals, all going through well-known places of the time. It was left to later cartographers to decide that a regular grid would work better for specifying locations in coordinates. His work was published in a three-volume set titled *Geography*—the first use of the word.

The Greeks and subsequently the Romans continued to refine the art of cartography. The maps of Eratosthenes and his successors were soon specialized to serve different needs. In the process, cartography evolved from a field of scientific research into something that began to resemble geospatial intelligence, as defined in chapter 1; that is, it increasingly drew on knowledge of the Earth, involving some type of human activity, and provided information to support decisions. For example, by 4 BCE, Egyptians were documenting property boundaries on maps. The Romans were especially interested in information that would help them in conquering and administrating an empire, so their cartographers responded. Maps of the time became centered upon the Mediterranean Sea and the lands surrounding it—what became the heart of the Roman Empire—and included administrative boundaries and road networks.

The Romans were fortunate to have Claudius Ptolemaeus (Ptolemy, c. 90–c. 168 CE)—the foremost cartographer, mathematician, and geographer of the time—to develop their maps and to describe how to properly prepare them. A Greek who held Roman citizenship, Ptolemy researched the Library of Alexandria. From its records, in about 150 CE he bested Eratosthenes by producing a detailed, eight-volume record of world geography, named, of course, *Geography*. In it, he explained how to create maps using mathematical principles, as well as provided the most complete maps of the time, stretching from the Atlantic Ocean to the middle of China. And in a service to later mapmakers, he established the equator as the point of zero degrees latitude. Ptolemy measured longitude eastward from zero degrees at the "Blessed Islands" (which he did not clearly identify; they could be Cape Verde, the Azores, Madeira, or the Canary Islands).

In his book, Ptolemy also argued for a spherical Earth, reaffirming Aristotle's arguments and adding one of his own. He observed that when a ship approaches a mountain from a distance, the mountain appears to rise from the sea, implying that the sea's surface is curved.

The International Cartographic Association today defines cartography as "the discipline dealing with the conception, production, dissemination, and study of maps."[4] Cartography is also the process of representing useful information in map form: gathering, evaluating, and processing information; determining how best to present it in graphical form; and drawing the map.

After Ptolemy's time, advances in cartography stalled—at least in Europe. From about 500–1500 CE, cartography and geography saw few advances in Europe. Mapmaking, in fact, regressed in one sense: under Roman rule, maps had contained only what their maker thought a user would need. The Catholic Church changed that. Under the Church's influence, mapmakers (who mostly worked in monasteries) included artistic decorations, usually with religious themes, on their maps. Illustrations of angels, biblical quotations, and locations of biblical objects such as Noah's Ark and the Tower of Babel were added. While these colorful supplements may have provided some divine guidance to a traveler, they weren't much help otherwise on a journey.

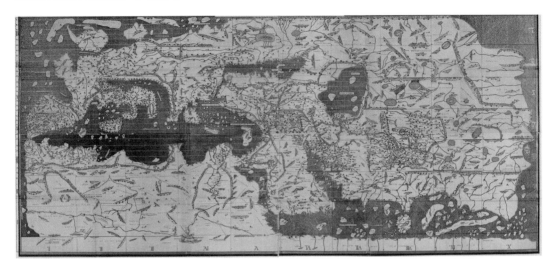

FIGURE 2.2 The 1154 *Tabula Rogeriana* map of the known world. *Source:* Wikimedia Commons. Note: al-Idrisi preferred to draw his maps with north downward and west rightward. This figure has been rotated so that north is at the top and west is to the left.

Things were quite different in the Islamic world. The Middle Ages were the golden age of art, science, and mathematics in the Middle East, and the disciplines of cartography and geography flourished. The outstanding cartographer of the period was Muhammad al-Idrisi, an Arab scholar and frequent traveler across North Africa and Europe. He apparently had good contacts with an extensive network of travelers, and he made productive use of them. His maps of Europe, Africa, and Asia drew on knowledge obtained by Islamic merchants and Norman seafarers. Al-Idrisi spent some time in the court of King Roger II at Palermo in Sicily. During that stay, in about 1154, he produced several maps and books with a geographic theme. His major work of the time, the *Tabula Rogeriana* (*Book of Roger*), featured a map of the known world that relied on Ptolemy's principles.[5] It was a remarkably accurate map for its time, illustrated in figure 2.2, and copies of it continued to be produced and used for the next three centuries.

During the fifteenth century, Ptolemy's *Geography* was translated into Latin and disseminated throughout Europe, thanks to a key invention of the time. Until then, each map had been created by painstaking manual labor using parchment and brushes. The result was that quality was variable, and maps weren't generally available. New technology, in the form of the printing press, changed all that. Maps could now be produced in quantity for users who previously could not afford them. The widespread publication of Ptolemy's works ignited a new interest in cartography. And in the same era, three other developments were making maps more accurate and useful: new types of map projections, better instruments for making maps, and a method of surveying called triangulation.

Map Projections

Do maps have a point of view? Yes, they do. Cartography is about telling users what they need to know in map form. But different people use maps for different purposes. One traveler may want to know the shortest distance between two points—for example, for navigation. Another person may wish to see the entire world on one map. A third might want to see an accurate picture of the geographical features of each area covered by the map. It's not possible to provide all three views in a single map. The solution? **Map projections** are methods for representing parts of the surface of the Earth on a plane surface. And there are many different types of projections, depending on how the map will be used.

In spite of the work of Aristotle, Eratosthenes, Ptolemy, and Al-Idrisi, if you had taken a worldwide poll about the shape of the Earth in the fifteenth century, you'd have gotten mixed results. The accepted belief in China was that the Earth was flat. Islamic scholars, relying on the work of Ptolemy, would have voted in favor of a spherical Earth. In Europe, opinion was mixed.

Unfortunately for cartographers, the Earth turned out to be round; Magellan's circumnavigation of the world from 1519 to 1522 established it beyond doubt. (Except of course, for members of the flat Earth societies around the world, who continue today to claim that the Earth is flat.)

For cartographers, the problem is that as the area covered by a map expands, distortions creep in around the edges. It's just not possible to map a curved surface onto a flat one (a paper map) without introducing distortion *somewhere*. Usually, the center of the map has the least deviation, being closest to the globe it is projected from. So maps of small areas are fairly accurate. But increasingly, governmental rulers and commercial interests wanted maps that displayed large areas, usually for conquest or commerce. So cartographers had to comply, and maps of larger areas had to be created—which meant introducing noticeable distortions.

Depending on the purpose of the map, some distortions are acceptable and others are not. Ptolemy recognized the problem, and developed three different types of map projections to deal with it. In doing so, he also developed the idea of latitude and longitude, allowing the mapmaker (and user) to specify locations in map coordinates.

Since Ptolemy's time, a plethora of map projections have been created, each preserving some features of the globe at the expense of other features. A few hundred different projections have been produced over the last 2,000-plus years, and there is no upper limit to the ones that are possible with current technology. Here are three of the most widely used: gnomonic, Mercator, and equal area.

Gnomonic. A Greek philosopher, mathematician, and astronomer named Thales reportedly developed, in the sixth century BCE, what is believed to be the oldest map projection. Called the **gnomonic projection**, it is still in use because of its chief advantage over others: a great circle (known as a **geodesic**)—the shortest distance between any two points on the Earth's surface—is represented by a straight line on the map.[6]

The projection is created by placing a tangent plane onto a point on the globe, and mapping each point of the Earth's surface onto the tangent plane in such a way that a straight line from the globe's center passes through the point on the surface and then onto the tangent plane. No distortion occurs at the tangent point, but distortion increases rapidly away from it. Figure 2.3 illustrates the problem. The tangent point here is the North Pole. By the time you reach the edge of the map in any direction—the southern United States, Japan, or North Africa, for example—the area depiction has become badly stretched.

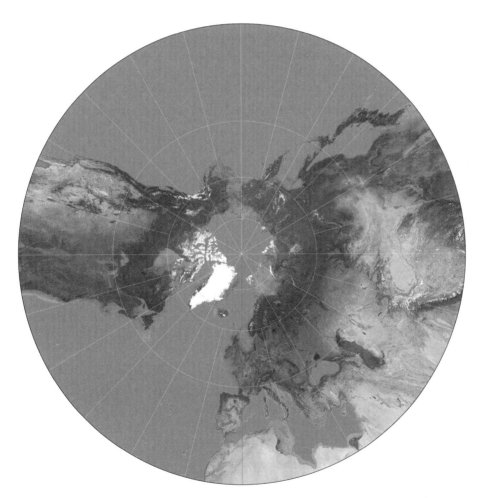

FIGURE 2.3 Gnomonic projection centered on the North Pole. *Source:* Strebe, Wikimedia Commons.

Mercator. When you see a map of the world in an atlas, it's likely to be a **Mercator projection**. It's also called a cylindrical map projection because it is created by (figuratively) wrapping a sheet of paper around the globe, touching at the equator, and projecting points on the globe's surface on a line from the center of the Earth, through the point, onto the paper. It was created by the Flemish geographer and cartographer Gerardus Mercator in 1569.[7] It soon became the standard map projection for nautical purposes because the distance between two points on the map was a line of constant course; the easiest way to navigate, after all, is to follow a single compass heading. But a constant compass course usually isn't a great circle route, so it isn't the shortest distance between two points. And again, the map becomes distorted as you move away from the tangent point (which is the equator shown in fig. 2.4).

Equal Area. Most maps distort the areas represented, but some do so significantly. The Mercator projection is one extreme example; distortion increases as you move farther from the equator. On it, the island of Greenland displays somewhat larger than South America, when in fact, Greenland is about one-sixth of South America's size.

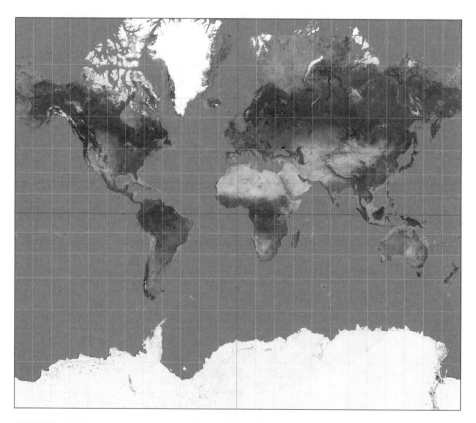

FIGURE 2.4 A Mercator projection of the world. *Source:* Strebe, Wikimedia Commons.

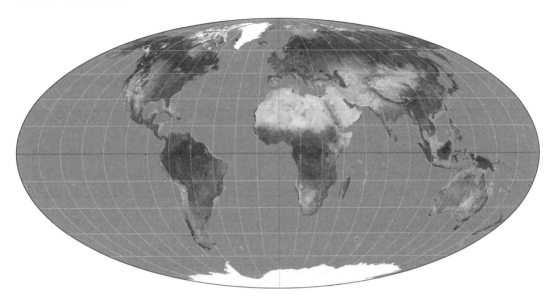

FIGURE 2.5 Equal area projection of the world. *Source:* Strebe, Wikimedia Commons.

Equal area projections—there are several, drawn using different techniques—all represent the size of any given area on Earth accurately. But to do so, they badly distort the *shape* of the area represented. Figure 2.5 illustrates the issue. Greenland is now the proper size, but is misshapen, compared with its more accurate representation in figure 2.3.

Locating the Prime Meridian. Once cartographers began using coordinates of latitude and longitude to determine location, they had to decide what line on a map to choose as the zero degree reference for both. For latitude, there was a natural reference point: the equator, a line around the world located halfway between the two poles.

For longitude, there was no such "natural" choice, though cartographers repeatedly tried to find one. Ptolemy located his prime meridian (0 degrees longitude) at the "Blessed Islands," an uncertain point west of Africa, and measured degrees eastward from there to avoid having to deal with negative degrees of longitude.

Later cartographers chose other points for the prime meridian. When Christopher Columbus reported that his compass pointed directly at true north somewhere in the mid-Atlantic, that line came into use as zero degrees longitude on some maps. Mercator drew the line through an island in the Canaries for his maps. Cape Verde was used in others. The British *Nautical Almanac*, first published in about 1767, used the Royal Observatory in Greenwich as its prime meridian, and it came into wide use. At the International Meridian Conference, held in Washington in 1884, Greenwich was adopted as the prime. The French objected to the choice and continued to use Paris as the prime. French mariners, though fiercely loyal to their country, weren't happy with the result. They found themselves at a definite disadvantage when they had to use non-French maps or depend on non-French coordinates, and the French finally adopted the Greenwich standard in 1911.

The standard geographic coordinate system was well suited for navigation at sea and, later, in the air. Not so much for militaries. By the time of World War II, ground force commanders needed to precisely calculate distances between points and define search areas. In the degrees-minutes-seconds format of the traditional system, such calculations were cumbersome and subject to error. That requirement led militaries to create a different type of map projection with a different coordinate system.

Universal Transverse Mercator. Militaries have long had a special interest in maps with specific features. Ground forces want maps with the lowest possible distortion, and nowadays they want the ability to clearly and accurately (i.e., to within 1 meter) identify the coordinates where something or some point on the Earth is located. That need was met by the **Universal Transverse Mercator projection**, or UTM.

The UTM projection was developed during the 1940s by either the German Wehrmacht or the US Army Corps of Engineers (or possibly by both, independently).[8] It is created by dividing the Earth into sixty north-south zones, along meridians beginning at the International Date Line (180° longitude) and proceeding eastward from there. Each zone, encompassing 6° of longitude, is then subdivided into 8° high lines of latitude, beginning at 80° south latitude. The resulting grids are numbered from 1 to 60 eastward, and lettered from C to X northward. Part of the resulting grid for Alaska and eastern Siberia is shown in figure 2.6.

Each grid zone then is mapped using a transverse Mercator projection. The grid zones are small enough that the resulting maps have relatively low distortion. And each grid zone uses a set of numbers to identify a specific location within the zone to within 1 meter (*not* the north-south, east-west coordinate system based on the equator and the Greenwich meridian). In the UTM coordinate system, for example, Anchorage would be located at 6V 344399 6786666. That pinpoints Anchorage as being in the 6V grid in figure 2.6, the 344399 meter number referring to location eastward on a UTM map and the 6786666 meter number referring to northward location. (Both numbers are quite large because the west and south edges of the map don't start at zero.)

Naval and airborne units typically don't need that level of precision and accuracy (except for targeting precision munitions). So they continue to use maps with the conventional E-W, N-S coordinate system.

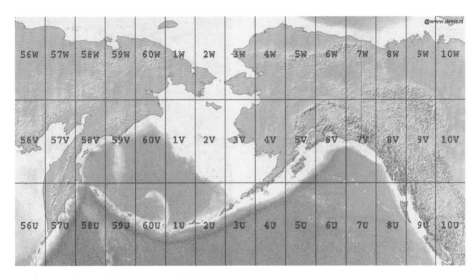

FIGURE 2.6 A Universal Transverse Mercator grid map of Alaska and eastern Siberia. *Source:* Demis (www.demis.nl), Wikimedia Commons.

Instruments of Cartography

Though al-Idrisi's *Tabula Rogeriana* was remarkably accurate, all maps (including al-Idrisi's) of that time suffered due to the difficulty of accurately measuring the coordinates of a point—and because cartographers had to estimate the direction and distance from one geographical location to the next in order to place both locations on a map. To make more accurate maps, you needed two instruments for measuring direction: one for determining north-south direction, and one for measuring the angle horizontally from north between two points. The compass (fig. 2.7a) and either the theodolite (fig. 2.7b) or transit (fig. 2.7c) served these two purposes. These instruments were developed and improved over a period of several centuries, and as the tools and the techniques for using them improved, the maps produced steadily became more accurate.

The Compass. Accuracy in mapmaking required determining the direction of true north. At night, you could use the stars, especially the North Star (in the Northern Hemisphere). But you needed at the same time to measure the direction to distant points, usually not possible at night. So a daytime reference was needed, and the position of the sun at local noon was the only real bet—until the invention of the compass.

China developed the first compasses somewhere between the third century BCE and first century CE. These were made of lodestone, an iron ore having magnetic properties. Between the ninth and eleventh centuries, the Chinese

FIGURE 2.7A (top left) The instruments of cartography: a dry magnetic compass. *Source:* Bios~commonswiki, Wikimedia Commons.

FIGURE 2.7B (right) The instruments of cartography: a German theodolite from 1851. *Source:* Bautsch, Helmholtz-Zentrum Potsdam–Deutsches GeoForschungsZentrum, Wikimedia Commons.

FIGURE 2.7C (bottom left) The instruments of cartography: a transit. *Source:* US Army Corps of Engineers, Office of History.

were using a navigational compass that relied on an iron needle, magnetized by stroking it with a lodestone, and floating it in a bowl of water. A major improvement appeared in about 1300 in the Islamic and European regions: the dry compass, made by balancing the needle on a pivot, enclosed in a glass case with angular markings and a star marking the eight major wind directions (known as the "compass rose" because of its flower-like appearance).

The compass was a great step forward, but of course, there was one shortcoming: it became unreliable near the magnetic poles. Because these were located in isolated parts of the world where people seldom ventured, that was usually not a serious problem; wars, commerce, and taxation weren't big issues at those two points of the globe. More of a problem was the fact that magnetic north seldom coincided with true north, so navigators had to correct for the compass deviation from true north.

The Theodolite. Invented in Germany during the sixteenth century, the theodolite initially comprised an open sight and a compass for measuring angles horizontally. It proved to be a valuable tool for creating maps. To use it, surveyors would use the highest point in an area, and often create a temporary tower to give them added height. From that known point, they could observe other points or towers and compute the coordinates of distant control points. In 1725, the theodolite's open sight was replaced with a sighting telescope to give it longer range and more accurate sightings. Over time, the design improved to allow surveyors to make increasingly precise angular measurements.

Today, theodolites have largely been replaced for mapmaking by aerial photogrammetry (described shortly). Yet they are still in use, albeit with many improvements to accuracy, the ability to measure angles vertically, and the capability to operate with laptop computers. New ones can measure distance as well as angle, using laser radar techniques.

The Transit. In 1831, the Philadelphia instrument maker William J. Young developed the surveyor's transit. This simplified version of the theodolite quickly became the most important surveying instrument in the United States. American surveyors didn't like the English theodolite; though more accurate, it was slow and inconvenient to use, and relatively fragile. In contrast, the transit was rugged, easy to use, and relatively inexpensive. The ruggedness was particularly important because American surveyors had to operate in difficult terrain, far from a repair shop.

First the theodolite, and later on the transit, were essential tools for the method of mapmaking that came into use in the seventeenth century: triangulation.

Triangulation

Surveying and mapmaking depend on **triangulation**—the process of determining the location of a point by forming triangles to it from known points. The concept was introduced by the Dutch cartographer Gemma Frisius in 1533, though it took nearly a century for the first serious use of triangulation to appear in mapmaking.[9]

Figure 2.8 illustrates how triangulation works. Starting from two known points 1 and 2, a surveyor measures the angular directions to a third point 3 from the line connecting 1 and 2. The location of point 3 is now known, and it can be used, along with point 2, to measure the angular directions to point 4. Then points 3 and 4 can be used to locate point 5, and so forth. Because the surveyor wanted the points to be as far apart as possible, a typical point was a hill or mountain peak or tall tower, if they existed. If not, you built your own observation tower, which made triangulation a slow process.

Ideally, the two known points used in a measurement will intersect with the new point at a 90-degree angle; the lines from points 4 and 5 would intersect

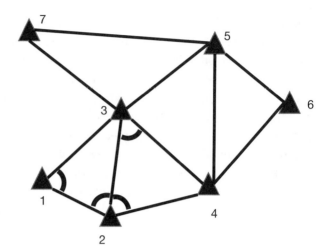

FIGURE 2.8 Triangulation.
Source: Author's collection.

with point 6 at 90 degrees, for example. That gives the most accurate location for point 6. If the intersection point is at a small angle (point 7), the location will not be as accurate, and the locations (and resulting map) will become more distorted as more points are added to the network. In triangulation, though, a 60-degree angle is about as good as you'll get.

The French priest and astronomer Jean Picard (perhaps the inspiration for the fictional *Star Trek* captain's name) carried out the first triangulation survey, starting in the seventeenth century. Using a series of thirteen triangles, he measured the distance of 1 degree of longitude from Sourdon, France, to a point near Paris. The astronomer and engineer Giovanni Cassini and his descendants continued the triangulation survey after Picard's death in about 1683, and by 1718 they had completed a map of France—the first country to be mapped by triangulation.[10]

A similar survey of Great Britain began more than a hundred years later. Major General William Roy directed a survey that began in 1784 from a point in what is now west London. Surveying by triangulation was of necessity a very slow process; the survey was completed in 1822.[11] Meanwhile, most countries in Europe were undertaking similar triangulation surveys, and by 1842 all of Europe had been mapped. An ambitious project to conduct a triangulation survey in India (the Great Trigonometrical Survey of India) had been started in 1802 with an initial baseline measurement in Madras; the entire project took nearly seventy years to complete.[12] The entire survey, shown in figure 2.9, illustrates how triangulation appeared on a survey map and also illustrates why the process was so slow; there were thousands of points, and each point might take days to establish.

Triangulation was the standard method of creating maps until the combination of photography and the airplane provided a better method to measure large areas—leading to the development of photogrammetry.

FIGURE 2.9 Map from the Great Trigonometrical Survey of India, 1870. *Source:* Survey of India, Wikimedia Commons.

Photogrammetry

Photogrammetry is the science of making measurements from photographs. The input to photogrammetry, obviously, is photographs. The output, in early photogrammetry, was a drawing, a measurement, or a three-dimensional model of some real-world object or scene using a photograph taken on the ground. After the invention of the airplane, photogrammetry was applied to creating maps from aerial photographs. The first photographs from an airplane for mapping purposes were taken by the Italian captain Cesare Tardivo in 1908, of a 50-kilometer stretch of the Tiber River.

During World War I, the photography from airplanes was used for battlefield reconnaissance rather than for mapping. But the cameras and photographic techniques developed were to pay dividends in mapping after the war.

The dominant figure in photogrammetry post–World War I was Sherman Mills Fairchild. Fairchild was one of those rare individuals who was good at both technology and business. He began as an inventor and later became a successful entrepreneur, founding more than seventy companies during his lifetime. But his first and lifelong love was building cameras. During his freshman year at Harvard, he invented the flash-synchronized camera. In 1918, he received a government contract to develop an improved aerial camera. Existing aerial cameras had slow shutter speeds, and the aircraft movement while the shutter was open caused major image distortion. His design largely overcame the distortion problem; it featured a fast rotary blade camera shutter mounted inside the lens, providing a sharper image and reducing distortion.

Fairchild, in the process, encountered two nontechnical problems that unfortunately are fairly common in government contracting. The first was a cost overrun of more than five times the contract budget; his father, a cofounder of IBM, paid the difference. The second was that the US government lost interest in the project. The war ended before the military accepted the camera, and the government's need for the camera also ended.[13]

Fairchild saw an opportunity after the war to use his camera for aerial surveying and cartography. In 1921, he formed Fairchild Aerial Surveys, and in 1923, Fairchild Aerial Surveys of Canada, Ltd. He purchased a World War I surplus Fokker D.VII biplane to carry the camera. His first photogrammetry project was the rather unglamorous aerial mapping of Newark. But a survey of Manhattan Island and photogrammetric surveys in Canada followed. The speed and reduced cost of these surveys, compared with conventional ground surveys, made Fairchild's project a commercial success; during the 1920s and 1930s, his company conducted aerial surveys of many US cities. His interest in aerial photography also led to an interest in the airplanes that carried them, and he went on to build aircraft designed to make good use of his cameras—a topic discussed in chapter 8.[14]

Photogrammetry at the time used simple mechanical devices to orient two photographs relative to each other and relative to the ground. The optical-mechanical device for doing that is called a stereoplotter. First developed in Europe before World War I, the stereoplotter became an indispensable tool for cartographers during the 1920s. It could nicely reconcile differences in scale between photographs and maps.[15]

Government interest in photogrammetry surged again with the outbreak of World War II. The demand for accurate and detailed maps during the war led to a dramatic increase in aerial photography, specifically to support mapmaking, and to major advances in the art of photogrammetry.

The greatest advances in what later would become geospatial intelligence were the result of the interactions between humans and machines. The stereoplotter was a beginning, but things really picked up in the late 1960s and early 1970s. Computers permitted cartographers to quickly and accurately orient photographs in producing maps. Precise numerical calculations, it turned out, provided better accuracy than the mechanical approximations of the stereoplotter. This method improved speed, efficiency, and accuracy, and it allowed a transition from paper-based maps to digital maps, a topic discussed in chapter 6.

Nautical Charts

Both triangulation and photogrammetry worked well for mapping land areas, where specific features could be identified and positioned on a map. Seafarers were primarily interested in the location of only a small part of the land: coastlines, harbors, ports, and hazards to navigation. And before the use of triangulation, ship captains had other mapping methods that had worked for thousands of years. Coastlines could be observed visually and—to the limits of a mariner's ability to determine location—mapped.

The Phoenicians were the first Western civilization known to have developed the art of navigation at sea. By about 1200 BCE, Phoenician sailors were navigating with the use of charts, using the positions of the sun and stars to determine approximate directions. In the same time frame, the massive Austronesian migration across the western Pacific was occurring, also aided by the sun and stars. But for centuries most navigators hugged the coast, moving from headland to headland. Once the magnetic compass became available, it was possible to proceed directly from one port to another across open sea with relative safety. After developing the compass, the Chinese used it for maritime navigation some time before the eleventh century; when the compass reached Europe, maritime charts, or sometimes pilot books that provided sailing directions, began to come into use during the thirteenth century.

The earliest known sea chart was prepared for the French king Louis IX, when he was preparing to join the Eighth Crusade in 1270. Such charts came to be known as portolans (pilot books) because they included ports, anchorages, and sailing courses.

In the United States, nautical charts came to be important as the country began to develop a robust overseas trade. On February 10, 1807, Thomas Jefferson established the Survey of the Coast—the primary objective being to improve the safety of navigation. In 1811, Ferdinand Hassler, a Swiss mathematician and surveyor, was selected to conduct the survey. Hassler's job was to determine the geographical location of points along the coast so that they could be used for a nautical survey and chart creation. He thereby became the first superintendent of what would become the US Coast and Geodetic Survey and is now part of the National Oceanic and Atmospheric Administration.[16]

Hassler's first field surveys were carried out in 1816–17 near New York City using triangulation surveys. But in 1818, Congress reassigned the survey to the US Army, Hassler's job ended, and progress basically stopped for the next fourteen years. In 1832, Congress removed the survey from both the army and navy, and Hassler was reappointed superintendent of the survey. From then until his death in 1843, Hassler completed triangulation surveys extending from New York eastward to Point Judith, Rhode Island, and southward to Cape Henlopen in Delaware.

The US government didn't treat Hassler very well. He was hired and fired from different jobs due to politics and was not paid what was owed him. He struggled to conduct a survey with limited appropriations. Congress badgered him about his results and interfered with his operations. On the other hand, Hassler probably didn't help his cause; he reportedly was irascible, impatient with those who didn't understand mathematics and surveying, and possessed an overbearing temperament. But he got his posthumous reward in 2012. That year, the National Oceanic and Atmospheric Administration named its newest coastal mapping vessel the *Ferdinand R. Hassler*.

Aeronautical Charts

In the early days of flight, pilots navigated using visible landmarks over short distances. By the 1930s, radio navigation aids made it possible for pilots to travel considerable distances through unfamiliar surroundings and with reduced visibility. But pilots needed to know the answers to questions such as, Where are the airports? How do I get to them? What obstacles and hazards exist en route? Aeronautical charts were developed to provide the answers to questions such as these, in map form. In 1941, an additional need—to be able to land an airplane during low visibility—was addressed by instrument approach and landing charts. Pilots now could use charts to help determine their location, the most direct route to their destination, the minimum safe altitude, navigation aids, and alternate airports for emergency landings. Later on, other features such as airspace boundaries were included—firing ranges and air defense zones, for example.

By this time, however, aeronautical chart users were encountering the same problem that bedeviled map users in the Middle Ages. Their charts were cluttered with all the "need to know" information—though a pilot seldom "needed" to know all of it on any given flight. Furthermore, different symbols and annotation standards were used in different countries, and that could create problems for international air travel. The first problem was solved when geographical information systems (discussed in chapter 6), allowed selective information displays. The second issue was dealt with by creating the United Nations International Civil Aviation Organization, with the mandate to standardize charts and specify an updating cycle, so all aircraft in the air would be working from the same data set.

Establishing Claims with Cartography

At the start of this book, we noted that the history of geospatial intelligence is the story of three things that have been with us since we organized into tribes: war, commerce, and taxes. These activities share a thread or construct in one form or another: governmental pressures. We also noted earlier in this chapter that maps have a point of view—a topic we'll be revisiting throughout this text. As mapmakers often work for leaders (as al-Idrisi and Captain Cook did), they encounter pressure regarding their product's point of view. Namely, the maps necessarily reflected the wishes of those in power. Even in modern times, cartographers often have to take into account the preferences of their national leaders, as we will see in the examples that follow. Maps found uses in geospatial intelligence long before the term had been coined. They have a history of use for deception, propaganda, and of course for establishing claims to territory. And because of this history, an astute observer can occasionally employ geospatial intelligence to anticipate future government actions, as geographers in one instance below could tell you.

The Tasmanian Deception

From 1768 to 1771, James Cook led the first of three Pacific Ocean voyages that he would eventually make from Britain as a captain in the Royal Navy. During his travels along the Australian coast, he found that the region known as Tasmania (previously discovered by the Dutch explorer Abel Tasman) was actually an island, separate from Australia. But Cook had been told by the British Admiralty that any islands found off the Australian coast should not appear on any maps that he produced. The British were already making plans to colonize the continent and did not want adjacent islands to be colonized by another country. Putting service to country ahead of charting integrity, Cook subsequently created a map of Australia (fig. 2.10), which shows, in place of the Bass Strait separating Tasmania from Australia, a land bridge connecting the two.[17] His cartographic deception helped to conceal the existence of the island for another three decades.

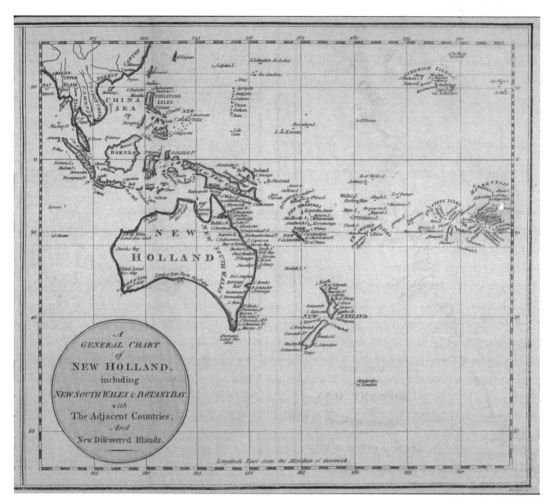

FIGURE 2.10 Captain Cook's deceptive map of Australia, "A General Chart of New Holland Including New South Wales & Botany Bay with the Adjacent Countries and New Discovered Lands." *Source: An Historical Narrative of the Discovery of New Holland and New South Wales* (London: Fielding & Stockdale, 1786), Wikimedia Commons.

The Northwest Territory

The settlement of the Americas resulted in many boundary disputes, some of which led to cartographic "warfare." One example involved the dispute between the United States and Great Britain over the Northwest Territory on the Pacific Coast. US maps in 1841 showed the 54°40' line as the boundary between the US and Canada. The British government had a different perspective, and the 1844 British maps of the region showed the Columbia River as the boundary. (Of course, the British claim had an economic motive; a boundary along the bottom edge of the "Disputed Area" in figure 2.11 would allow the Hudson's Bay

Company to continue its lucrative fur trade along the Columbia River.) The US claim was encapsulated in a Democratic Party political slogan "Fifty-Four Forty or Fight," but it fortunately never came to that. The dispute was resolved peacefully in 1846 by the two countries agreeing to extend the existing boundary westward (with an adjustment to keep Vancouver Island as part of Canada).[18]

Iraqi Cartographic Claim on Kuwait

On August 2, 1990, Iraq invaded and occupied its neighbor Kuwait. The invasion came as a surprise to most observers, in spite of many clues that existed about Iraqi intent. Iraqis had long regarded Kuwait as a part of their country that had been illegally carved off by the British subsequent to a 1913 treaty. One important clue about possible Iraqi intent to invade came to the attention of

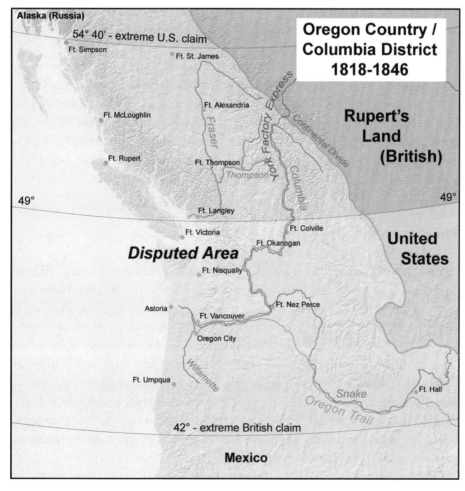

FIGURE 2.11 British and US claims in the Northwest Territory. *Source:* Kmusser, Wikimedia Commons.

geographers but was not widely noticed otherwise before the invasion. A 1990 map produced by the Iraqi government showed Kuwait as the nineteenth province of Iraq.[19] The occupation continued for seven months, until a United Nations–mandated military force led by the US expelled the Iraqis in one of the most lopsided victories in history.

China

The 1990 Iraqi map noticed by geographers has a conceptual name. It is an example of what the geographer Harm de Blij calls "cartographic aggression"—a practice at which China shows considerable skill.[20] Chinese maps since at least 1986 have indicated that parts of India—almost all of the state of Arunachal Pradesh and part of the state of Assam—are Chinese territory.[21] Chinese maps currently show most of the South China Sea as part of its territorial domain for fishing, exploration, and exploiting crude oil and natural gas beneath the seabed. These cartographic claims have been vigorously contested by all countries bordering the South China Sea.

Even without a boundary dispute, the names used on maps can evoke nationalistic tension. The body of water between the Korean Peninsula and Japan is called, on most maps, the Sea of Japan. But Koreans object to that appropriation, and Korean maps title it the East Sea. And what has historically been called the Persian Gulf is commonly called, within the Arab world, the Arabian Gulf, provoking strong objections from the Iranians (Iran being long called Persia).[22]

Chapter Summary

The history of cartography is a catalog of tenacious innovators from different fields (including librarians, priests, astronomers, and entrepreneurs) who were set on solving "just one problem" in order to improve accuracy or production of maps. And then their sights were set on the next "just one problem." From the earliest times to today, cartography has been aided by technological advances from countries around the globe—including inventions such as the printing press and the tools of surveying, later the aerial photograph that permitted use of photogrammetry in mapping, and most recently the technologies that enabled computers, visual displays, and the easy communication of geospatial information to the user. The computer has provided cartographers with unprecedented control over the mapping process—a subject we'll revisit in chapter 6.

With this introduction to maps and charts, let's turn our attention to the third dimension of a map and yet another problem for cartographers: terrain.

Notes

1. Joshua J. Mark, "Silk Road," in *Ancient History Encyclopedia*, March 28, 2014, www.ancient.eu/Silk_Road/.
2. Aristotle, *On the Heavens*, 350 BCE, translation by J. L. Stocks, http://classics.mit.edu/Aristotle/heavens.1.i.html.
3. Cristian Violatti, "Eratosthenes," in *Ancient History Encyclopedia*, April 5, 2013, www.ancient.eu/Eratosthenes/.
4. Canadian Cartographic Association, "What Is Cartography?" http://cca-acc.org/resources/what-is-cartography/.
5. Paul B. Sturtevant, "A Wonder of the Multicultural Medieval World: The Tabula Rogeriana," *Public Medievalist*, March 9, 2017, www.publicmedievalist.com/greatest-medieval-map/.
6. Timothy J. Freeman, *Portraits of the Earth: A Mathematician Looks at Maps* (Providence, RI: American Mathematical Society, 2002), 41.
7. Nicholas Crane, *Mercator: The Man Who Mapped the Planet* (London: Henry Holt, 2014).
8. Manfred F. Buchroithner and René Pfahlbusch, "Geodetic Grids in Authoritative Maps: New Findings about the Origin of the UTM Grid," *Cartography and Geographic Information Science*, 2016, DOI: 10.1080/15230406.2015.1128851.
9. Beau Riffenburgh, *Mapping the World: The Story of Cartography* (London: Carlton Books, 2014), 71.
10. Riffenburgh, 68.
11. Joseph F. Dracup, "Geodetic Surveys in the United States: The Beginning and the Next One Hundred Years," n.d., www.ngs.noaa.gov/PUBS_LIB/geodetic_survey_1807.html.
12. Riffenburgh, *Mapping*, 104.
13. Farnsworth Fowle, "Sherman Mills Fairchild Is Dead at 74; IBM Heir Invented an Aerial Camera," *New York Times*, March 29, 1971, www.nytimes.com/1971/03/29/archives/sherman-mills-fairchild-is-dead-at-74-i-b-m-heir-invented-an-aerial.html.
14. Leon T. Eliel, "The Story of Fairchild Aerial Surveys, Inc.," *Photogrammetic Engineering*, September 1942, www.asprs.org/wp-content/uploads/pers/1942journal/sep/1942_sep_163-170.pdf.
15. G. Petrie, "A Short History of British Stereoplotting Instrument Design," *Photogrammetric Record* 9, no. 50 (1977): 213–38, http://petriefied.info/Petrie_British_Stereo-Plotting_Instrument_Design_fulltext.pdf.
16. "Ferdinand Rudolph Hassler," n.d., https://westpoint.edu/sites/default/files/pdfs/Math/Hassler.pdf.
17. Margaret Cameron-Ash, *Lying for the Admiralty* (Kensworth, Australia: Rosenberg, 2018).
18. Henry Commager, "England and Oregon Treaty of 1846," *Oregon Historical Quarterly* 28, no. 1 (1927): 18–38.
19. Harm de Blij, *Why Geography Matters* (New York: Oxford University Press, 2005), 45.
20. De Blij.
21. De Blij.
22. De Blij, 38.

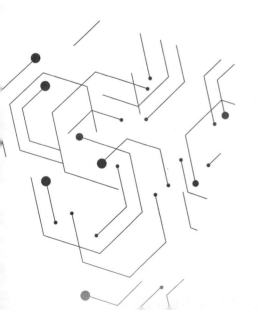

Terrain

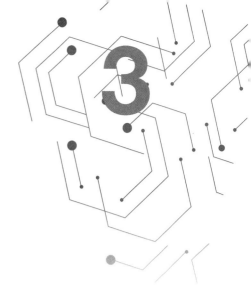

Paper maps are flat. The Earth isn't. That inconvenient fact gave cartographers heartburn from the time that they recognized the Earth was a sphere, and representing it on a flat surface introduced distortion. But their woes didn't end there. Earth's surface is neither smooth nor uniform. The distance between two points in eastern Kansas may be the same as the distance between two points in western Colorado. But the time it takes to travel that distance is likely to be quite different. Travel over land between any two locations differs, as any traveler knows, depending on whether you must move over flat desert, through brackish swamps, or up mountainous paths. And while the oceans can be represented as flat, they have depth; and exactly what that depth is can be rather important for seafarers who wish to arrive at their destination while keeping their craft in one piece. In air travel, the height of both natural terrain and human-made objects is critical information. And most importantly, the composition of all the Earth's surface is important for exploiting its resources—agriculture and mineral exploration, in particular.

This chapter is about the third dimension of cartography—the vertical dimension. **Topography** involves studying the shape and features of the Earth's surface. It typically requires making records of relief or terrain, capturing the three-dimensional quality of the surface, and identifying specific natural features. In a broad sense, it includes representation of artificial features as well. A topographic map is primarily concerned with the description of all the physical features (natural and human-made) of an area.

From the earliest times, topographers encountered two problems: how to measure the height of terrain, and how to represent it on a map.

Measuring and Representing Terrain

The same instrument and technique used for mapping in the horizontal plane—the theodolite and triangulation—could be used to determine height as well.

And for centuries, that was the accepted method for doing so. Not until the development of airborne and spaceborne radar did we see dramatic improvements in measuring terrain elevation.

Even with the elevation measurement issue resolved, the problem of accurately representing terrain features remained. Luckily, there are many possible solutions. Terrain or relief can be shown in a variety of ways, including (especially in the twentieth and twenty-first centuries) by using contour lines to show elevation. Deserts, swamps, snow, and forests can be represented by appropriate shading, for example. As usual, however, there's still one problem. The resulting maps tend to become extremely cluttered, making it difficult for a user to pull out the needed information. For centuries, the solution to that was to make a series of maps of the same region, each containing different information. Pick up any atlas produced in the twentieth century, and it will include at least two maps of the same region. One will contain political boundaries, and one will show relief. The replacement solution—map layers—had to await the development of a new technological breakthrough—the geographic information system (described in chapter 6).

Once it became possible to capture accurate data about the elevation of every point on Earth, the next step was to create a detailed elevation model of the Earth—both for representing it in map displays and for analyzing problems that depended on such detailed knowledge as the use of topography for navigation and targeting, discussed below. That led to the creation of digital elevation models.

Digital Elevation Models

A **digital elevation model** (DEM) is a matrix of numbers. Each entry in the matrix represents a specific point on the surface, and the number attached to that entry is the altitude of that point above some reference (usually sea level, for DEMs of the Earth). DEMs have been created for the Earth, its moon, and several planets.

DEMs date back to at least the 1970s. They originally were created by digitizing the contours from existing topographic maps, and then by using stereographic imagery. Those models were rather crude by today's standards. Far more accurate models of the Earth became possible after airborne and spaceborne radar imagery became available (discussed in chapter 12). The National Aeronautics and Space Administration (NASA) Shuttle Radar Topography Mission, flown in 2000, resulted in a topographic map of about 80 percent of the Earth's surface.[1] A subsequent joint US-Japanese effort using NASA's Terra satellite created the Global Digital Elevation Model, the most complete DEM of the Earth ever made (covering 99 percent of Earth's surface) in 2009.[2] Worldwide, many DEMs now exist; many national mapping agencies produce their own versions.

The US Geological Survey, for example, produces its own version, called the National Elevation Dataset.

So what are DEMs good for? First, a DEM can be used to create a topographic map of the Earth. It is widely used to rectify aerial and satellite imagery for mapping (i.e., making corrections to ensure that what is displayed will be free of distortion). Also, three-dimensional physical models (relief maps) of the Earth have long been used for military purposes, and DEMs provide the accurate physical models that modern militaries need.

On the civilian side, DEMs find use in applications such as flood modeling, mineral exploration, modeling water flows, doing groundwater studies (where to dig wells), and ensuring flight safety. Scientific studies such as archaeology and **geomorphology** (study of the physical, chemical, and biological features of the Earth) make use of DEMs. So does the engineering of large structures such as dams and bridges. DEMs support more accurate satellite navigation and precision farming. GIS (in chapter 6) depends on the availability of DEMs.

With the development of maps depicting terrain features, planners and decision-makers had a tool that they could use to produce geospatial intelligence for military and civil uses. The subsequent availability of DEMs made those tools more powerful. Let's look first at military applications of terrain mapping, then go on to civil applications, as they illustrate our three themes in the story of geospatial intelligence history: war, commerce, and taxes.

Military Use of Terrain

Militaries through history have needed to move masses of troops, and in later times to move heavy equipment, though difficult terrain. Topography is a factor to consider in all warfare. The Chinese general and military strategist Sun Tzu, in his classic book *The Art of War*, characterized it this way: "We may distinguish six kinds of terrain, to wit: (1) accessible ground; (2) entangling ground; (3) temporizing ground; (4) narrow passes; (5) precipitous heights; (6) positions at a great distance from the enemy."[3]

Consider Switzerland. It has the advantage of mountainous terrain that has obstructed any form of modern warfare. The basic Swiss defense plan has always been to retreat to the mountains if attacked. The terrain favors small rifle units that have accurate shooters, and it gives the Swiss a level of self-confidence that is reflected in an incident that reportedly occurred in 1912. German Kaiser Wilhelm II was the guest of the Swiss government and while there observed Swiss military maneuvers. According to a story that appeared subsequently on a Swiss postcard, the Kaiser asked a Swiss militiaman: "You are half a million and you shoot well, but if we Germans attack with a million men, what will you do?' The soldier replied: 'We will shoot twice and go home.'"[4]

Switzerland is an example of what Sun Tzu called "precipitous heights," where the best defensive strategy is to occupy the high ground and wait for the opponent to attack—a subject to which we'll return in chapter 8.

Of the many battles in history where terrain made an important contribution, a few of the best known follow here.

Thermopylae

In 480 BCE the Greeks had a very good understanding of the Persian Army. They knew, for example, that it couldn't stay in one place long because of supply logistics. It had to attack or retreat. They knew how the Persians used cavalry effectively. And, they knew Persian battle tactics.

The Greeks also had a great understanding of terrain, and they used it. The pass at Thermopylae was the only known way into Greece (an example of Sun Tzu's "narrow passes" terrain). The Greek phalanx could block the narrow pass with ease, without risk of being outflanked by cavalry. In the pass, the phalanx would be difficult to assault for the more lightly armed Persian infantry.

The vastly outnumbered Greeks held the pass for seven days, exacting heavy casualties on the Persians. The battle would have gone their way if it had not been for a local Greek traitor who led the Persians on a little-known path that led behind the Greek line—a great example of the use of HUMINT to provide geospatial intelligence. The trapped Greeks, led by the Spartan commander Leonidas, fought on to the last man. The battle also featured one of the greatest one-liners in the history of warfare. In response to a pre-battle offer by the Persian commander Xerxes to allow the Greeks to lay down their arms without bloodshed, Leonidas replied, "Come and get them."[5]

Agincourt

During the Hundred Years' War, the English achieved one of the most lopsided victories in history, and against a considerably superior force. A French army had trapped the English and their king, Henry V, near the French village of Agincourt. On October 25, 1415, the French attacked.

The English forces, totaling fewer than 10,000 men, were mostly unarmored English and Welsh archers equipped with the English longbow. The French forces comprised as many as 50,000 men, a large portion of whom were mounted knights and foot soldiers.

However, the terrain most heavily favored the defenders. It was a relatively narrow 700-meter stretch of plowed land, hemmed in on either side by dense forest, which was muddy due to recent rains. Over this field the French knights and footmen charged—and sank in mud up to their knees. Any armored knight who fell had great difficulty getting back up. Some drowned in the mud, and many were killed by the English archers where they lay, struggling to rise.

The French cavalry also charged, but could not get past a row of sharpened stakes that the archers had set up; and the horses, panicking upon being hit by arrows, fled and trampled the advancing French forces. The exhausted French foot soldiers who succeeded in engaging the English in hand-to-hand combat found themselves at a disadvantage against the more lightly armored bowmen in the thick mud. It was a perfect example of what Sun Tzu called "temporizing ground"; the side that attacked first was at a disadvantage.[6]

While the number of casualties on each side have been difficult to determine, most estimates are that the French lost ten (killed or captured) for each English casualty.

The Ardennes

The Ardennes Forest has long been thought to be unsuitable for large-scale military operations due to its difficult terrain and narrow lanes for road movement. However, three times—once in World War I and twice in World War II—Germany successfully moved its military units through the Ardennes to attack a lightly defended part of France.

During World War I, the Battle of the Ardennes was fought from August 21 to 23, 1914. The French expected few German forces to be in the area because of the difficult terrain. They were surprised by a full-scale German offensive through the area, and were forced to retreat in the first major battle on the Western Front.

The French did not learn the lesson. During the Battle of France in May 1940, Germany's First Panzer Division pushed through the Ardennes, again surprising the French. The German advance was slow at first because of the large number of tanks and other vehicles trying to force their way along a poor road network. But once they were through the forest, the German panzers cut off and surrounded the British, Belgian, and French forces in Belgium.

After these two defeats, the French understood that the Ardennes, difficult territory though it was and an exemplar of Sun Tzu's "entangling ground," could still be an effective route for a surprise attack. The Americans were to learn this lesson in December 1944. On December 16, three panzer armies moved through the Ardennes, pushing aside the thin American defenses. But the American army units tenaciously held the critical road junctions at Bastogne and Elsenborn Ridge long enough for reinforcements to arrive and win what we know today as the Battle of the Bulge.

Operation Desert Storm

Trafficability of terrain will always be a factor in war. In 1415 at Agincourt, terrain trafficability was a critical factor in the English success. Almost six centuries later, the same issue became important in the lead-up to the international coalition attack to retake Kuwait from the Iraqis—Operation Desert Storm, in 1991. The difference between Agincourt and Desert Storm was that one commander obtained geospatial intelligence about the terrain before attacking.

In 1990 the coalition was planning the routes for its forces to take in the proposed attack. The force commander, General Norman Schwarzkopf, planned to send airborne units and armor into the desert west of Kuwait as part of his famous "left hook" maneuver, which raised the question: Just what conditions would those units encounter on the ground in the desert? The American army needed to know if their M-1 Abrams tanks could traverse the planned invasion route across desert terrain.

Intelligence agencies were asked for all they could obtain on the trafficability of the deserts that the "left hook" would go through. There wasn't much available; no records in intelligence community holdings had any soil condition data for the region. But an unlikely source turned out to be the key. The Library of Congress had archaeological records from early in the twentieth century, recorded in the dairies of an archaeological team that had crossed the area on camelback. The diaries gave intelligence officers the terrain information that Schwarzkopf needed to execute his decisive maneuver of that brief conflict—areas of soft sand, and the locations of ravines that would need to be avoided or bridged.[7]

This was an example of GEOINT derived primarily if not entirely from non-imagery sources. Since 1990, imaging technology has advanced considerably; imagery would likely be a key part of a similar terrain assessment today.

Using Topography for Targeting

Before there were maps, hunters and warriors used terrain landmarks for navigation overland. The electronic age made it possible to use topography for the precise navigation needed for missile targeting. As modern militaries developed cruise missiles for land attack, they needed highly accurate navigation systems to guide the missiles to their targets. They found a solution in the use of terrain models.

If you draw a straight line between any two points on the Earth, the terrain elevation changes constantly along the line (unless you are over water). These changes form a contour pattern that is *unique for any given path*. Given accurate models of the terrain contour between a missile's launch point and its target, the missile can measure its altitude above the terrain in-flight using an altimeter, compare the readings with a stored terrain map, and use the comparison to find out where it is and to correct its course. The technique—called terrain contour matching, or TERCOM—was successfully used to guide cruise missiles such as the US Tomahawk until it was replaced by a global positioning system receiver in 1993.[8]

Civil Use of Terrain

The terrain beneath us has to be explored for many reasons having nothing to do with military operations. The earliest such exploration probably was for the simple purpose of obtaining drinkable water. Wells were being dug in antiquity. And crops had to be planted. Later on, mines were dug in likely places, or where there was evidence of mineral wealth to promote commerce. But as governments were established and developed a need for revenue, characterizing terrain became their key. The most obvious way to tax citizenry is through property.

Property taxes were one of the earliest known forms of taxation. Rulers needed to assess what land was worth to know where best to look for revenue, and that depended on how much wealth could be extracted from the land by agriculture, mining, or other types of exploitation. Taxable values were determined according to whether the land was richly agricultural; consisted of barren hills, forest, swamp; or had been developed. Thus, leaders took a keen interest in the topographic details about the terrain over which they ruled. Land became a source of wealth and power, and therefore of income to the state. To learn what rulers needed to know, they turned to surveyors to produce cadastral maps.

Surveys and Cadastral Mapping

A cadastre is a form of registering land and people. From the age of agriculture to feudalism, human beings were physically tied to the land. It was the primary symbol and source of wealth. In this phase, the cadastre publicly recorded ownership for fiscal purposes. Surveys of land for taxation were being conducted as early as about 3000 BCE in ancient Egypt and 700 BCE in China. The Romans, however, gave the practice more structure and a name for the product.[9]

The Roman Empire, as it expanded, found it convenient to pay its soldiers in conquered land, and over time this meant that large parcels of land wound up in private hands. Later, emperors, annoyed at the loss of what had been public property, ordered a series of cadastres (derived from the Latin word for a poll tax register) across the empire. The cadastre was a record of land locations and boundaries, ownership, tenure, and property value. Using the information in the cadastres, Roman emperors were able to impose taxes on all landholders.

Later on, England's first king made further improvements. In 1085 King William the Conqueror commissioned the great survey of England. The result is known as the Domesday Book. The complete book was a roll of all English landholders, along with a description of their holdings. It included details about each parcel—dimensions, suitability for agriculture, and property improvements such as buildings, mills and fishponds. Most importantly for the king, it listed property values in pounds, so that the holders could be properly taxed. For that reason, the survey was not popular in England, and the name it was appropriately given is Middle English for "Doomsday Book." The decisions of assessors about property value were final; like the Last Judgment, they could not be appealed.

But, it was France's most famous military leader who established the foundations of modern cadastral systems. In 1807, Napoleon established a comprehensive cadastral system for France that is regarded as the forerunner of most modern versions around the world. The French cadastre made more use of maps than had previous cadastres, and provided more details about land use.[10]

Surveying and cadastral mapping were essential as states acquired new territory that had to be managed—as happened in the settlement of North and South America. The US acquired large increments of territory at the end of the American Revolutionary War; later via the Louisiana Purchase, as a result of the Mexican-American War in 1846; and finally after the purchase of Alaska in 1867. The US government over time needed to distribute these lands, either for money or as a reward for service to the nation. But in order to do so, the lands had to be topographically surveyed. Canada faced a similar problem after the Dominion of Canada was created in 1867, and accordingly it began the Dominion Land Survey in 1871.

During the Industrial Revolution land became more important in commerce: it was a tradable commodity and the primary source of capital. This step in cadastral evolution gave birth to land markets, and so the cadastre took on another focus—a tool for transferring land.

After World War II, reconstruction and an increasingly mobile, growing population led to another view: land as a scarce resource that may not be sufficient for the world's needs. The result was an interest in urban and regional planning, an important new application for cadastres. Later on, the focus on civil use of terrain expanded to include issues such as environmental degradation, sustainable development, and social equity. Along the way, a wide range of specific concerns developed; let's take a brief look at a few of them.

Agriculture

The Bible tells the story of Abraham and Lot, who decided to divide up the Promised Land in order to settle a dispute between their shepherds. Given the right to choose, Lot made the choice that seemed to be better terrain for his flocks (it turned out to be a bad decision, but the lush green grass had nothing to do with that): "Lot looked around and saw that the whole plain of the Jordan toward Zoar was well watered, like the garden of the LORD, like the land of Egypt.... So Lot chose for himself the whole plain of the Jordan and set out toward the east."[11]

Much later, Abraham's descendants were returning from a long stay in Egypt, and had much the same concerns about terrain. Moses's guidance to his spies was this: "Go up from here through the Negev, then ascend to the hill country. See what the land is like.... Look to see whether the land where they live is good or bad.... Examine the farmland, whether it's fertile or barren, and see if there are fruit-bearing trees in it or not."[12]

Terrain establishes whether an area can grow crops, and if so, what kind. Is it flat and well-watered enough to grow rice? Is it suitable only for forage? Or should it be left in forests? Those have been questions documented through history in both secular writings and ancient religious texts.

Over time, as our ability to assess the quality of terrain has improved, we have applied the tools of geospatial intelligence to determine specifically what crops are most suitable for an area. That long had been a matter of personal observation, experience, and soil sampling. Late in the twentieth century, agricultural assessments of large areas became possible. Aircraft, and later satellites, were equipped with sensors that could "see" outside the visible part of the spectrum. These imagers allowed researchers, by measuring the health of vegetation, to infer the quality of underlying soils. In 2017, in combination with onsite measurements, researchers made an alarming discovery: Earth has experienced a significant decline in agricultural soil quality worldwide due to the expansion and intensification of agriculture. The loss is forecasted to continue—which likely will result in food shortages and additional conflicts, such as those occurring in Chad and Sudan.[13]

Construction Planning

As commerce between villages and cities developed in ancient times, topography became an important factor in where to place ports, roads, and bridges—and, much later, where to build railroads. Some terrain features were easily observed and avoided: floodplains, swamps, and steep rocky areas, for example. Some were less obvious and required careful investigation over time—such as sinkholes and harbors subject to silting.

When architects and engineers began to build tall structures, subterranean features became important. One can, of course, find the subsurface conditions by drilling a borehole down to bedrock, and that was long the common practice. Today, a technique called seismic tomography is used to give a more complete picture. It involves drilling boreholes and emplacing sensitive instruments in them to monitor seismic activity. Then, either by waiting for the seismic shock from earthquakes or by setting off underground explosions to create **seismic waves**, we can get a three-dimensional image of the geological structures, assess soil and rock properties, and map cavities. Geophysicists and engineers use it to investigate the foundation and underlying rock for constructing buildings and bridges. And they should. Bad outcomes can occur when the subsurface terrain isn't well considered in construction planning. The result can be either a tourist attraction or a catastrophe. The Leaning Tower of Pisa in Italy is one of the few examples of such a tourist attraction. The Millennium Tower in San Francisco is an example of a catastrophe.

The Millennium Tower is a luxury residential high-rise building that was completed in 2008. Since then, it has sunk 17 inches and tilted 14 inches. Residents who paid anywhere from $1 to $10 million for their condominium homes now face steep losses—though finding a buyer might be problematic. While the reason for the subsidence is being disputed, one factor is the builder's decision to anchor the building 80 feet into packed sand rather than going all the way to bedrock, 200 feet down. In the event of an earthquake, the sand that the tower rests on could experience liquefaction, in which the sand and silt behave like a liquid as a result of the increased water pressure due to seismic activity—in which case, the building could sink rapidly or collapse. One of the two engineers who had originally concluded that the tower's design was structurally sound subsequently concluded that the builders should have hired a geotechnical expert to look at the soil's building strength—that is, they should have drawn on geospatial intelligence.[14]

Oil and Natural Gas

Oil was first discovered in China in about 600 BCE. The first oil wells were drilled there in about 347 CE using bamboo poles with drill bits attached. Early oil exploration was often a matter of drilling a well and hoping to get lucky. Sometimes there were clues about where to drill. Surface features could indicate that the underlying geology had attributes that would cause petroleum to accumulate. Oil seepage to the surface was usually a good clue. But such exploration was limited to examining surface characteristics for a sign. If an area appeared likely, then you could drill an exploratory well—which often turned out to be a waste of money.

In the last several decades, exploration companies have developed a rich set of geospatial intelligence tools to more precisely locate the drill point. High-resolution aerial cameras and imaging radars permit detailed digital terrain maps to be created. Imaging cameras that work in the infrared bands—often the same types as those used in agricultural assessments—give important details about surface rock formations. Laser radars can detect the presence of ethane or methane, indicating seepage of natural gas.

Armed with a better knowledge of where to look, oil and natural gas companies today have a wealth of tools for geophysical exploration that can be applied onsite before digging a well. They can search for small variations in the Earth's gravity field or magnetic field and passively survey for seismic vibrations that indicate the presence of oil-bearing rock. After they identify features of interest (called *leads*), the next step is to use active seismic surveys to

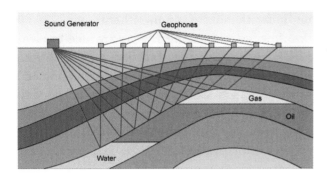

FIGURE 3.1 Layout for a seismic survey. *Source:* Author's collection.

produce a three-dimensional profile of the rock structure. An active survey uses a sound generator (e.g., an explosion) with a set of ground-emplaced microphones, called **geophones**, scattered around the area to sense reflections from geological formations beneath the surface, as shown in figure 3.1. Then, only after a promising area has been identified and evaluated, an oil company drills an exploratory well to determine the presence of oil or gas.

Mineral Exploration

Finding valuable minerals originally was much like finding oil and gas: you looked for surface evidence of what was beneath the surface. Over time, certain geological features became known indicators of the presence of minerals. Rock types that might contain the sought-after minerals were identified. Assays determined the metallic content of a sample of rock. Precious metals (gold and silver) and gemstones always were high on the search list, followed by base metals (copper, nickel, zinc, and lead), iron, and coal.

As more complete terrain information became publicly available, the first stop in exploration was to examine public records. Mining companies would begin by looking at existing mineral maps, academic geological reports, and local, state, and national geological reports. They would then consult property assays and well-drilling logs, supplemented with knowledge obtained from locals. Once a company had identified a likely geological formation, it would drill a borehole and conduct an assay.

Technology has largely replaced the need to perform laborious field searches or depend on luck to find knowledgeable locals. In fact, mineral exploration today is a field that employs GEOINT analysts, both in government and in the commercial sectors. Their searches typically begin with a detailed look at satellite and airborne imagery. Satellites have been providing images for mineral exploration since 1972. In that year, NASA put into orbit the first of the Landsat Earth imaging satellites. Landsat remains today the longest-running program for Earth imaging; its series of satellites have captured tens of millions of images contributing to our understanding of, among other things, agriculture, geology, and forestry worldwide. Its imaging sensors are able to identify geological features that indicate the existence of mineral deposits.[15] And Landsat now is only one of many space-based imagers being used in mineral searches. Aircraft are also used for mineral exploration, some equipped with **magnetometers** (sensors that detect slight changes in the Earth's magnetic field) to detect iron deposits. Most recently, **unmanned aerial vehicles** (UAVs), commonly called drones, have been similarly equipped and used to perform the surveys at lower cost.

Flood Modeling

Like digging oil wells or drilling mines, determining how to avoid building on a floodplain long was a matter of observing over time or guesswork. Today, terrain models are used to estimate the likelihood and effects of flooding. In the US, government agencies such as the US Army Corps of Engineers and the US Geological Survey develop such models for, among others, the Federal Emergency Management Agency. The models typically are maps that depict the spatial extent and depth of flooding at specific water level intervals in areas that might be subject to flooding. The maps are created using hydraulic and topographic modeling techniques. They are then used, along with National Weather Service forecasts, to provide flood warning. The modeling products include the Federal Emergency Management Agency's 100- and 500-year flood maps, showing areas that are subject to flooding once every 100 or 500 years, respectively.

The models aren't perfect; no models are. The residents of Houston can tell you that. Three times in three years—2015, 2016, and 2017—Houston was hit with 500-year floods. That's statistically a very unlikely occurrence. The last of the three, Hurricane Harvey in 2017, was a 1,000 year flood. The models may have accurately characterized the terrain, but they appear not to have taken into account climate changes.[16] Rainfall on a scale not seen before, rather than inaccurate terrain modeling, was the culprit in the case of Harvey. Meteorologists were searching for words to describe it as the storm neared Houston: "epic flood catastrophe"; "a demarcation in people's lives"; "we've never forecast this much before." Meteorologists had to tweak their rainfall maps as the totals exceeded the upper bounds of their color legends.[17]

Oceanographic Terrain

Many people—including commercial mariners, navies, and scientific researchers—are concerned with getting an accurate picture of ocean conditions. Depth is important, but features such as surface wave conditions, currents, and the nature of the ocean floor are also significant. The study of those terrain features is called hydrography.

Hydrography

Hydrography is a scientific discipline concerned with measuring and describing the physical features of bodies of water and the land areas adjacent to them. It includes the measurement and description of the features of oceans, seas, coastal areas, lakes, and rivers. It also predicts their change over time. Hydrography has many customers; its primary concern for millennia has been to contribute to the safety of navigation. Now it also has a role in scientific research, environmental issues, economic development, and defense, among others.

Bathymetry

Bathymetry, a subdiscipline of hydrography, is the study of underwater depth of river, lake, or ocean floors. So bathymetry is the underwater equivalent to topography on land. It's been practiced to assist in navigation since at least 500 BCE. Around that time, Herodotus reportedly provided in his sailing instructions that, when the ocean's depth reached 11 fathoms in approaching Egypt, you were still a day's sail from the coast.[18]

Hydrographic charts, also called nautical charts, are more familiar to navigators. They typically show features such as water depth, nature of the sea or river bottom, contours of the coastline and bottom, and tides and currents. They are created based on hydrographic surveys. Bathymetric charts include detailed representations of the seabed depth, and look just like topographic maps. Figure 3.2 (a. and b.) illustrates the difference.

Depth measurements (known as *soundings*) have long been important for mariners—often a matter of survival. In Herodotus's time, the common practice was to tie a stone—later replaced by a lead weight—to a rope, throw it overboard, and measure the length of rope it took to reach the bottom. Over time, the rope was marked for every fathom (6 feet) of length, and the sailor handling the rope would call out the mark number to the helmsman. (The fathom measurement originated from the fact that a sailor, with arms outstretched, could hold about 6 feet of rope.) For Mississippi River steamboats, the safe depth was the second mark on the rope, indicating a depth of 2 fathoms, or 12 feet. And the call for that was "mark twain" (mark two)—a term that author Samuel Clemens liked so much that he adopted it as his pen name.

The lead weight sounding method was both inefficient and inaccurate (since the rope could not easily be made perfectly vertical). In 1913 a solution was found; that year, a German inventor received a patent for an echo sounder, also known as a depth sounder. It transmitted a pulse of sound downward from a

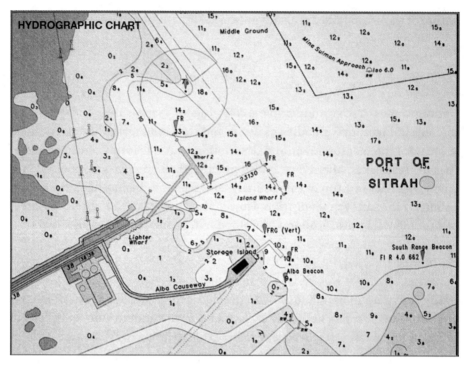

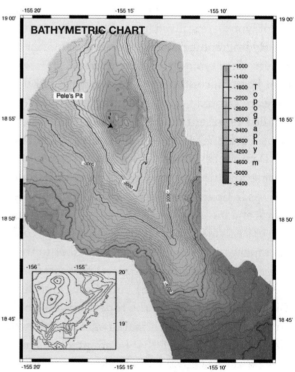

FIGURE 3.2A Comparison of hydrographic and bathymetric charts. *Source:* MapXpert, Wikimedia Commons.

FIGURE 3.2B Comparison of hydrographic and bathymetric charts. *Source:* National Oceanic and Atmospheric Administration.

ship and measured the time for a return of the echo from the ocean bottom. Since the speed of sound in water is fairly predictable, the echo time could be converted to depth. Starting in the early 1930s, depth sounders were used to make bathymetry maps of the ocean floor. Much later, they would become important for finding schools of fish.

Bathymetry became important for a different type of geospatial intelligence during World War II in both the Pacific and Atlantic oceans: planning for amphibious assaults. In the Pacific, one US Navy hydrographic survey ship was available; the USS *Sumner* spent much of 1942 conducting surveys of potential fleet bases in the southwest Pacific. By 1943, several US Coast and Geodetic Survey ships, complete with crews, had been transferred to the navy. Additional survey ships, often converted minesweepers, joined the Pacific Fleet in 1944 and 1945.[19]

The survey teams' most hazardous job was conducting pre-invasion surveys off hostile shores. The teams usually conducted their surveys at night in rubber boats, measuring water depth, shore conditions, and tides. They relied on lead-lines, pressure gauges, and luminous compasses to make observations of the tides. At night, there was almost no opportunity to record the depth and position data as they were collected. The information had to be committed to memory.[20]

Before the Normandy invasion during World War II, the Allies needed detailed information about the condition of the Normandy beaches and the water depth near the beaches. Water depth was particularly important for planning the troop landings, and for selecting locations for the large floating docks that were to be emplaced after the landings. The UK Hydrographic Office assigned a team to conduct the soundings. In late 1943 and early 1944, the team clandestinely collected beach samples and completed the depth measurements.

During the Cold War, when ballistic missile submarines were developed, the products of bathymetry became important for their navigation. Submarines on patrol could not risk discovery by routinely surfacing to determine their position. But terrain maps of the ocean floor were sufficiently detailed to be used for accurate navigation by ballistic missile submarines—much the same technique, in a different medium, as the terrain contour matching used by cruise missiles.

Hydrographic and bathymetric surveys also are conducted for commercial purposes. The marine dredging, oil exploration, and drilling industries all rely on such surveys. Telecommunications companies survey their routes for laying submarine communications cables, making use of military-class acoustic imagers.

Chapter Summary

Terrain—surface or subsurface, on land or sea—is a factor in a wide range of human endeavors. Armies, from earliest history, have considered terrain in their planning and movements. In civil matters, terrain has influenced taxation, resource exploitation, and construction, among others. The third dimension of a map has always been an essential element of geospatial intelligence. It has shaped the history of navigation, the subject of chapter 4.

Notes

1. Tom G. Farr and Mike Kobrick, "The Shuttle Radar Topography Mission," n.d., https://trs.jpl.nasa.gov/bitstream/handle/2014/15891/00-1678.pdf?sequence=1.
2. NASA, "ASTER's Global Digital Elevation Model (GDEM)," June 29, 2009, www.jpl.nasa.gov/spaceimages/details.php?id=PIA12090.
3. Jessica Hagy, "Sun Tzu's *The Art of War, Illustrated* (Chapter 10: Terrain)," November 5, 2013, www.forbes.com/sites/jessicahagy/2013/11/05/sun-tzus-the-art-of-war-illustrated-chapter-10-terrain/#e52f25d55332.
4. Stephen P. Halbrook, "Target Switzerland," n.d., www.davekopel.org/2A/OthWr/Target_Switzerland.htm.
5. Mark Cartwright, "Thermopylae," in *Ancient History Encyclopedia*, April 16, 2013, www.ancient.eu/thermopylae/.
6. Robert McCrum, "Agincourt Was a Battle like No Other . . . but How Do the French Remember It?" *The Guardian*, September 26, 2015, www.theguardian.com/world/2015/sep/26/agincourt-600th-anniversary-how-french-remember-it.
7. "Intelligence Successes and Failures in Operations Desert Shield/Storm," report to Committee on Armed Services, US House of Representatives, August 16, 1993, www.dtic.mil/dtic/tr/fulltext/u2/a338886.pdf.
8. "TERCOM, System and Symbol: Part One," October 15, 2015, https://devilofhistory.wordpress.com/2015/10/15/tercom-system-and-symbol-part-one/.
9. International Federation of Surveyors, "History of Cadastral Systems," n.d., www.fig.net/organisation/perm/hsm/history_of/cadastre.asp.
10. International Federation of Surveyors.
11. Genesis 13:10–11.
12. Numbers 13:17–20.
13. Jonathan Watts, "Third of Earth's Soil Is Acutely Degraded Due to Agriculture," *The Guardian*, September 12, 2017, www.theguardian.com/environment/2017/sep/12/third-of-earths-soil-acutely-degraded-due-to-agriculture-study.
14. Riley McDermid, "Engineer Who Signed Off on Millennium Tower Says Builders Needed One More Expert to Certify Soil," *San Francisco Business Times*, February 3, 2017, www.bizjournals.com/sanfrancisco/news/2017/02/03/engineer-millennium-tower.html.
15. NASA, "Landsat Science," n.d., https://landsat.gsfc.nasa.gov/about/history/.
16. Jason Samenow, "Harvey Is a 1,000-Year Flood Event Unprecedented in Scale," *Washington Post*, August 31, 2017, www.washingtonpost.com/news/capital-weather-gang/wp/2017/08/31/harvey-is-a-1000-year-flood-event-unprecedented-in-scale/?utm_term=.fa560f28cfd5.
17. Jonathan Erdman, "'We've Never Forecast This Much Before': Hurricane Harvey's Ominous Forecast and How Meteorologists Reacted a Year Ago," Weather Channel, August 23, 2018, https://weather.com/storms/hurricane/news/2018-08-22-hurricane-harvey-diary-meteorologists-reactions.
18. A. J. Grant, *Herodotus*, vol. 1 (New York: Charles Scribner's Sons, 1897), 131.
19. Hydro International, "Charting the Pacific War," n.d., www.hydro-international.com/content/article/world-war-ii-charting-the-pacific-war?output=pdf.
20. Hydro International.

Navigation

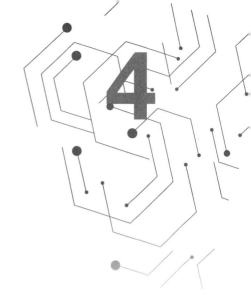

Navigation probably was the earliest known application of geospatial intelligence. We humans almost daily have always wanted the answer to questions such as, Where am I? Where do I want to be? What path do I take to get there? How long will it take? What hazards exist along the way? Those questions were asked at least a few millennia ago by traders moving goods along the Silk Road as well as military generals facing combat. We still ask the questions, but now we have a display on our mobile phone that can answer them—and "hazards" may refer to traffic congestion.

Cartography provided two dimensions of a map; terrain measurements provided the third dimension. But to find out where you are on a map, or how long it will take to get to where you want to be, you need to work within the fourth dimension: time.

Throughout history, knowing time has been important in navigation and geolocation. In early times, if you were on land and in possession of a detailed map or had knowledge of the local terrain, you usually could figure out where you were without referring to time. But without a map, or when traveling on the ocean with no landmarks, the job was more difficult. Most early mariners around the Eurasian coasts consequently preferred to hug the coast, staying in sight of land and reading the wind patterns.

The earliest and simplest form of navigation was (and still is) called dead reckoning. You could calculate your present location by knowing where you had been, the direction you had been going, and how fast you had been moving since your last known location. And from that point, determining how long it would take to get to a desired future location was easy. Not very accurate, perhaps. But we still tend to use the method to make a quick mental estimate of how long it will take us to get somewhere—by ship, airplane, or automobile.

Ocean travelers, however, early on found that the positions and motions of celestial bodies could help them navigate—and, if they knew what time it was, navigate accurately.

Celestial Navigation

As they gained seafaring experience, mariners began to rely more heavily on the sun or stars for open ocean navigation. The invention of the compass allowed them to sail directly from one port to another, but ascertaining progress along the way required some way of finding latitude and longitude. Both depended on determining the time.

Latitude, it turned out, was easier to figure out. During daytime, you measured the angular elevation of the sun near local noontime. When the sun stopped rising and started to go down, you were at local noon and the maximum elevation reading could be used, with a navigator's almanac, to determine latitude. At night, much the same method worked, using the moon or a navigational star (i.e., one listed in the almanac). The first instrument used to measure a celestial body's elevation with some degree of accuracy was the mariner's astrolabe, which was in use by the end of the fifteenth century. In about 1730, the marine sextant (another instrument for measuring the elevation of the sun, moon, or stars above the horizon) was invented, and quickly replaced the astrolabe because of its better accuracy.[1]

The issue lay in determining longitude. That required measuring the elevation of the sun, moon, or a navigational star when it was positioned to the east or west of your position. And you had to know the time precisely. Of course, there was just one problem: for every 4 seconds you were off in measuring time, the measured location would be wrong by up to 1 nautical mile at the equator. That doesn't sound terribly bad. Except that timepieces of the day had accuracies measured in minutes, not seconds; that meant you could be mistaken by 100 miles or more.

The problem got the attention of the British Parliament in 1707. On October 22 of that year, the Royal Navy fleet encountered a rock outcropping during severe weather off the Isles of Scilly. Four vessels sank on the rocks, and 1,550 sailors lost their lives. Today, it continues to rank as one of the worst maritime disasters in British naval history. The tragedy has been attributed to the navigators' inability to accurately calculate their longitude.[2]

After the Scilly incident, the British Parliament passed the 1714 Longitude Act. The act offered financial rewards of up to £20,000 (equivalent to over £2.5 million today) for a method of determining longitude that was accurate to half a degree (30 miles at the equator). Leading scientists and astronomers of the time took on the challenge—but they concentrated on finding an approach to determining longitude without knowing the exact time, using various tricks of astronomical observation.

An English carpenter turned clockmaker took a different approach. John Harrison decided to build a clock that would keep accurate time at sea: a device he called the marine chronometer. Twenty-one years after the passage of the act, Harrison's first watch, the H1, was completed in 1735. It performed well on a sea trial, and he began developing improved versions—first the H2, then the H3, and, finally, his successful H4 "sea watch," which took him six years of work. The Board of Longitude directed a sea trial from Portsmouth, England, to Kingston, Jamaica. The H4 was tested on a voyage by the HMS *Deptford* in 1761. On arrival in Kingston, the watch was found to be 5 seconds slow after correcting for its known time loss—an error of about 1 nautical mile.

The board, claiming that the outcome might have been just luck, ordered another sea trial. Meanwhile, Harrison had been working on another improved sea watch, the H5. The trial was again a success, giving a longitude error of less than 10 nautical miles. However, the board again declared that the result could have been just luck.

At this point, Harrison had had enough. He appealed to King George III for a hearing, and the king decided to test the newest watch, H5, for himself at the palace. Over a ten-week period in 1772, the watch was determined to be accurate to a third of a second per day—a remarkable level of accuracy for that period. The king, impressed, told Harrison to petition Parliament for the full prize. Harrison did, and in 1773, three years before his death, he received an award of £8,750 from Parliament.[3]

While Harrison never received the official award of £20,000, he wasn't treated quite as badly as it may seem; he had received a total of over £23,000 for his work in progress payments from the board and the final award from Parliament. He died a wealthy man.

The Harrison chronometers were slow to be adopted at first, because they were difficult to make and accordingly expensive. The initial versions cost almost one-third of the cost of the ship that carried them. But they represented a significant step forward, not only in marine navigation but also in mapmaking. During Captain James Cook's voyages (1768–79), he made three expeditions and created more charts of the Pacific Ocean than any previous explorers. More to the point, his charts were the most accurate yet produced because he had two important instruments: the recently developed sextant and Harrison's H4 chronometer.[4]

By the early 1800s, the chronometer's cost had dropped sufficiently that most ship owners could afford it; in fact, ocean navigation without one was considered to be foolhardy and dangerous. It became the standard tool for celestial navigation at sea until the arrival of radio navigation in the 1900s.

Radio Navigation

Once radios were developed, the idea of using them for navigation followed quickly. Radio waves were known to travel outward in a straight line from their transmitter. So if you could determine the directional bearing to a known transmitter, it would be possible to draw a line of position (LOP) on a map; you had to be somewhere on that line. If you could then get a second bearing to another transmitter that was located in a different direction, the intersection of the two LOPs determined your location.

The Telefunken Kompass Sender

The first radio navigation system to be put into use depended on accurate timing. Developed in 1908 by the German electronics firm Telefunken, it was called the Kompass Sender—an appropriate name for its method of operation.

A Kompass Sender station sent out a narrow radio beam that steered continuously through 360 degrees. The station transmitted its identifier in Morse code, then began to electronically sweep its beam around, starting from true north. An aircraft or ship could listen for the identifier, then start a stopwatch and listen to the signal strength. When the signal reached its maximum, the radio beam was pointing at the receiver, and using the elapsed time, a navigator could determine the direction to the transmitter. Navigators quickly found that the minimum signal was easier to identify, and they used it instead to get the station bearing.

To geolocate yourself, however, you needed two stations to get two lines of position. The intersection of those two lines on a map marked your location, within some margin of error. Telefunken built two Kompass Sender stations: one at Kleve near the Dutch border, and one at a German Zeppelin base in what is now part of Denmark. The two stations provided coverage of the North Sea and the English Channel. They were used primarily for long-distance navigation by Zeppelin airships (discussed in chapter 8) before and during World War I.[5]

The Radio Direction Finder

A more accurate way of finding location was to have the receiver on a ship or aircraft determine the direction to a transmitter without the need for the ground station to transmit a special signal. Then any station—a commercial radio station or a beacon set up for the purpose—could be used. This required that the ship or aircraft have a directional antenna, usually in the form of a loop, and that it be turned until the strongest (or weakest) signal was received. Soon after World War I, antennae were developed that could do that; and during the 1930s, ships and aircraft were fitted with radio direction finders (RDFs). Low-power beacons were set up at airports and harbors to provide the needed signal.

RDF is the exact inverse of the triangulation method used in surveying and mapmaking. In triangulation, you knew where you were, and you determined the location of a distant point using the bearings to it from two known points.

FIGURE 4.1 Amelia Earhart's Lockheed Electra. *Source:* US Air Force.

In RDF, you didn't know where you were, and you used the bearings to two known points to find your location.

The best-known RDF system of its time was that carried by Amelia Earhart in her ill-fated attempt to become the first woman to circumnavigate the globe. Figure 4.1 shows the Lockheed Electra that was used for the flight. The loop antenna used by the RDF system is visible on top of the aircraft. Earhart departed Miami on June 1, 1937, and after stops—in South America, Africa, India, Southeast Asia, and Australia—arrived at Lae, New Guinea, on June 29. She departed Lae on July 2 for Howland Island, in the Pacific. This was the longest leg of the flight, and near the maximum range of the Electra.

What happened next has been the subject of controversy and conspiracy theories ever since, but Earhart and her navigator, Fred Noonan, never reached the island. The US Coast Guard Cutter *Itasca* was anchored at Howland, ready to assist Earhart's approach. The *Itasca* was able to receive Earhart's communications, but apparently Earhart did not receive *Itasca*'s responses. For some reason, the RDF system failed to guide the Electra to Howland. One theory is that the frequency of the radio beacon (7,500 kilohertz) was too high to allow the RDF system to get a bearing; another is that Earhart did not understand how to use the new system. Repeated searches by the Coast Guard and US Navy were unable to locate the aircraft's wreckage.[6] Many theories (including one where she survived the crash) have been proposed since then about what happened to Earhart and Noonan. A detailed 2018 forensic study presented a convincing argument that bones found in 1940 on the Pacific island of Nikumaroro belong to Earhart.[7]

While other radio navigation systems were developed and used during the twentieth century, most were abandoned as satellite navigation came into use. However, one system (that did not use RDF) remained in use until well into the twenty-first century: Loran-C.

Loran-C

Introduced in 1957, Loran-C (LORAN is an acronym for long-range navigation) had several advantages. It required that the user (on a ship or aircraft) receive a low-frequency radio signal from at least two land-based transmitters. Based on the timing of pulses received, a receiver could establish its location. The signal could be received at very long ranges, and it gave accurate locations—less than a quarter of a mile. But the receivers originally were complex and expensive, so the system was mostly used by military forces at first. Later, solid state receivers became available at reasonable cost, and began to be used by civilian aircraft and ships. In 2009 Loran-C was declared obsolete by the US president, and most stations worldwide were shut down by 2015; the **Global Positioning System** (GPS) had become the gold standard navigation method. But GPS signals can be jammed, and North Korea reportedly has accomplished that on several occasions. An updated version of Loran is being planned for areas (e.g., the Korean Peninsula) where GPS jamming is a threat.[8]

Satellite Navigation

For ships at sea and distant from shore stations, location accuracy wasn't absolutely critical because hazards to navigation weren't common in mid-ocean. However, when ballistic missile submarines started patrols, they did need fairly precise location information—in case a missile launch became necessary.

In the early 1960s, precise location at sea became possible. The US Navy recognized that satellites could provide the needed position accuracies for their ballistic missile subs. It accordingly funded the development of the first navigation satellite system, called Transit.

Transit

Transit, also known as NAVSAT, was the first satellite navigation system to become operational. It owed its design to a feature of the first satellite ever launched—the Soviet Sputnik 1. Launched in October 1957, Sputnik 1 transmitted radio beacon signals at 20 and 40 megahertz. Scientific observers exploited the Doppler effect of the radio signal as the satellite passed by to obtain a rough position fix.[9] The US Navy quickly realized that the Doppler effect could be used to give precise locations.

Originally developed to provide accurate location information to Polaris ballistic missile submarines, Transit later was used by both naval and commercial ships and for hydrographic surveys. It went into operation in 1964 and continued in use until 1996. While in operation, the system normally used five active satellites in polar orbits at about 600 nautical miles altitude to provide global coverage; five satellites, properly spaced, meant that at least one satellite would be visible at any time.[10]

Each satellite broadcast two signals, on 150 and 400 megahertz. The broadcasts contained information about the satellite's orbit and the precise time. Because the satellite's position was predictable, the Doppler shift could be used to locate a receiver. A satellite would be above the horizon for about 15 minutes. During that time, a computer-equipped receiver could measure the changes in Doppler shift as the satellite moved, and from that shift it would obtain a position fix. For US ballistic missile submarines, the shift was measured for only 2 minutes because that was the amount of time the receiver antenna was allowed to be above the sea surface to protect the submarine from being detected. The fix had an accuracy of about 100 meters, but US submarines had a special encrypted mode that allowed accuracies of less than 20 meters.[11]

Satellite Navigation Systems

While Transit was an improvement over other navigation systems, the Doppler shift method of measurement had a definite accuracy limit; so the search began for a more accurate satellite-based system that would provide worldwide coverage. A system that could precisely measure the distance to multiple satellites appeared capable of high accuracy. But a globally available replacement system that could do that depended on the development of two things.

First, engineers needed a computer that could be placed in a small receiver package for submarine use. Because satellites are in orbit and normally move with respect to the receiver, the computer had to be able to calculate the position of each satellite to within a few meters. And it had to be a manageable size because there wasn't much room to spare on a Polaris submarine. By the late 1960s, minicomputers were generally available, and their size was shrinking each year.

Second, engineers needed an accurate time standard. The accuracy of satellite navigation depends on the accuracy of the clock used. In Transit, time originally was measured using a quartz crystal oscillator in a temperature-controlled oven located at the ground station. More accurate clocks were needed for the new generation of navigation satellites. Atomic clocks could provide the needed accuracy; but for the new generation that was envisioned, the clocks had to be placed onboard the satellite.

This problem was a bit thornier. The most accurate atomic clocks rely on the cesium atom; these clocks were invented at the UK's National Physical Laboratory in 1955. However, cesium clocks originally were bulky, needed cooling, and consumed lots of power; though they were used to maintain a standard time at most national laboratories worldwide, they weren't suitable for satellite use. Clocks using rubidium for timing are compact and less expensive. So they are widely used to control the frequency of television stations, cell phone base stations, and test equipment. And by the 1970s, rubidium clocks had shrunk enough in size and power requirements to be placed on a satellite.[12]

Two advancements—miniature computers and highly accurate atomic clocks that could be placed on satellites and updated as needed—allowed the development of what became known as satellite navigation, or SATNAV, systems. These comprised a constellation of satellites that could provide autonomous geospatial positioning. Receivers with computers and the proper software could determine their location (longitude, latitude, and altitude/elevation) to high precision (within a few meters), using the signals transmitted by multiple satellites. The first such system was the GPS.

The Global Positioning System

As happens with many revolutionary technological advances, the GPS developers had no idea that they were starting a revolution. They simply intended to create a system for military personnel and mobile units (ships, submarines, and aircraft) to use for precise geolocation. It was intended to be just a more accurate military navigational aid. The US Defense Department had been studying the concept of a GPS since the 1960s, but—astonishing as it may seem today—almost no one thought it would be useful. The project struggled along under a series of Air Force managers during the 1960s, lacking funding or what the military often refers to as "top cover"—that is, high-level support.

That changed in late 1972, when Colonel Brad Parkinson took charge of the program. Parkinson saw the potential where others had not. He became a zealot for the concept, and as he later observed, "Zealots are not easily discouraged." He pushed for support both within and outside the military, and in 1973, he got it. Late that year, Congress appropriated funding to begin development. Parkinson pushed ahead and also upped the ante: he asserted that not just the military, but civilians too, should have access.[13] That argument continued well after the first GPS launch in 1978. And then, five years later, a tragic incident settled the argument.

On September 1, 1983, Korean Air Lines Flight 007, a Boeing 747, was en route from New York to Seoul. A navigational error caused it to stray into the USSR's prohibited airspace. The Soviets sent interceptors aloft and shot it down, killing all 269 people aboard. President Ronald Reagan subsequently issued a directive making the GPS freely available for civilian use once it was sufficiently developed.

That wasn't the end of the story, however. Out of an abundance of caution, the military restricted civilian users to receiving a lower level of accuracy in location data to prevent GPS use by adversaries for targeting purposes. It took almost two decades, but in 2000, President Bill Clinton changed that policy to allow civilian users to have the same GPS accuracy that the military enjoyed.[14]

Twenty-four GPS satellites are in medium Earth orbit, at an altitude of 12,000 miles. At that altitude, each satellite completes two orbits in less than 24 hours. The satellites rely primarily on solar energy, with battery backup for times when a satellite is in the Earth's shadow. Each satellite is built to last about ten years, and replacements are regularly built and launched into orbit.

Today, at least four GPS satellites are always visible to an observer anywhere on Earth. Each satellite regularly transmits its location and the current time. When these signals arrive at a GPS receiver, the receiver compares the arrival time with the time when the satellite transmitted the signal. The difference in arrival time tells how far the receiver is from the satellite—giving a line of position on the Earth's surface. The signal from a second satellite gives a second LOP, which intersects the first LOP at two points on the Earth's surface. A third satellite's signal determines which point is the receiver's position on the surface. A fourth satellite's signal is used to determine position vertically—the altitude of an aircraft, for example.[15]

A GPS receiver calculates position, but it can do much more. It can provide speed for moving objects, and navigation or guidance information. It allows tracking the position of an object fitted with a receiver (satellite tracking). It calculates the current local time to high precision. GPS time is theoretically accurate to about 14 nanoseconds, enabling a geographically dispersed network of stations to synchronize their timing. The US electrical power grid is synchronized and kept running by using GPS clocks. GPS-based timing is essential to the functioning of worldwide seismic monitoring networks. Traders in the stock market rely on it to avoid damaging losses in timing purchases and sales.

Global Navigation Satellites

The United States' GPS success led other countries to develop similar systems, as table 4.1 indicates. Russia developed its GLONASS. The European Union developed Galileo, which can work cooperatively with the US system to provide better position accuracy. Several countries are developing systems that provide regional rather than global coverage—China, India, France, and Japan, for example. China is expanding its regional BeiDou Navigation Satellite System into the global BeiDou-2. All the satellites in these systems operate at medium Earth orbit, about 10,000 miles in altitude.[16]

TABLE 4.1 Global Navigation Satellite Position Accuracy

BeiDou	Galileo	GLONASS	GPS
2.5m	1m	4.5m–7m	< 3.5m

Countering Global Navigation Satellites

Global navigation satellites now are essential in both military and civil applications, and are also used by terrorist groups and criminal organizations. So it's no surprise that countermeasures against them have come into being.

On January 12, 2016, two US Navy patrol boats were seized by Iran's Islamic Revolutionary Guard Corps Navy after they entered Iranian territorial waters near Iran's Farsi Island in the Persian Gulf. It was later determined that the two boats entered Iranian waters because of navigational errors. Some observers speculated at the time that Iran had used the deception of GPS signals to lure the boats off course. The hypothesis was based in part on a precedent that had been set five years earlier. In 2011 Iran had claimed that it had captured a US RQ-170 Sentinel drone by deceiving its GPS receiver to lure it from Afghanistan into Iranian airspace.

Both the hypothesis and the Iranian claims are open to doubt. The US military uses a different set of codes than do civilian GPS systems to receive broadcasts from GPS satellites. The military codes are encrypted and therefore difficult to deceive. But even encrypted GPS signals can be jammed and result in navigational errors—which could have happened with both the US drone and the patrol boat incident.

Because nonmilitary global navigation satellite systems are not encrypted, it is possible to deceive a commercial receiver about its location. In June 2013 a Cornell University researcher demonstrated the ability to deceive the GPS navigation system on a private yacht in the Mediterranean, luring the ship 1 kilometer off course.[17]

Russia in particular has a history of jamming navigation satellite signals, with the goal of protecting VIP visits and strategic installations from attack. Russian activity since at least 2016 has included more than simple denial-of-service jamming; it has included transmitting false position information to ships and drones (spoofing). Russian spoofing has been especially prevalent in locations such as Syria, Kaliningrad, and the Crimea. A 2019 report on the Russian activity noted that "the tools and methodologies for conducting similar activities are readily available to nonstate actors."[18]

Chapter Summary

All navigation begins by answering one question: Where are we? The answer, whether determined by celestial observation or radio-based positioning, has long depended on accurate time measurements. It still does. That question today can be easily and definitively answered, thanks to global navigation satellites. They are now an essential part of knowing where we are, how long it will take to get someplace, the exact time, and much more. Navigation satellites are the driving force behind many aspects of GEOINT, for example, by providing the timing that is essential for geolocating radiofrequency emitters (described in chapter 7).

Notes

1. Andrew K. Johnston, Roger D. Connor, Carlene E. Stephens, and Paul E. Cerruzzi, *Time and Navigation* (Washington, DC: Smithsonian Books, 2015), 25.
2. Johnston et al., 32.
3. Johnston et al., 30–31.
4. Beau Riffenburgh, *Mapping the World: The Story of Cartography* (London: Carlton Books, 2014), 76–77.
5. Rafael Saraiva Campos and Lisandro Lovisolo, *RF Positioning* (London: Artech House, 2015), 7.
6. Dorothy Cochrane, "Amelia Earhart: Missing for 80 Years but Not Forgotten," Smithsonian Air and Space Museum, June 29, 2017, https://airandspace.si.edu/stories/editorial/amelia-earhart-missing-80-years-not-forgotten.
7. Richard L. Jantz, "Amelia Earhart and the Nikumaroro Bones: A 1941 Analysis versus Modern Quantitative Techniques," *Forensic Anthropology* 1, no. 2 (2018): 83–98.
8. Sean Gallagher, "Radio Navigation Set to Make Global Return as GPS Backup," *Ars Technica*, August 7, 2017, https://arstechnica.com/gadgets/2017/08/radio-navigation-set-to-make-global-return-as-gps-backup-because-cyber/.
9. The Doppler effect, or Doppler shift, occurs when a source of sound or radiofrequency energy is moving with respect to an observer. The observer senses a higher frequency as the source approaches, and a lower frequency as the source recedes. It is most commonly observed in listening to a passing automobile horn or siren.
10. Johnston et al., *Time*, 142–47.
11. Johnston et al.
12. Norman F. Ramsey, "History of Early Atomic Clocks," *Metrologia* 42 (2005): S1–S3, http://geodesy.unr.edu/hanspeterplag/library/geodesy/time/met5_3_S01.pdf.
13. Bjorn Carey, "Stanford Engineer Bradford Parkinson, the 'Father of GPS,' Wins Prestigious Marconi Prize," *Stanford News*, May 16, 2016, https://news.stanford.edu/2016/05/16/stanford-engineer-bradford-parkinson-father-gps-wins-prestigious-marconi-prize/.
14. Johnston et al., *Time*, 156–64.
15. Johnston et al.
16. "Other Global Navigation Satellite Systems (GNSS)," GPS.Gov, www.gps.gov/systems/gnss/.
17. Mark L. Psiaki and Todd E. Humphreys, "GPS Lies," *IEEE Spectrum*, August 2016, 26.
18. C4ADS, "Above Us Only Stars," March 26, 2019, https://static1.squarespace.com/static/566ef8b4d8af107232d5358a/t/5c99488beb39314c45e782da/1553549492554/Above+Us+Only+Stars.pdf.

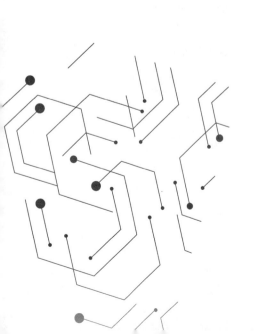

Geopolitics

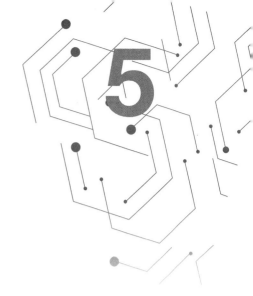

Geopolitics may not immediately come to mind as a part of geospatial intelligence. But it fits the definition introduced at the end of chapter 1. It

- is anticipatory,
- involves some type of human activity,
- draws on knowledge of the Earth, and
- provides information to support decision-making.

From antiquity, countries have had to be concerned about their geographical location in relation to possible opponents. Geography can confer advantages—consider the history of island nations England and Japan—or disadvantages; for example, Poland, sandwiched between traditional opponents. And the physical shape of a country's boundaries can either help to unify a nation or present problems. Each type of shape—compact, elongated, or geographically separated—has benefits, along with drawbacks.

Geopolitics is the study of the advantages and disadvantages offered by geography in international relations. It allows researchers to understand, explain, and predict likely international political behavior by examining geographical variables. It is geospatial intelligence from a strategic perspective. Those who are in the field develop theories of international politics in terms of geography—that is, the location, size, and resources of a country. They describe relationships between geographic space, resources, and foreign policy.

Several geopolitical theories were developed in the nineteenth and twentieth centuries. They arguably shaped the most significant events of the twentieth century, as they were used to justify imperialism and wars of aggression. The two most influential theories were proposed by Alfred Thayer Mahan and Halford John Mackinder, who are widely regarded as founding fathers of geopolitics and geostrategic thinking.

Mahan's Sea Power Theory

Alfred Thayer Mahan served in the US Navy during the Civil War, and afterward, he became a lecturer in naval history and tactics at the Naval War College. His studies on the influence of sea power led him to conclude that national power was tied to a country's ability to control the sea, and thereby to control international commerce. His most influential work on the subject was *The Influence of Sea Power upon History: 1660–1783*.[1]

Mahan's book discusses the factors that lead to supremacy of the seas, using the example of Great Britain's rise to a dominant position in the seventeenth and eighteenth centuries. It examines the importance of geography in the global balance of power and the role of sea power in national security policy. Dominance of the sea, in his view, meant control of commerce, leading to control of the land areas around the sea. An army could occupy territory for a while, but would eventually be starved for supplies and defeated by a naval blockade.

Mahan's influence on geostrategic thinking was profound. Kaiser Wilhelm II made Mahan's book required reading for his naval officers, and that led to Germany's construction of a navy capable of threatening British sea supremacy—provoking a naval arms race in Europe.

Mahan's book also was translated into Japanese and became a textbook for the Imperial Japanese Navy. It shaped Japan's thinking about how to contain Russian naval expansion in the Far East, resulting in the Russo-Japanese War of 1904–5, which ended favorably for Japan.[2]

Mackinder's Heartland Theory

Halford John Mackinder was an English geographer who rose to become the director of the London School of Economics and Political Science and later became a politician and a member of Parliament. Mackinder was a dominant influence in the establishment of geography as a major academic discipline in the United Kingdom. But his fame today rests on his development of the heartland theory of geopolitics, first proposed in a 1904 article titled "The Geographical Pivot of History."

He divided the globe into two parts. One part he described as the "World Island," comprising Europe, Asia, and Africa. The remaining continents were an afterthought, in his view; he called them the "outlying islands." Much of Eurasia formed the "heartland," which, he argued, contained the resources and internal communications needed to control the rest of the world, in part due to economic and industrial development of southern Siberia. Figure 5.1 shows

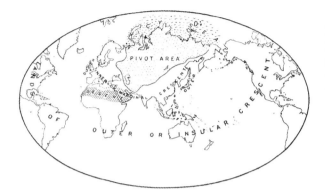

FIGURE 5.1 Halford Mackinder's 1904 map of Heartland Theory—"The Geographical Pivot of History," *Geographical Journal* 23, no. 4 (April 1904): 435.

his concept of the heartland, which he describes as the "pivot area." His theory is summed up in three lines:

> Who rules East Europe commands the Heartland;
> Who rules the Heartland commands the World Island;
> Who rules the World Island commands the world.[3]

Mackinder's theory initially received little notice outside academic circles. But it, along with Mahan's theory, would later shape a German variant of geopolitics.

German Geopolitik

Mahan's theory caught the attention of a German geographer and ethnographer named Friedrich Ratzel. A student of zoology and admirer of Charles Darwin, Ratzel developed a biological view of geography. In his concept, a state is a living organism that must grow or decline. Healthy states grow by expanding their borders; sick ones decline and shrink. Building on Mahan's theory, Ratzel argued that sea power could sustain itself. The profits from international trade would pay the cost of a powerful navy that could protect a country's merchant marine. Among his many published papers was a 1901 essay called "Lebensraum"—a word that would later become associated with Nazi Germany.[4]

Ratzel was a close friend of an economics professor named Max Haushofer and of Haushofer's son, Karl. As a result, Ratzel shaped the thinking of a man who was to become the dominant figure in the German school of *Geopolitik*. Karl Haushofer, a general in World War I, believed that Germany had lost the war due to a flawed view of political geography. He enthusiastically accepted both Mahan's and Mackinder's theories as part of his *Geopolitik*. And because his ideas fit well with the inclinations of Adolf Hitler, his theory later became a part of the Nazi regime's strategy for seizing control of the heartland—in its interpretation, the USSR and Eastern Europe.[5]

Spykman's Rimland Theory

Nearly forty years after Mackinder's article, a professor of international relations at Yale University agreed with Mackinder on two points, but had a different perspective on the critical geographical regions. The political scientist Nicholas Spykman concurred with Mackinder that geography was the primary factor to consider in developing foreign policy. He wrote in 1944 that geography was "the most fundamentally conditioning factor because of its relative permanence."[6] And, like Mackinder, he believed that the domination of the world island meant eventual domination of the world. But that is where they parted ways.

Spykman asserted that Eurasia's rimland, shown in figure 5.1 as the "inner or marginal crescent," was more important than Mackinder's heartland. The rimland surrounds the heartland. Whoever could control the rimland would in time control the heartland, and the world island as well. In place of Mackinder's three points, he developed his own set of two:

> Who controls the rimland rules Eurasia;
> Who rules Eurasia controls the destinies of the world.[7]

After World War II, Spykman's theory gained attention among diplomats in Washington and Western Europe. The Soviet Union had control of Mackinder's heartland. The West needed a policy to prevent the USSR from extending its control to the surrounding rimland. The fear was that if they succeeded, the Soviets would control the heartland, the rimland, and the world island.

The result of that concern was the *containment* theory, first proposed by the US diplomat George Kennan in 1947. It relied on Spykman's revision to Mackinder's heartland theory. Containment became a successful geopolitical strategy pursued by the US and its NATO allies for over forty years, until the end of the Cold War made it no longer necessary. Spykman is sometimes referred to as the "godfather of containment."

The Continuing Influence of Geopolitical Theories

The groundbreaking theories of Mahan, Mackinder, and Spykman continue to be studied and, as in the twentieth century, are often used to justify expansionist policies today. China and India, strengthened by their economic development, appear to be pursuing geopolitical strategies.

China's national security long has been focused on internal and land-based threats. Geographically speaking, China lives in a tough neighborhood. But although threats continue to shape its strategic thinking, China's focus has shifted toward control of natural resources and becoming a powerful player on the world stage.

Chapter 2 introduced the original Silk Road—a land and maritime trade route between China and regions to its west. In 2013, Chinese president Xi Jinping proposed an updated version that seems to draw on Spykman's rimland theory.[8] Originally it was called "One Belt, One Road," but in 2015 the Chinese began describing it variously as the "Silk Road Economic Belt" and the "21st-Century Maritime Silk Road," before settling on the "Belt and Road Initiative." The proposed road(s) follow both the overland and maritime routes of the original Silk Road, and connect the countries of the rimland. The Chinese government has described the project as "a bid to enhance regional connectivity and embrace a brighter future."[9] Others view it as a move to strengthen China's position in Asian, African, and European affairs through a trade network.

But the country has many strengths, and its economy depends on trade, causing Chinese leadership to also pursue Mahan's dictum of the need for a strong navy.[10] Mahan's theory also likely is used to rationalize China's ongoing attempt to seize control of the Spratly Islands in the South China Sea.

India, like China, also has been historically concerned about internal and land-based threats—primarily from Pakistan and China. And, increasingly, India views the Indian Ocean as its preserve, embracing Mahanian concepts about naval power projection, control of sea routes, and access to bases.[11]

Thematic Cartography

If asked, most people would say that maps are intended as a reference. Atlases, topographic maps, and navigation charts typically present maps that are tailored to the needs of the user who has a straightforward question about a location. But during the nineteenth century, maps also found another purpose: to convey a message to a broad audience. Geopolitical thinkers such as Mahan, Mackinder, Haushofer, and Spykman relied on maps to present their arguments because maps provided a medium for conveying complex ideas in a way that the general population could easily understand. These products have a name: **thematic cartography**.

Thematic cartography emphasizes spatial attributes to present a visual theme, explain a concept, or send a political message. Its goal is to capture the

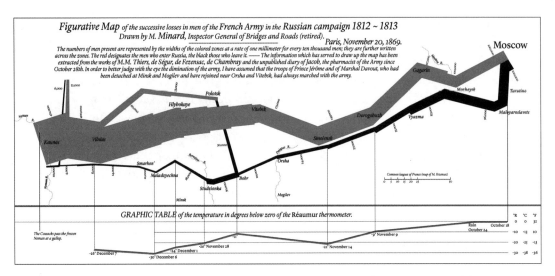

FIGURE 5.2 A modern redrawing of Charles Joseph Minard's figurative map of the 1812 French invasion of Russia, including a table of temperatures translated from Réaumur to Fahrenheit and Celsius. *Source:* DkEgy, Wikimedia Commons.

viewer's attention. One early use was to present statistical information in an elegant and compelling fashion.[12]

One of the most compelling examples of the genre was created a century and a half ago, and since then it has repeatedly been used to illustrate the power of brilliant graphical design. In 1869 Charles Minard created a map graphically showing the history of Napoleon's 1812 Russian campaign. Figure 5.2 presents several types of data in geospatial form, but the most dramatic is the line showing Napoleon's advance toward Moscow, with nearly 500,000 troops (red shaded lines), followed by his retreat from Moscow (black shaded lines), until he reached his starting point with some 10,000 surviving troops. The line thickness corresponds to the number of troops Napoleon commanded as he first advanced, then retreated after the Battle of Borodino. Minard's theme was the terrible human cost of Napoleon's venture.

Thematic cartography became the favored way to present geopolitical ideas or make dry numbers more interesting. Making the abstract and unfamiliar concrete was an application that took hold during World War II. The war provided an incentive for people to understand what was happening in distant parts of the globe. Maps suddenly became of interest to everyone; they gave meaning to news reporting. They could visually explain why Great Britain was sending troops and supplies to Egypt or why the US was invading a small island in the south Pacific called Guadalcanal. At the same time, a novel approach to mapmaking began to emerge, drawing on the principles of thematic cartography. Maps that were more than simple providers of geospatial information became popular.

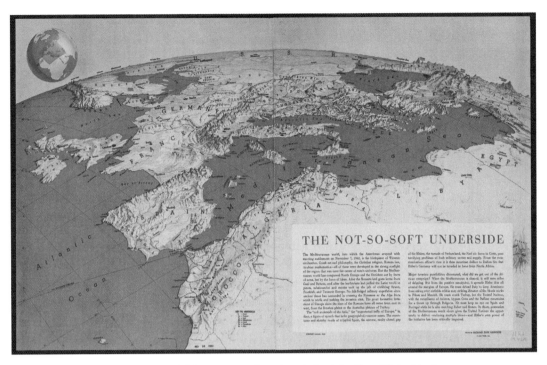

FIGURE 5.3 Richard Edes Harrison's "the not-so-soft underside," originally published in *Fortune* in January 1943. *Source:* Yale University, Beinecke Rare Book and Manuscript Library. Used by permission of the Estate of Richard Edes Harrison.

The pioneer at the forefront was not a cartographer; he was an artist. Richard Edes Harrison created maps that didn't follow the conventions of cartography and often relied on no known type of projection. He employed techniques (beyond those of Minard) that are part of an artist's tool kit: color, perspective, and lighting (or shading). His maps carried visual appeal. They emphasized geospatial relationships that revealed what was at stake in the various campaigns of the war. They were, in fact, excellent examples of thematic cartography.[13]

An example of Harrison's skill is shown in figure 5.3. During the Casablanca Conference in January 1943, Churchill and Roosevelt agreed to invade Europe across the Mediterranean Sea, after they had driven the Germans and Italians out of North Africa. Churchill believed that an attack through the "soft underbelly" of Europe would be more effective than a direct assault on northwestern Europe. Harrison thought differently, and he created this map and published it in *Fortune* in 1943. It sends a powerful message: the "underbelly" is anything but soft; the terrain would be rough for an invading army. The Allies apparently weren't convinced, and invaded Sicily and subsequently mainland Italy anyway. It proved to be as tough a campaign as Harrison's map suggested it would be.

Throughout the war, Harrison produced a series of maps that were published in *Fortune* magazine. Later, in 1944, they were published in an atlas that became an immediate bestseller.[14] Harrison's innovative maps led to a new era

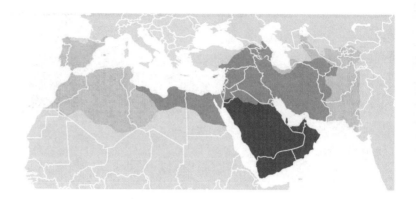

FIGURE 5.4 Territory controlled by the caliphates, 622–750. *Source:* DieBuche, Wikimedia Commons.

of cartography—one in which maps would be engaging enough to shape public opinion and change how we view geography. A Mercator projection leads the viewer to think of the world in East-West terms. A gnomonic projection centered on the North Pole emphasizes the closeness of North America and North Eurasia. Harrison's maps showed the contrasting views that different world leaders might take. His 1944 map "Europe from the East," for example, portrayed the world as it might be viewed from the USSR: as a region ripe for the taking.[15]

Since Harrison's time, maps have been used during many critical times, to send messages, establish territorial claims, and mobilize public opinion. During the Cold War, all parties used geospatial representations to convey the impact of the threats they faced.

Most recently, radical Islamist groups have employed maps to remind Muslims that in the time of the caliphates, Islam dominated a large swath of Asia, Europe, and Africa. Maps such as figure 5.4 transmit two messages to recruits and followers alike: a sense of loss over what Islam once had, and the idea that "we did this once in little more than a century; we can do it again." Other maps remind all Muslims that under the Ottoman Caliphate, they also controlled Turkey and the Balkans. Such maps are widely shared in social media to repeatedly suggest to Muslims that "this should be our land."

Since Minard's time, thematic cartography has developed into a valuable way for intelligence organizations to provide geospatial intelligence—especially statistical information—in a form that intelligence customers can readily comprehend. The Central Intelligence Agency's map library for decades has provided many such maps to present the results of analysis.

Geopolitical Strategy

Geopolitical theories tend to stress the importance of factors that are contained or implied within geography. Mahan emphasized the economic power enabled by control of commerce via the sea. Mackinder focused on infrastructure and economic strengths (resources and transportation) of his "heartland." Extremist recruiting materials rely on social factors (ethnicity and religion) to unite followers within the realm of the historical homeland. These theories have the implicit use of force behind them, but in all cases there is something more.

The PMESII View

Geopolitics is the study of conflicts and shifts of power around the world as they relate to geography. But important factors that intersect with geography obviously must be considered. Military planners have identified six such factors or perspectives: political, military, economic, social, infrastructure, and information—abbreviated *PMESII*.[16] Here is a brief explanation of what each factor encompasses:

- **Political:** Describes the distribution of responsibility and power at all levels of governance—formally constituted authorities, as well as informal or covert political powers.
- **Military:** Explores the military and/or paramilitary capabilities or other ability to exercise force of all relevant actors (enemy, friendly, and neutral) in a given region or for a given issue.
- **Economic:** Deals with individual and group behaviors related to producing, distributing, and consuming resources.
- **Social:** Describes the cultural, religious, and ethnic makeup within an area and the beliefs, values, customs, and behaviors of society members.
- **Infrastructure:** Details the composition of the basic facilities, services, and installations needed for the functioning of a community, business enterprise, or society in an area.
- **Information:** Explains the nature, scope, characteristics, and effects of individuals, organizations, and systems that collect, process, disseminate, or act on information.

Four of these—political, military, economic, and information—are often considered instruments of national power. They are the "levers" that leaders can pull to influence world events.

At about the same time that the PMESII construct was developed within the US military, a similar geopolitical framework was developed in the Middle East.

Meta-Geopolitics

In 2009, the Saudi neuroscientist and geostrategist Nayef Al-Rodhan proposed a framework for statecraft he called *meta-geopolitics* that is similar to the PMESII

TABLE 5.1 Comparison of Factors of State Power

PMESII Construct	Meta-Geopolitics
Political	Domestic politics
	International diplomacy
Military	Military and security issues
Economic	Economics
Social	Social and health issues
	Science and human potential
Infrastructural	The environment
	Science and human potential
Informational	International diplomacy

view.[17] He argued that seven factors shape a state's power, as shown in table 5.1. The two formulations are close but not an exact match; Al-Rodhan's factors of "international diplomacy" and "science and human potential" each fall into two different PMESII factors.

Both frameworks emphasize the same conclusion: state power is multidimensional, and though geography is inextricably related, it is only a starting point for assessments. The instruments of national power include both hard- and soft-power aspects. In 2012, Al-Rodhan extended his framework to include outer space as a major geopolitical factor in state power.[18]

Whether you prefer the six PMESII factors or the seven meta-geopolitical factors, all are important in assessing national strength. In the view of the geopolitical thinker Samuel Huntington, the social factor is dominant.

The Clash of Civilizations

Samuel Huntington was an American political scientist who worked briefly for the US government and was employed for fifty years at Harvard University, many of them as director of the Center for International Affairs. In 1993 Huntington proposed a geopolitical theory that argues for the dominance of the social factor in a post–Cold War world order. He elaborated on that theory in his 1996 book, *The Clash of Civilizations and the Remaking of World Order*. Huntington's main thesis is that "the most important distinctions among peoples are not ideological, political, or economic. They are cultural."[19] In his view, future wars would be fought not between countries but between cultures (social factors). Figure 5.5 presents the fissures in cultures where he saw conflict as being most likely to develop. Finally, he asserted that following this reasoning, Islamic extremism would become the biggest threat to world peace.

Huntington proposed his theory just after the fall of the Berlin Wall and the dissolution of the Soviet Union, when superpower conflicts appeared to be a thing of the past. As with most bold theories, his was partially right on some

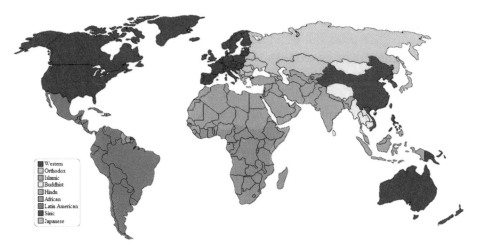

FIGURE 5.5 A depiction of Samuel Huntington's view of the different cultural blocs and fissures from *The Clash of Civilizations and the Remaking of World Order*. *Source:* Kyle Cronan and Olahus, Wikimedia Commons.

fronts. In fact, new superpower confrontations have gradually developed—Russia and China versus the West, for example. But, many of the conflicts have been *within* the separate regions outlined in the figure—Russia versus Ukraine, and Iran versus Saudi Arabia, for instance. And though Islamic extremism has not become the centerpiece of global clashes as he predicted it would, it has emerged over the past thirty years as a major source of conflicts worldwide.

During his career, Huntington drew several conclusions about international relations that provoked controversy, including,

- As societies modernize, they become more complex and disordered.
- Western values are not universal.
- Any attempt to create a multicultural state is doomed to failure—such states are inevitably riven with internal conflict.

Time will tell, but it's likely that parts of these theories will be borne out as valid. Certainly, he was correct that the social factor has been significant in shaping twenty-first-century geopolitics; however, the other factors in the PMESII and meta-geopolitics constructs continue to be relevant.

Chapter Summary

Geopolitical theories, aided by maps, convey an understanding of national and international power in a form that leaders can understand and act upon. They shaped the thinking behind every major war, hot and cold, of the twentieth century, and they continue to be proffered to influence the thinking of leaders throughout the world. The trend in such geopolitical frameworks is to view state (and nonstate) powers through a multidimensional lens. Geography is the starting point, but political, military, economic, social, infrastructure, and information factors have major roles to play.

During the nineteenth century, thematic cartography came to the fore as an effective means of political messaging for leaders and the general public alike. The wide availability of maps that contain geopolitical themes since then has encouraged the mobilization of popular opinion for either noble or less-than-honorable causes.

Notes

1. Alfred Thayer Mahan, *The Influence of Sea Power upon History: 1660–1783* (Boston: Little, Brown, 1890).
2. Toshi Yoshihara and James R. Holmes, "Japanese Maritime Thought: If Not Mahan, Who?" *Naval War College Review*, Summer 2006, https://digital-commons.usnwc.edu/cgi/viewcontent.cgi?referer=https://www.google.com/&httpsredir=1&article=1930&context=nwc-review.
3. H. J. Mackinder, *Democratic Ideals and Reality* (London: Henry Holt, 1919).
4. Woodruff D. Smith, "Friedrich Ratzel and the Origins of Lebensraum," *German Studies Review*, February 1980, 51–58, www.jstor.org/stable/1429483?seq=1#page_scan_tab_contents.
5. Colin Flint and Peter J. Taylor, *Political Geography*, 7th ed. (New York: Routledge, 2018), 2.
6. Nicholas J. Spykman, *The Geography of the Peace* (New York: Harcourt, Brace, 1944).
7. Nicholas J. Spykman, *America's Strategy in World Politics: The United States and the Balance of Power* (New York: Harcourt, Brace, 1942).
8. BAOLUO, "The 21st Century Strategic Pivot, The Rimland," *Geopolitics*, September 17, 2016, https://gpindex.org/2016/09/17/the-21st-century-strategic-pivot-the-rimland/.
9. Xinhua News Agency, "China Unveils Action Plan on Belt and Road Initiative," March 28, 2015, http://english.gov.cn/news/top_news/2015/03/28/content_281475079055789.htm.
10. Francis P. Sempa, "The Geopolitical Vision of Alfred Thayer Mahan," *The Diplomat*, December 30, 2014, https://thediplomat.com/2014/12/the-geopolitical-vision-of-alfred-thayer-mahan/.
11. David Scott, "India's 'Grand Strategy' for the Indian Ocean: Mahanian Visions," *Asia-Pacific Review* 13, no. 2 (2006): 97–129.
12. Kenneth Field, "Compelling Thematic Cartography," *ArcUser*, Summer 2012, www.ESRI.com/news/arcuser/0612/compelling-thematic-cartography.html.
13. Susan Schulten, "World War II Led to a Revolution in Cartography—These Amazing Maps Are Its Legacy," *New Republic*, May 20, 2014, https://newrepublic.com/article/117835/richard-edes-harrison-reinvented-mapmaking-world-war-2-americans.
14. Schulten.
15. Timothy Barney, "Richard Edes Harrison and the Cartographic Perspective of Modern Internationalism," *Rhetoric & Public History* 15, no. 3 (2012): 397–434, doi:10.1353/rap.2012.0037.
16. R. Hillson, "The DIME/PMESII Model Suite Requirements Project," *NRL Review*, 2009, 235–39, www.dtic.mil/cgi-bin/GetTRDoc?AD=ADA525056.
17. Nayef Al-Rodhan, *Neo-Statecraft and Meta-Geopolitics* (Geneva: Geneva Centre for Security Policy, 2009), www.gcsp.ch/News-Knowledge/Publications/Neo-statecraft-and-Meta-geopolitics.

18. Nayef Al-Rodhan, *Meta-Geopolitics of Outer Space: An Analysis of Space Power, Security, and Governance* (Geneva: Geneva Centre for Security Policy, 2012), www.gcsp.ch/News-Knowledge/Publications/META-GEOPOLITICS-OF-OUTER-SPACE.
19. Samuel P. Huntington, *The Clash of Civilizations and the Remaking of World Order* (New York: Simon & Schuster, 1996), 21.

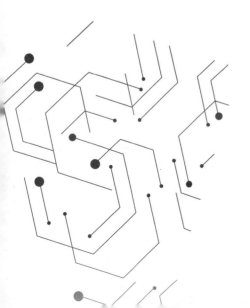

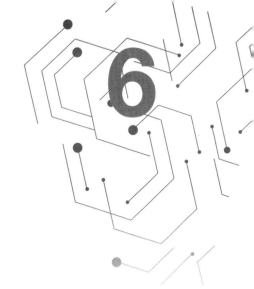

Geographic Information Systems

In 1854, a severe cholera outbreak hit residents living near Broad Street in London's Soho district. In the space of ten days, 500 people died of the disease, and most residents had fled the area.[1]

At the time, it was widely believed that "bad air" was the cause of cholera. Dr. John Snow thought otherwise. Snow was an English surgeon, a student of respiratory diseases, and at the time widely recognized for his pioneering work in anesthesia. He did not understand how cholera was transmitted (Louis Pasteur had not yet proposed the germ theory), but suspected that some agent in the water was responsible. Snow noted that the cases seemed to be located around a public water pump on Broad Street.

Though he could not identify evidence of cholera in the pump water, Snow convinced the authorities to remove the handle from the pump, and the outbreak ended soon thereafter. Snow, however, did not stop there. He subsequently identified all the cholera victims and placed dots on a map locating their places of residence. The dots of this "map layer" clustered around the water pump.[2] Dr. Snow had made perhaps the first analytic use of what is known today as a **geographic information system** (GIS) layer, and in the process he had created a new medical discipline: **epidemiology**—the study of the spread of disease. He is recognized today as the "father of epidemiology."[3] Less well known is his role as a pioneer in GIS, the subject of this chapter. What he actually did, using the definition from chapter 1, was produce geospatial intelligence.

The Cluttered Map

Ancient maps were (and still are) often difficult to read, because their makers tried to include everything that their customers might want to see. Eventually, they simply gave up on that approach; the maps were becoming too dense with information to be usable. In today's terms, if you are a weary traveler searching for the nearest Starbucks, your smartphone display becomes useless if the location is hidden among icons for grocery stores, drug stores, car washes, dry cleaners, horse stables, movie theatres, libraries, nail salons, restaurants . . . and so on. We all want our maps and charts (and smartphone displays) to show us the things we need to know and not to clutter the display with things we don't need to know. For a long time, that's why mapmakers not only used different projections but also crafted multiple maps of the same area, displaying different things on each copy.

As cadastral maps developed through history, recording more and more details about parcels of land, they captured the basic idea of a GIS. A key feature of England's Domesday Book, for example, is that it portrayed geographical areas in several different ways: *layers* of information about each area existed, and the book recorded them. Map layers could be used for purposes other than taxation: to solve real-world problems, which is exactly what Dr. Snow did in 1854. But Snow really used only one layer: the locations of cholera victims (not counting the water pump as a separate layer). The next step was to make use of multiple layers, and one professional group was in the forefront of this advance.

Hard Copy Layers

Landscape architects played a major role in promoting the concept of map layers that eventually led to GIS. A pioneer in the field was Warren H. Manning, the cofounder of the American Society of Landscape Architects. Manning thought of landscape design from an environmental perspective. He viewed any project as part of a regional system, in which soil, water, terrain, existing structures, and vegetative cover needed to be considered. His attitude naturally led to the development of a set of map layers for planning landscape architectural designs. In 1912 Manning didn't have computer displays to use, but he did have a light table—a drawing table with a translucent glass top that could be illuminated from beneath that architects normally used for design tracings. (It later would find use in photo interpretation.) Manning used his light table with a series of overlays in what became known as geodesign. His most ambitious map-layering effort took the form of a landscape plan for the entire United States that he completed in 1912 and published in *Landscape Architecture* in June 1923. The plan was intended to help the US government manage the exploitation of the country's natural resources.[4]

Still, traditional cartography relied on paper maps to be both the database of geographic information and the means for displaying that information. But a database, its cartographic representation and display, and the spatial analysis of the data are three distinct things. Though interrelated, they need to be created, managed, and stored separately. A GIS does that. As usual, however, there was just one problem: a GIS required information to be in digital form. And the road to digitization was not short. It began with a chance 1961 meeting of two people on an airline flight.

Roger Tomlinson, the Father of GIS

In 1961, a mere forty years after Manning's *Landscape Architecture* article, Canada enacted the Agricultural Rehabilitation and Development Act. It required an inventory of land use and usability across Canada. The result was the creation of the Canada Land Inventory, and Lee Pratt was appointed to head it. Pratt's assignment was to create a map of the southern part of Canada, covering 1 million square miles, which would characterize its agricultural land, forests, wildlife, and areas suitable for tourism.

Pratt's job was to in effect create a modern-day version of the Domesday Book (or Manning's US landscape plan) for Canada. But far more detail was needed: soil types, drainage and climate characteristics, suitability for specific crops or for grazing or recreational use, forested areas, and much more. While Pratt was considering just how to go about creating and analyzing the massive number of maps that would be needed, he boarded a flight and took a seat next to a British geographer named Roger Tomlinson. The two men discussed Pratt's problem, and Tomlinson proposed a solution: use computerized spatial data to create the Canada Land Inventory, and then subsets of it could be selectively placed on a map. Pratt listened, and then acted. Soon thereafter, Tomlinson was in charge of building what he called the Canadian Geographic Information System—the first use of the term "GIS."[5] The project was a success, and Tomlinson became known as the "Father of GIS."

The introduction to a subsequent article about Tomlinson's work describes the key concept of GIS:

> At the heart of the innovations that led to the Canada Geographic Information System was the fundamental idea of using computers to ask questions of maps and to render useful information from them. To do this, maps had to be in digital form. This led to the idea that many digital maps could be stitched together to represent the whole of Canada and that the maps could be linked intelligently to digital databases of statistics, such as the Census.[6]

Tomlinson's team was the first to implement the concept, but it wasn't the only group working on GIS. A team at Harvard did some pioneering work in GIS that became the incubator for the most successful commercial company in the field.

The Harvard Connection

In 1963 a University of Washington faculty team presented a workshop on computer-based mapping in Chicago. One attendee, a landscape architect named Howard Fisher, thought that he could design a better software package than the one demonstrated. With the aid of a Ford Foundation grant, he began developing a computer mapping program called SYMAP. In the 1960s, his alma mater, Harvard University's Graduate School of Design, had a landscape architecture faculty, and Fisher persuaded them that computer technology could be used to further Manning's layering concept. Harvard consequently set up its Laboratory for Computer Graphics and Spatial Analysis, with Fisher as its head. In the period between 1965 and 1975, the laboratory improved SYMAP, and in the process pioneered in the application of computers to cartography. Its most important product was the Odyssey project: a GIS that marked a milestone in the creation of integrated mapping systems.[7] Odyssey was not a commercial success, but it became the template for subsequent GIS software that was successful, thanks to the vision of two participants in the project.

ESRI and Intergraph

One member of the Odyssey project was a landscape architect and environmental scientist named Jack Dangermond. Dangermond worked as a research assistant at the laboratory during 1968 and 1969. He quickly recognized the commercial potential of Odyssey, and in 1969 he left the project to found the Environmental Systems Research Institute (ESRI). Eleven years later, Scott Morehouse, the lead developer for the Odyssey project, joined Dangermond at ESRI. The two launched the first commercial GIS, named ArcInfo, in 1982. Morehouse since then has been the lead architect of ESRI's software systems. Its flagship product today, ArcGIS, traces its heritage back to the two men's work on Harvard's Odyssey project.[8]

ESRI soon had competitors offering GIS, including agencies of the US government that independently developed their own software. But by the mid-1980s Intergraph Corporation was ESRI's primary competitor.

Founded at about the same time as ESRI, Intergraph from the beginning had been a hardware and software company aimed at the government market. One of its projects required building an interactive graphics system for the design and layout of printed circuit boards under a NASA contract. Circuit boards are designed in layers, so Intergraph built software that relied on the same layering principles as GIS. Intergraph subsequently landed a contract with the city of Nashville to build an interactive mapping system based on its circuit board software.[9]

The contract launched Intergraph into the business of managing geospatial data, and during the next thirty years it delivered several billion dollars' worth of GIS technology worldwide. It delivered mapping software to the oil and natural gas industry for managing exploration and production data, and it provided classified GIS to the US government.

ESRI and Intergraph became fierce competitors during the 1980s, each striving to dominate the GIS market. The two companies went into the competition with different perspectives:

- ESRI's founders had a landscape and land resource management background. The company's early mission was to help land planners and land resource managers make well-informed environmental decisions.
- Intergraph's founders had grown their company on government contracts, and were comfortable in the defense industry.

In the end, it came down to which firm had the best sales strategy. Early on, ESRI provided its GIS software free to universities. By 2014, the company was giving 1 billion dollars' worth of software annually to US schools. It proved to be a highly effective strategic approach. Students learned proficiency with ESRI's GIS, and when they graduated and moved into the workforce, they didn't want to switch to a different operating system. ESRI came to dominate the field, and today is the largest GIS software provider in the world.

Interactive Maps and Charts

Well before the end of the twentieth century, the computer had become an indispensable tool for all cartography, and GIS was the tool of choice. From the beginning, GIS was developed around the idea of layering information: points of interest, areas, infrastructure, and political and social information, among others. The secret to GIS was to allow features to be selected interactively for display to meet specific needs. Figure 6.1 illustrates a few examples in a display. Interactive, computerized maps now are commercially available, allowing users to select the desired information and zoom in (increase the scale) or zoom out (decrease the scale) in the process, often by replacing one map with another of different scale that was centered on the same point.

Layers in a geospatial display are familiar to everyone today. If you've used the features of Google Earth or the map display on any electronic device, you've relied on GIS and its layers. Who hasn't used it when on the go—to find places to eat, to have a cup of tea, or to spend the night? Staying with the example of everyday conveniences, GIS's computational power also allows you to create queries, such as "What is the best route to avoid highways and traffic?"

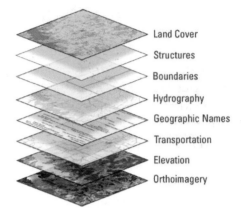

FIGURE 6.1 Examples of map layers used in a geographic information system. *Source: US Geological Survey.*

The mapping capabilities have continued to become more robust—allowing the user to select roads, boundaries, points of interest, and photographs of an area that people have taken over the years—with just a tap or click. The need for multiple different maps has disappeared; one electronic map display handles all of it. Simply choose the layers that contain what you need to see, and a specific map projection if necessary. The ability to select layers was a critical step—a revolutionary one, in fact—in the development of GEOINT for everyone.

In addition to Google Earth, several geographic information systems are readily available online—NASA Worldwind, OpenStreetMap, OpenWebGlobe, and Bing Maps, among others. Their common feature is a virtual globe. The advantage that it has over a conventional (physical) globe is that a user can choose any of many different projections (points of view) of an area. Using a cursor (or a finger on a touchscreen), you are able to easily move around in this virtual environment by changing the location and viewing angle. And the features to be viewed are optional—to include, for example, natural geographical features or human-made features such as roads and buildings.

But why stop there? The next step in making maps interactive meant turning to the tools of augmented reality and virtual reality—two related but different things. In **augmented reality**, the real-world map or picture that you see is augmented with additional details created by the software.

Today, tools for augmented reality travel are more sophisticated and have become widely available. Such features are now common in, for example, some of the navigation displays on mobile phones and in automobiles. You can gather information about an area by simply aiming the camera lens on your smartphone and calling up details about its history.

Virtual reality takes interactive mapping one step further by immersing the user in the scene. Anyone can create a virtual tour of a location by taking 360-degree pictures and adding the resulting scenes to an existing GIS tool such as Google Street View.[10] Then a virtual reality headset allows you to, in effect, tour the area.

The GIS Choice: Raster or Vector?

Some of the bitterest fights in the information technology world occur between proponents of different tools or approaches. The history of GIS has been no different. There are two basic ways to create, store, and display map layers: raster or vector, depicted in figure 6.2. Each has its strong and weak points, and in the early years each had its evangelical corps.

Raster models are simply electronic images of paper maps or charts and as such provide no more information than that available on their paper counterparts. Electronic images created by television cameras or aerial cameras are produced in raster form; they consist of a regular grid of tiny picture elements that we all know as pixels. Viewed closely, they appear as dots having different colors and shades.

Vector models, in contrast, contain "smart" dots. Vector maps are created using points, lines, and polygons. Each such entry on a vector map has an associated data file that can contain extensive geospatial information. So vector GIS includes a set of tools for collecting, processing, and displaying geographic information, and a separate database containing information about each point, line, or area.

Both technologies have advantages and drawbacks. In the case of raster, each pixel contains no information other than a geospatial (locational) reference.

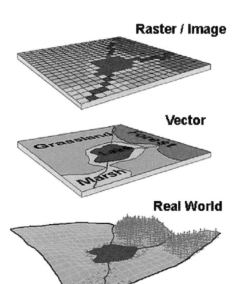

FIGURE 6.2 Raster and vector depictions of a scene. *Source:* Defense Mapping School, National Imagery and Mapping Agency.

But that also translates into the chief advantage. Raster maps and charts are cheaper and faster to create: a raster map can be created by simply scanning a paper map. Also, aerial photography naturally arrives in raster form.[11]

A *vector* map, in contrast, has additional information associated with each object. Simply click on a map element, such as a building or geographical area (a polygon), and data about it appears: what is inside the building or area, its ownership, age, or whatever else has been placed in the database. You have options and control; you can select the map or chart layers to display.[12] The major disadvantage is that a vector chart must be created from a raster map or image—an extra, and in the early years of GIS, expensive, step. The difficulty is that each object in the scene has to be georeferenced; that is, it must have a location assigned using map coordinates.

During the 1980s, in the fierce rivalry that developed between the two GIS companies, the raster/vector issue was central. ESRI had introduced ArcInfo in 1982 using the raster approach. In 1989, Intergraph released its Modular GIS Environment. Because it was developed using CAD/CAM software, Intergraph relied on the vector approach. GIS users were polarized; depending on what features they considered important, users praised one approach and vehemently criticized the other.

Of course, battles gradually died out as information technology advanced and allowed using the best features of each. Today, most GIS software—for example, products by ESRI, Google, Apple, and Bing—runs both types. If you call up Google satellite imagery of an area, you'll get a raster display. If you ask it to display borders and buildings, you'll get a vector display. It was natural to make use of both. (Because imagery is created in raster form, it is still the fastest way to update maps with new imagery.)

The Power of GIS

It is not easy for the average person to see the importance of GIS, because most of us routinely use it without knowing that we're doing so—it operates in the background. You can of course interact directly with a GIS. Its power then comes from the fact that almost any location-associated data can be placed in a file and then combined with other files (other layers) in a geographic display for research, analysis, and decision-making. One author has noted that "a trained GIS specialist is able to create maps that are capable of describing amazingly complex problems and the space these problems occupy with an impressive level of detail. As any tool or system used by analysts, the production of GIS products is rarely the end of a workflow, it requires an analytical mind-set to determine what the results are actually showing and what new data may be required for future analysis."[13]

Let's illustrate what that means by looking at two of many applications where you choose, manipulate, and analyze layers while working directly with a GIS.

Thematic Cartography

Thematic cartography, of course, predates GIS. And the maps used in geopolitics typically apply thematic cartography. But GIS gave *everyone* the tools to create thematic maps. The technology has, in the words of the ESRI senior research cartographer Kenneth Field, "opened up the world of mapmaking, supporting anyone to author and publish thematic web maps in interesting ways on an unlimited array of topics."[14] In a 2012 ESRI article, he pointed out an interesting survey result concerning recent maps that had been cited for their superlative design qualities. Three-quarters of them had been created by people without a background in cartography.[15] Today, GIS puts that power into the hands of a broad spectrum of users—who now create thematic maps on almost any subject—in effect, telling a story or sending a message with a map.

Financial Crime Analysis

GIS also has proved to be a powerful tool in analyzing a number of financial crimes—which inevitably have a geographic pattern. Probably the financial crime of most concern to governments is money laundering. It supports the financing of kleptocracies such as Putin's Russia, terrorist groups such as Boko Haram, drug cartels such as Sinaloa, and numerous black market operations. Estimates are that between $1.5 trillion and $2.85 trillion are laundered every year.[16]

Law enforcement organizations therefore put a priority on identifying countries and financial institutions that are most likely to be involved in money laundering. This calls for the sort of anticipatory analysis that was described in chapter 1. Using the power of GIS, analysts are able to identify by country the data sets that describe factors such as poverty, quality of life, corruption, and access to the political system. In the GIS, these then become "layers" that can be analyzed and combined to give an overall picture of the risk that a given region or financial institution is engaged in money laundering.[17]

The Explosion of GIS Applications

The previous examples illustrate the value of interacting directly with the GIS. But it's likely that you use a GIS daily and never realize it; you may be selecting layers and interacting with them through some application—an interactive map, for example. *Gisgeography* keeps an article on its website titled "1,000 GIS Applications & Uses—How GIS Is Changing the World," though the title may need to be changed since the August 2019 update included 1,002 entries. What GIS has done in providing interactive access to layers relevant to the user finds application in over fifty industries and fields across world.[18] Three examples, with brief historical context, include the aeronautical, nautical, and, interestingly, entertainment and recreational industries.

Aeronautical Charts

In the early years of aviation, flights were short in duration and conducted during clear weather. The pilot typically navigated by flying along a road or railroad, or sometimes using a compass when heading to a known destination. Air travel was limited by visibility. By the 1930s radio direction finding (discussed in chapter 4) made it possible to navigate over greater distances, in unfamiliar areas, and in poor visibility. That drove a need for a type of chart designed specifically for aviators. These charts provided information on airport locations, radio aids to navigation, terrain height, and the location of obstacles such as towers and power lines. In 1941 the first instrument approach and landing charts were developed, serving pilots who needed to land in low visibility.[19] During World War II, the demand for aerial navigation charts boomed among major participants. But each country applied its own standards for chart design and content.

After the war, standardized charts with agreed-upon marking conventions gradually developed to support international air travel. The United Nations International Civil Aviation Organization standards for civil aviation became accepted generally for such charts.[20]

Things have changed a lot in the last three decades. Aircraft navigation now is mostly handled electronically by a flight management system rather than using traditional paper charts. Electronic systems today function much like the map and GPS displays common in automobiles, including, of course, augmented reality features. The systems integrate all the available navigation aids to display the position of the aircraft, and the pilot can select the layers of information to display.

Nautical Charts

Electronic displays are the new generation of GIS-based nautical charts. Charts are available in both raster and vector formats and are referred to as, obviously enough, electronic navigation charts. Figures 6.3 and 6.4 show the two types with identical coverage of the harbor at Vancouver; they illustrate the differences discussed earlier.[21] Raster charts are simply duplicates of paper nautical charts suited for electronic display, and all the available information is displayed

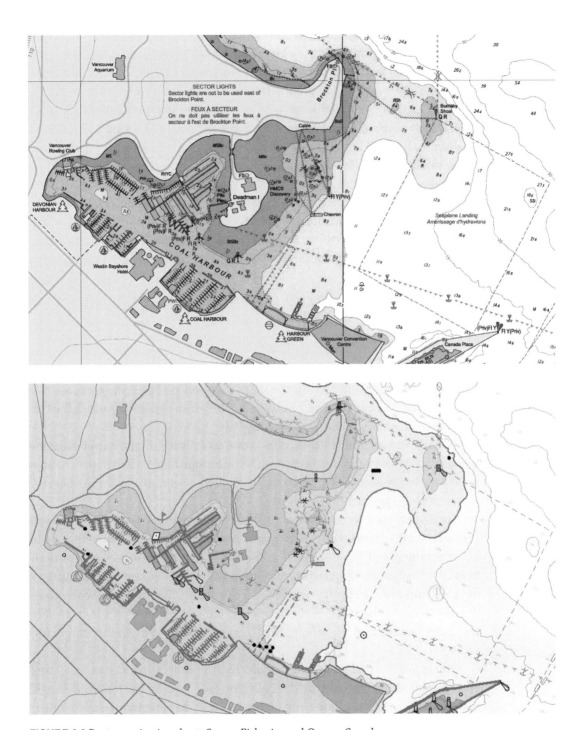

FIGURE 6.3 Raster navigation chart. *Source:* Fisheries and Oceans Canada.

FIGURE 6.4 Electronic or vector navigation chart. *Source:* Fisheries and Oceans Canada.

on the chart. The vector version of the same chart is less cluttered, but has additional information about features such as lights, wharfs, and buoys that are displayed by clicking on the object or area.

Electronic navigation charts function much like the airplane's flight management system, only in an ocean environment. They automatically integrate a ship's location (obtained from GPS or other navigation aids) with a chart display. The ship's navigator has a complete picture of the ship's position in relation to the shoreline and shore installations, charted dangers to navigation, and often to nearby vessels as well.

In the United States, the National Oceanic and Atmospheric Administration provides (and updates) both types of navigational charts. Increasingly, they are replacing the traditional paper charts long used by ship navigators. But there are many areas of the world yet to be covered by adequate charts. Many mariners carry paper charts for such areas or in case the electronic charting system fails.

Recreation and Entertainment

In Wilmington, North Carolina, a couple drives a sport-utility vehicle up to an empty parking lot in the Mayfaire Town Centre. A woman gets out, walks over to a light pole, and lifts up the lid that covers its base. She removes two packages from inside the base, photographs the contents, and replaces the two items with an object. What is happening?

A casual observer might conclude, as one actually did, that the event was the unloading/loading of a "dead drop"—the technique long used by spies to pass on secret documents. The photography, though, was the tipoff: the observer was watching a geocaching event.

Geocaching is a recreational activity in the form of outdoor hide-and-seek carried out on a global scale. Participants use a GPS receiver or smartphone to hide containers called "geocaches" at specific locations identified only by their map coordinates. The geocache typically is a waterproof container containing a log book and some "treasure" in the form of trinkets. Geocachers use the coordinates to locate the stash, record their finds in the logbook, and then replace the cache (sometimes exchanging the treasure for items of similar or greater value.)

The availability of smartphones has led to many other recreational and entertainment applications of GIS. Pokémon Go is one for the books. Released in July 2016 as a smartphone app, it immediately became immensely popular worldwide. The real world is the setting. As a person moves around with the app open, the smartphone will vibrate and a small cute or curious-looking creature appears on the screen. You electronically throw a Poké ball with a swipe of your finger on the screen, and with luck you will have "captured" the creature for your collection. This addictive game is credited with popularizing location-based and augmented reality technology but also with something not normally associated with online games: promoting physical activity. It has people out on the streets and in parks, walking around to catch wild Pokémon. The analytics firm Superdata reported that more people played Pokémon Go in May 2018 than at any point since the game's launch. The game garnered $104 million, a 174 percent jump from the previous year, and active users numbered at 147 million.[22]

Are Paper Maps Obsolete?

It might seem so. In fact, US states—a primary source of maps for motorists—have backed off on producing paper maps. Some states now print new maps every two years rather than on the traditional yearly cycle. Others now publish maps in five-year intervals. A few states no longer publish maps at all.

Nevertheless, paper maps continue to be in demand for many reasons. Rand McNally continues to produce its popular road atlas. Michelin maps are in use across Europe. After almost a decade of steady decline, paper map sales in the UK began to increase in 2014. The Ordnance Survey, Britain's mapping agency, reported a 7 percent increase in paper map sales from 2014 to 2015.[23] When taking unfamiliar routes, truckers rely on paper maps to verify items that are important for them—tolls and underpass clearances, for example. People who do not have paper maps can find themselves stranded in "dead spaces" where there is no cell tower/satellite service.

Then there are the survivalists, also known as "doomsday preppers"—those people preparing for the supposedly imminent catastrophic collapse that will end civilization. They include paper maps as an essential part of the preparation. As one prepper notes, "Maps take some skill to use properly, but unlike a GPS they don't need batteries. They won't stop working if some mad dictator hits us with an EMP [electromagnetic pulse] attack, the military decides to scramble the signal for unencrypted (i.e., all nonmilitary) receivers, or you just drop it on pavement. They're robust, low-tech, and, in a crisis, invaluable. If you want to be prepared, and your plans involve anything more than sitting quietly at home, you need to have maps."[24]

And, so long as the sport of orienteering remains popular, detailed maps will be in demand. The sport originated in late-nineteenth-century Sweden. It requires crossing unfamiliar terrain aided only by a paper map and a compass. Electronic aids such as GPS are not permitted. From its origin as a military training exercise, it evolved into a competitive sport practiced by civilians around the world.

For the military, at least, digital maps solved a major problem: that of how to produce and distribute millions of maps to military units around the world and how to keep them up to date. Updating a digital map is a very quick process. There was some resistance to putting laptop computers in the field in the early part of the digital age, before rugged versions came along. For example, a possibly apocryphal story tells of a senior US Army officer putting a bullet through a map, then pointing out that, unlike a laptop, the map still worked just fine.

And sometimes, no matter how good the digital map support is, military services find paper maps useful—especially when another country made the map. During World War II, General George Patton relied extensively on Michelin maps as he led his armored forces in the breakout from Normandy. Much later, as US planners began preparation for the invasion of Afghanistan after 9/11, they discovered that their best available maps lacked critical detail. But Russian maps of the region, made during the Soviet occupation of Afghanistan in the 1980s, had just what was needed: locations of mosques, schools, and hospitals, water wells, passable terrain markings, and more. As Thom Kaye, a cartographer at the National Imagery and Mapping Agency, observed, "We relied on the Russian maps because they had incredible detail about the things you could only have known if you had people on the ground."[25]

GIS and GEOINT

GIS has been concisely described as enabling "a dialogue between the map and the map user."[26] It's important to distinguish GIS from GEOINT. They are two different things, though closely related and therefore often confused with each other. In brief, "a GIS is a *tool* comprising hardware, software, and data used for capturing, managing, and displaying all forms of geographically referenced information. As a tool, it often enables GEOINT. . . . GEOINT is both a process, carried out by trained analysts, and the geospatial product of that process. The process most often makes use of a GIS."

GEOINT relies on the application of intelligence **tradecraft**. The word "tradecraft" has a long history in the field. Some in the intelligence community describe it as techniques developed over years of experience; others will argue that it is an art. For GEOINT analysts, it encompasses the skill at evaluating and using intelligence sources, topics covered in chapters 7 through 12.

Those who are unfamiliar with the intelligence business might hear the use of "trade" and think of occupations such as carpentry or masonry, often referred to as "trades." Rather than wave it away, we can make use of that analogy. GIS, then, is a tool that is used to produce geospatial intelligence by people versed in the tradecraft of intelligence. Consider GIS as much like a battle tank—the weapon that revolutionized ground warfare in the twentieth century. A tank, standing alone, might as well be a large hunk of scrap metal. Only when paired with a well-trained crew—that is, one knowledgeable in the tradecraft of tank warfare—does it become a fearsome combat weapon. In the same fashion, a GIS package, in the hands of an intelligence professional with the proper analytical tradecraft, becomes a powerful tool for producing geospatial intelligence.

Chapter Summary

Map users long were bedeviled by the problem of a cluttered map; too much unnecessary detail made it difficult to identify the important features. That problem was solved first by designing maps to fit specific needs, and later by creating layers of information. But the layers had to be applied manually until computers provided the tools to handle them electronically. The result was GIS, first applied as the Canadian Geographic Information System. Commercial versions of GIS software quickly followed. A controversy over raster versus vector display was gradually resolved as it became apparent that both were necessary. But the real explosion in GIS usage came with the spread of the internet.

In pre-internet days, you typically had to visit a brick-and-mortar library when doing research. You might have pulled ten volumes from shelves from which to draw information on any spatially related topic. GIS is the electronic version of that shelving and those books; it provides different aspects, in graphic form, about your research topic, in a far less excruciating process. How you use that information is the domain of geospatial intelligence, or GEOINT. If you analyze the information in service to decision-making, that process and your resulting product is GEOINT. Again, GIS enables GEOINT—a topic we'll revisit in chapter 14.

Notes

1. John Snow, *On the Mode of Communication of Cholera*, 2nd ed. (London: John Churchill, 1855).
2. Snow.
3. "The Remarkable History of GIS," *GISGeography*, n.d., http://gisgeography.com/history-of-gis/.
4. Robin Karson, Jane Roy Brown, and Sarah Allaback, eds., *Warren H. Manning: Landscape Architect and Environmental Planner* (Amherst, MA: Library of American Landscape History, 2017).
5. "Remarkable History."
6. Roger Tomlinson, "Origins of the Canada Geographic Information System," *ArcNews*, Fall 2012, www.ESRI.com/news/arcnews/fall12articles/origins-of-the-canada-geographic-information-system.html.
7. Nicholas R. Chrisman, "Academic Origins of GIS," in *The History of Geographic Information Systems*, ed. Timothy W. Foresman (Upper Saddle River, NJ: Prentice Hall, 1998).
8. Nick Chrisman, *Charting the Unknown: How Computer Mapping at Harvard Became GIS* (Redlands, CA: ESRI Press, 2006).
9. David E. Weisberg, *The Engineering Design Revolution*, chapter 14 (self-published, 2008), www.cadhistory.net/14%20Intergraph.pdf.

10. Jacob Kleinman, "How to Make Your Own Virtual Reality Tours with Google Street View," Lifehacker, May 10, 2018, https://lifehacker.com/how-to-make-your-own-virtual-reality-tours-with-google-1825929630.
11. Government of Canada, "What Is the Difference between a Raster Chart and a Vector Chart?" www.charts.gc.ca/charts-cartes/digital-electronique/raster-enc-eng.asp.
12. Government of Canada.
13. Michael Wakeman, "Exploring Potential Uses of Geographic Information Systems and Predictive Analysis in AML/CTF Investigations," ACAMS white paper, www.acams.org/aml-white-paper-geographic-intelligence/.
14. Kenneth Field, "Compelling Thematic Cartography," *ESRI ArcUser*, Summer 2012, www.esri.com/news/arcuser/0612/compelling-thematic-cartography.html.
15. Field.
16. Wakeman, "Exploring Potential Uses."
17. Wakeman.
18. "1000 GIS Applications & Uses: How GIS Is Changing the World," August 3, 2019, https://gisgeography.com/gis-applications-uses/.
19. ESRI, "A Brief History of Aeronautical Charting," n.d., www.ESRI.com/news/arcnews/summer07articles/a-brief-history.html.
20. ESRI.
21. Government of Canada, "What Is the Difference?"
22. Paul Tass, "'Pokémon GO' Is More Popular than It's Been at Any Point since Launch in 2016," *Forbes*, June 27, 2018, www.forbes.com/sites/insertcoin/2018/06/27/pokemon-go-is-more-popular-than-its-been-at-any-point-since-launch-in-2016/#3b492d99cfd2.
23. S. J. Velasquez, "Why Paper Road Maps Won't Die," BBC, October 7, 2016, www.bbc.com/autos/story/20161007-why-paper-road-maps-wont-die.
24. Fergus Mason, "5 Maps You Should Have at Home," *Ask a Prepper*, October 12, 2018, www.askaprepper.com/5-maps-you-should-have-at-home/.
25. Betsy Mason and Greg Miller, *All Over the Map: A Cartographic Odyssey* (Washington, DC: National Geographic Press, 2018).
26. Harm de Blij, *Why Geography Matters* (New York: Oxford University Press, 2005), 48.

Geolocation

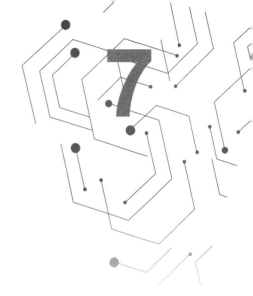

On August 2, 1990, Iraq invaded and essentially annexed its tiny neighbor Kuwait, in the process seizing about 20 percent of the world's oil reserves and a large chunk of Persian Gulf coastline. Six months later, after the Iraqi dictator Saddam Hussein ignored an ultimatum to leave, a US-led coalition began Operation Desert Storm, the mission to expel Iraqi forces and return the Kuwaiti government to its people.

Two days after Desert Storm began, the coalition encountered an unexpected challenge. The Iraqis began to deploy and fire their Scud missiles from mobile launchers, targeting Israel, Saudi Arabia, and Bahrain. Strictly speaking, the missiles weren't much of a threat; they were inaccurate and unreliable. But in targeting Israel, Hussein had a strategy that might have worked. If Israel could be provoked into responding militarily—standard Israeli policy when attacked—the result would likely have been a broken coalition. The coalition's Arab members could not afford to be seen as fighting another Arab state alongside Israel. The Scuds had to be located and destroyed.

Destroying a Scud wasn't difficult, given the coalition's immense advantage in airpower. *Finding* the mobile missile transporter-erector-launchers (TELs) was the problem. The Iraqis hid their Scuds in hardened shelters both at airfields and in cities; in the countryside, they made good use of the terrain, hiding the TELs in culverts, gullies, and highway underpasses. From their hiding places, the launch crews would move out quickly to a launch point and use the "shoot and scoot" tactic: fire the missile, then move the TEL immediately back into concealment and reload.

Against these tactics, the coalition's standard approach of using radar or optical sensors to locate a target was not working. The Iraqi units weren't staying in one place long enough to be attacked, and the TELs proved to be difficult to locate amid all the clutter that the sensors picked up. In the meantime, the Israelis were threatening to begin attacking the Scuds in western Iraq. Because of the imperative to keep Israel out of the war, the Americans and British resorted to a different approach: they sent their Special Operations Forces (SOF) units in to locate the TELs and call in airstrikes. The SOF units divided up responsibility for searching the part of western Iraq that became known as "Scud Alley." The teams would patrol during the night and hide during the day in wadis or gullies.[1]

The SOF were predictably excellent at a form of geospatial intelligence, much like that practiced by Lieutenant O'Neil in the *GI Jane* movie described in chapter 1. As one SOF member explained it,

> Scuds were usually launched at night and gave a huge signature, a great big ball of light. You could see the fireball at the base of the [Scud's] motor from thirty miles away across flat open desert, and that gave us an indication of where to look. The launcher would be moved immediately after firing, but if you looked at the layout of the roads and interpreted it intelligently, you could generally pick up where the launcher was going to be.[2]

Since Operation Desert Storm, the geolocation process has speeded up substantially, thanks to improved sensors and a faster process for transmitting target locations to the combat units that can use them.

Geolocation Basics

Geospatial intelligence almost inevitably involves **geolocation**. And we're concerned with two basic types of geolocation: locating things that don't move, and things that do (e.g., vehicles and people).

Geolocating fixed points of interest on the Earth's surface—that is, establishing their coordinates—is known as **georeferencing**, a process used in GIS, as noted in the previous chapter. That's what we do in the mapping process. In its most general sense, it also includes locating a fixed target *under* the Earth or ocean—creating bathymetric charts or mapping oil reserves, for example.

The other type of geolocation requires locating *and tracking* moving targets in all environments—on the ground, in air and space, on and under the water, and sometimes even underground. That is a more pressing concern in GEOINT, because it includes the temporal element. The precision weaponry used by militaries today requires that their targets be located accurately and precisely, down to a few meters. And the location needs to be established quickly during conflicts. There is ample time to geolocate targets that don't move. But combat units in the twenty-first century are mobile, and, as with the Iraqi TELs, if the location process takes too long, the unit won't be there when your side is ready to do something about it. Geolocation, especially locating moving targets, is also an important enabler in commerce and social interaction—the subjects of chapters 15 and 19.

Using Imagery

After the invention of the airplane, mapmakers and intelligence officers alike immediately saw its potential for geolocating objects—both for mapping and for intelligence reporting. The original military use envisioned for aircraft, after all, was for aerial observation. And indeed most geolocation today, military or civilian, relies on imagery produced using aircraft, unmanned aerial vehicles, or satellites as the imaging platform. Imagery is the most straightforward way to locate natural and human-made features—the fixed points of interest—for mapping.

With imagery, the method of determining location is much like that of surveying: triangulation. Using two or more images of the same scene, an imagery analyst marks known ground points on both (or all) photographs, then locates the coordinates of an unknown point using geometry. When digital imagery became available, the triangulation could be conducted electronically using software.

Static imagery usually works for locating fixed objects, but it gives only a snapshot of where a moving target is at that moment. Video imagery can keep track of moving targets, though it has limits imposed by clouds and the need to be reasonably close to the target. But neither static nor video imagery helps when you need to track them under all weather conditions and at long ranges. That required the development of radiofrequency geolocation.

Radiofrequency Geolocation

Soon after radio was invented, people wanted to find out where a radio signal was coming from—first for military purposes, later for law enforcement and for search and rescue, and more recently to locate missing automobiles or pets. This is the inverse of the navigation problem described in chapter 4. There, you knew where a radiofrequency (RF) emitter was located, and you used that information to locate your position. Here, we know our location, and we use the same techniques to locate an emitter.

Several different techniques have been developed to geolocate RF emitters. The following are the most widely used methods.

Angle of Arrival

The oldest RF geolocation technique is to determine the direction that a signal is coming from—a method called radio direction finding (RDF), or simply **direction finding** (D/F). Because electromagnetic (EM) waves from a transmitter travel in a straight line, the direction of arrival of the signal is the direction in which the

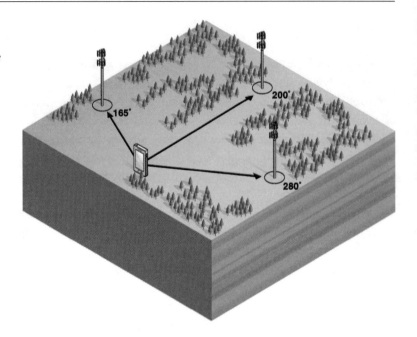

FIGURE 7.1 Geolocation by the angle of arrival. *Source:* Robert M. Clark, *Intelligence Collection* (Washington, DC: CQ Press/Sage, 2014). Used by permission.

transmitter lies. The method is called **angle of arrival** (AOA). Two AOA antennas, well separated, can locate a target on the Earth. Three can locate a target in the air or in space. Figure 7.1 shows a cell phone being geolocated by AOA, but the same idea works with any RF signal. Again, it's the same technique as the radio direction finders used for navigation (as in the story of Amelia Earhart, as told in chapter 4) but it's the inverse problem. Earhart didn't know where she was, but she knew where the Howland Island radio beacon was. Here, you know where you are, and you want to know where the radio transmitter is located.

During World War I, the British built a number of AOA systems to operate against German high-frequency radio transmitters. Soon after the war's outbreak, the British put two of these D/F stations in France. Initially, they allowed the British to identify German wireless positions in the trenches and observe their movements; later, they were used to track Zeppelin dirigibles (described in chapter 8) and German aircraft.

The D/F stations in France proved to be so successful that the British set up a chain of stations in England to track the transmissions from German submarines, surface vessels, and Zeppelins. By May 1915, the Admiralty was able to track German submarines crossing the North Sea. By 1916 the coastlines of Britain were covered by networks of D/F stations. And in May 1916 these stations made a critical discovery. German naval transmissions from the fleet port at Wilhelmshaven changed in direction by 1.5 degrees on May 30, and signal activity increased. The Admiralty reasoned that the German fleet had put to sea, and the British fleet was ordered to intercept the German fleet. The next day the Battle of Jutland was fought—resulting in a British strategic victory.[3]

Since World War I, several countries have built AOA-type D/F systems. Perhaps the best of its time—copied by both the US and the USSR—was called the Wullenweber. This D/F station was developed for the German Navy during the early years of World War II and was used to track Allied ship and troop movements. It featured a large circular antenna array (over 100 meters in diameter). After the war, the Soviets built several copies of it, called "Krug" (Russian for "circle"). The US built its own substantially larger versions of the Wullenweber—some so large that they acquired the name "elephant cage." As high-frequency radio declined in importance, these arrays were gradually closed down. AOA-type D/F stations, for higher frequencies and using other designs, continue to be used in military forces around the world.[4]

Soon after satellites began to collect RF signals, AOA went into space. Beginning in 1967, the USSR launched a series of satellites called Tselina, which used AOA to locate radar installations.[5]

Occasionally, the AOA method makes it into the movies, as it did in the 2013 film *Zero Dark Thirty*—the story of the hunt for Osama bin Laden. In the film, the CIA signals technicians fan out in Peshawar and Rawalpindi, two cities in Pakistan where bin Laden's courier has appeared in cell phone communications. They drive through streets teeming with cars, people, and donkeys, using an AOA device with eavesdropping equipment to locate the courier's cell signal. And the courier leads them to the Abbottabad compound where bin Laden is hiding—an example of GEOINT without imagery, though imagery subsequently was used (in the movie and in reality) to obtain the details of the compound and to persuade US leaders to conduct the raid that killed bin Laden.[6]

AOA is inherently limited in accuracy at longer ranges, however. We not only want to know where the emitter is; increasingly, we want to locate it to within a few meters. The AOA technique had a definite upper limit on geolocation accuracy, based on how wide the direction-finding antenna beam was; the wider the beam, the less accurate the geolocation. And the further the emitter is from the AOA receiver, the worse the accuracy gets. So more accurate methods were needed, and three major ones have been developed.

Time Difference of Arrival

Another process for geolocating a signal depends on the fact that signals travel at a defined velocity—acoustic signals at the speed of sound, electromagnetic signals (and optical signals) at the speed of light. And because the speed of electromagnetic energy is a constant value, it can be used to geolocate the source of a signal.

The geolocation technique is called **time difference of arrival** (TDOA). It has been the preferred technology for high-accuracy geolocation since the

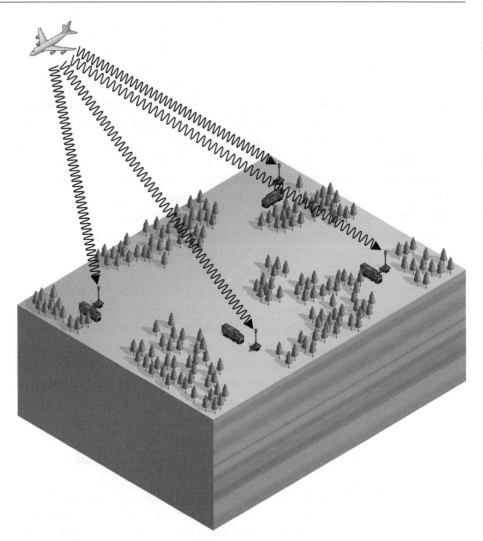

FIGURE 7.2 Geolocation using the time difference of arrival. *Source:* Robert M. Clark, *Intelligence Collection* (Washington, DC: CQ Press/Sage, 2014). Used by permission.

1930s. In its simplest form, several widely spaced receivers receive a signal transmitted from an RF emitter, as indicated in figure 7.2. Each of the sites has an accurate timing source—usually an atomic clock of the sort discussed in chapter 4. The signal reception is time-stamped at all the receiver sites, and the times are sent to a base station for comparison. As the figure suggests, the path lengths to each receiver are different, so the arrival times will be different. A comparison of the differences identifies the only possible location the signal could have come from.

The GPS uses TDOA, but everything is reversed: there is only one receiver (e.g., the smartphone in your hand) and at least four transmitters whose locations are known (the ones on the GPS satellites). The satellites transmit a time-stamped pulse. The pulses are received at different times, and the same comparison is made to locate the GPS receiver position.

Radio Frequency Identification Tags

The idea of using a **radio frequency identification** (RFID) tag to identify and track a moving vehicle dates back to World War II. As radar-controlled anti-aircraft guns came into use, both the British and the Germans had the same problem: you couldn't be sure you were shooting at an enemy aircraft, because you couldn't identify it visually. So combat aircraft and antiaircraft units both began to carry a special receiver/transmitter device. The ground unit would transmit an interrogation signal toward a suspicious aircraft. When it received the signal, the aircraft's device would automatically transmit back a signal that allowed the radar controller to identify the aircraft as friendly. No return signal meant that the aircraft was an enemy. The unit naturally acquired the name identification-friend-or-foe system, and became standard on military aircraft after the war. Also after the war, the tags became useful in a number of other fields, both governmental and commercial.

The nuclear arms buildup after the 1950s, and the increasing use of nuclear material in civil applications, meant that substantial quantities of nuclear material were being produced, moved around, and stored. The governments of both the US and USSR became concerned about keeping close track of it, especially when it was transported between storage sites. In response, during the 1970s engineers at Los Alamos National Laboratory developed an RFID tag to track trucks carrying nuclear materials. Some of the engineers, recognizing a commercial opportunity, left the lab to develop tags for toll payment systems. These tags, which are now commonplace, are mounted somewhere on a vehicle's front windshield, containing detailed information about the car. They are electronically interrogated by a card reader instantly at toll road booths, once known as the scourge of all commuters and travelers as each person waited to interact with "operators" who took cash and made change; the traffic backups were miles long. These RFID tags subsequently became inexpensive to build, and they are now used for all sorts of vehicle interactions, for example, to enter and exit secure or paid parking lots. As smart cities are becoming popular, the integrated tags are already being used in law enforcement intelligence. Meanwhile, the US national laboratories have continued to improve the tags, so that now they monitor the condition of nuclear materials and any physical changes to the packages, both during transportation and storage.[7]

RFID tags can be either active or passive. An active tag is battery powered; it periodically transmits an identification signal. A passive tag may have a small battery, but it transmits only when it senses the presence of an RFID reader;

or it can be made cheaper and smaller by eliminating the battery and simply reflecting back the radio energy transmitted by an RFID reader, modified with a tag identification code.

One of the earliest applications of passive RFID tags also was developed by the Los Alamos National Laboratory, for what has to have been at the opposite extreme from tracking nuclear material: identifying and tracking cows. Cows are given hormones and medicines when they are ill. But it's difficult to ensure that each cow gets only the right dosage and not a double dosage. A passive tag with a unique cow identifier allows ranchers to keep track of dosage and hunt down cows who need a dose.[8]

RFID chipping is best known today for its widespread use in pets—specifically, in helping owners find their beloved pets who have gone missing. The first question anyone asks today when a frantic owner has lost Fluffy is—"Well, is Fluffy chipped?" The reference is to a passive RFID microchip about the size of a grain of rice, injected through a needle, typically between the shoulder blades. The chip can be scanned at any veterinarian or animal control location in the US, and the unique identifier is matched with owner information. It may be the only instance that RFID is known to produce joy in humans. It was only a matter of time, of course, before someone made the leap from pets to people. Some companies now encourage employees to have a microchip emplaced in one hand between thumb and forefinger, allowing an employee to purchase items and to access buildings and computers, all with a wave of the hand near a scanner.[9] Such small devices don't currently work with GPS, but going back to the pet industry (about $70 billion in America), there are now dog and cat collars that will. They claim to pinpoint Fluffy's location anywhere in the US using the same principles as RF tagging, which is discussed next.

RF tagging and tracking of mobile or transportable objects is less well known in popular culture, but it is widely used in intelligence and law enforcement. In those professions, you have to clandestinely tag the object to be tracked, which can be difficult to do if it must be tagged while in unfriendly hands. And the tag has to be very difficult to find by physical inspection or SIGINT.

Vehicle tags are widely used by long-haul trucking companies for fleet management: fleet tracking, routing, dispatching, on-board information, and security. They are also installed in some passenger cars as a theft prevention, monitoring, and retrieval device. These concealed tags include a GPS receiver that tracks the automobile's position. When notified of a stolen car, police search for the signal emitted by the tracking system and locate the stolen vehicle. Tags also go to sea, as the next two examples illustrate.

The Marine Asset Tag Tracking System. The US Department of Homeland Security's Marine Asset Tag Tracking System (MATTS) is used to track shipping containers. It's a sophisticated tag in a small package: a sensor, computer, GPS tracking system, and radio transceiver, all packaged in a waterproof container. These tags are fixed to shipping containers and transmit a signal that jumps from one container tag to another until it finds a path to the ship's communications system. The tag also records its location history and automatically reports in to the Department of Homeland Security via cell phone or internet when it comes within range of a cell phone tower or internet wireless connection. Its GPS chip can estimate its location history if it loses access to a GPS signal. If anything unusual has happened during the container's journey—for example, if the container has been tampered with—the Homeland Security authorities are alerted to inspect that container.[10]

The Automatic Identification System. Many oceangoing commercial vessels are required to be fitted with an electronic tag that reports the ship name, position, course and speed, classification, call sign, and registration number. Called the automatic identification system (AIS), the tag is designed to provide safety of navigation; nearby ships receive and display the information when within radio range. AIS also allows maritime authorities to track vessel movements. Vessels fitted with AIS can be tracked through a network of satellites that are fitted with special AIS receivers. The AIS global display of ship traffic is available online (https://www.marinetraffic.com/). (Try it—you can see a great deal of information about a ship anywhere in the world as well as its location at that moment.) International agreements require that AIS be fitted aboard ships with 300 or more gross tonnage traveling internationally, and all passenger ships regardless of size. But because of the danger that pirates can use AIS to track ships for attack, regulations allow the AIS to be shut off in high-risk areas, such as off the coast of Somalia.

Mobile Phones

Mobile phones can be geolocated and tracked in several different ways. The phones do not need to be in use. They simply must be turned on, as happened with the bin Laden courier. The phone emits a roaming signal periodically when turned on, and the signal strength at nearby cell phone towers can be measured to give an approximate location for the phone. More accurate location comes from using TDOA—timing the arrival of the signal at different cell phone towers.

Smartphones—most of which carry GPS—are even easier to geolocate. In fact, they might be considered another type of RFID tag; they contain both location information and the identity of the user. People every day willingly allow their phones (and therefore themselves) to be tracked with apps like Find My Friends.

Acoustic Geolocation

Acoustic geolocation works in much the same way as radiofrequency geolocation. The difference is that we're dealing with sound waves, not radio waves; and unlike radio waves, sound waves don't travel at a constant speed. Predictability is the worst problem with underground sound. In soil, sound waves travel at about 500 meters per second; in solid rock, the speed increases to about 5,000 meters per second. Sound in the atmosphere is more predictable, but it still varies with temperature, atmospheric pressure, and humidity. Underwater sound speed is the most predictable, at about 1,500 meters per second; but it also varies, increasing with rising pressure, temperature, and salinity. So identifying the source of a sound based on time of arrival can be a challenge.

Two techniques are commonly used to geolocate any source of sound: measuring the direction of arrival, and measuring the time difference of arrival. Both depend on having at least two widely separated sensors to do the geolocation, and both resemble the geolocation techniques used for RF signal geolocation.

During World War I, several European countries developed acoustic arrays to measure the angle of arrival of sound from artillery fire.[11] Two such arrays that were well separated could triangulate on the source of the sound, much as the angle-of-arrival measurement does for RF geolocation.

The other geolocation method depends on accurate measurement of the time of arrival of the sound, again at widely spaced sensors. The time delays measured by such a sensor array allow you to triangulate on the source and pinpoint the location of the noise. The technique is similar to the RF time difference of arrival (TDOA) method, with one very important difference: The speed of EM signals can be assumed to be the speed of light; as noted above, it is not so simple with sound waves.

Acoustic geolocation, like radio frequency geolocation, requires highly sensitive receivers. Most acoustic receivers use microphones that are optimized to work in a specific environment (air, underground, underwater) and in a specific part of the acoustic spectrum. Typically, several microphones are used for geolocation rather than a single microphone; they may be spaced some distance apart, as in radio frequency geolocation, or placed close together in an array.

The next three subsections describe a bit more of the history and uses of atmospheric (air), underwater, and underground acoustic sensing. But first, let's take a quick glance back at chapter 1 to recall the concept of signatures, which are critical to the production of GEOINT. Signatures are the distinctive features of any entity that identify or characterize it. Signatures appear in many forms: chemical, optical, radiofrequency, and, of course, acoustic. (We return to the topic of signatures in chapter 10 in our discussion of visible imagery analysis.) But for now, let's look at acoustic signatures, which have many GEOINT applications.

Sound in Air

The sensing of atmospheric sound for military purposes is as old as militaries are; but until the cannon-fire locators of World War I, it relied on the use of our unaided ears. Since then, battlefield intelligence has made increasing use of short-range sound or infrasound collection not only to locate but also to identify sounds. Land and air vehicles such as trucks, tanks, helicopters, and unmanned aerial vehicles have unique acoustic signatures (e.g., from the noise of a gasoline or diesel engine and the sound of rotating tires or tank treads). Most of these acoustic signatures are detectable only at short ranges, on the order of a few kilometers or less.

Acoustic geolocation has found increasing use in law enforcement in recent decades. Urban police forces now use gunshot detection and location systems to pinpoint the location of gunfire. One of the most widely used gunshot detection systems is called ShotSpotter. Most such systems rely on a network of acoustic sensors located around the city. When a gun is fired, the network uses TDOA on the arrival of the sound at different sensors to identify the source of the gunshot. The nature of the sound can sometimes even be used to identify the type of weapon fired.[12]

A challenge, of course, is to distinguish a gunshot from fireworks or automobiles backfiring. But again, each such sound has a unique acoustic signature, and current detection systems employ signature recognition to distinguish gunshots from false alarms.

Underwater Sound

Underwater sound can be used for communication and to detect, identify, and track entities. Sea creatures such as whales and dolphins have done this for millions of years. Its use by humans has a much shorter history, and, like sea creatures, we use it both actively and passively.

Active use is called sonar (shorthand for "sound navigation and ranging"); it was first used during World War I to detect submarines. Sonar relies on the same principles as radar (i.e., transmitting a signal and timing the interval between transmission and the reception of an echo); but, as the name suggests, sonar transmits a short pulse of sound (known as a "ping") and listens for the echo from an underwater object such as a submarine. The listening device is called a **hydrophone** (essentially a microphone that is designed to operate effectively underwater).

Passive use, often misnamed passive sonar, depends on the fact that ocean-going vehicles generate underwater noise; and that noise can be detected and tracked at ranges of many kilometers. Naval intelligence organizations have used passive acoustic systems for years to identify and track both surface ships and submarines in open ocean areas. Underwater sound can be used to detect, track, and even to identify the type of vessel. Fine-grained measurement of the acoustic spectrum allows a vessel to be "tagged" by its unique signature, in the same fashion that land vehicles are uniquely identified by their configuration in imagery.

A ship or submarine generates several types of sound from different sources: machinery vibration, water flow over the vessel's hull, and propeller rotation all combine to generate a unique sound signature. And sound propagates farther, and about five times faster, in water than in air. So underwater sound can be detected at long ranges.

Concerned about the threat from Soviet submarines during the Cold War, the US Navy installed an underwater surveillance system that relied on underwater sound for submarine tracking. The sound surveillance system (known as SOSUS) is a network of hydrophone arrays in the Atlantic and Pacific oceans. These hydrophone arrays listen to the low-frequency sounds in the sound channel, record them, and transmit the data via undersea cables back to shore stations for analysis.[13] Those original arrays were mounted on the seafloor. Later on, an array was designed to be towed behind a ship or submarine instead of being fixed to the ocean bottom. Figure 7.3 illustrates how a towed hydrophone array works. The sound from a submarine arrives at each hydrophone at different times, depending on the direction from the array to the submarine. Processing the signals allows the direction to be determined. As the figure suggests,

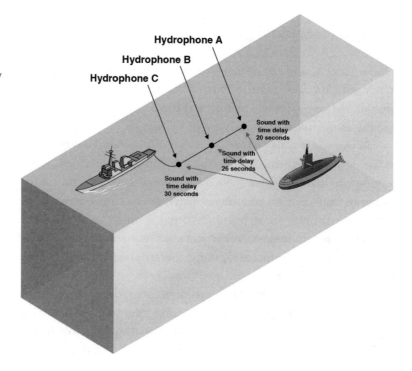

FIGURE 7.3 Tracking a submarine with a hydrophone array. *Source:* Robert M. Clark, *Intelligence Collection* (Washington, DC: CQ Press/Sage, 2014). Used by permission.

acoustic processing involves dealing with seconds (or fractions thereof) in the time scale; this contrasts to radiofrequency geolocation, where time differences are measured to nanoseconds. So it operates much the same as radiofrequency TDOA, with time scale being the big difference.

Seismic Geolocation

Detecting and locating seismic events—specifically, earthquakes—has a long history. The earliest known device for detecting earthquakes dates to 132 CE. In that year, a Chinese inventor created a bronze vessel that was about 2 meters in diameter. It had eight dragon heads around its top, and each dragon was holding a bronze ball in its mouth. When an earthquake occurred in the region, one mouth would open and drop its ball onto a bronze toad below. That dragon was the one located in the earthquake's direction.[14]

Seismic sensing has become considerably more accurate since the time of the Han Dynasty and its dragon sensor, and the stakes have become higher. We're still interested in locating earthquakes, but even more so in underground explosions, because nuclear weapons tests are conducted underground.

Both earthquakes and underground explosions can be detected at very long ranges. They create what is called a **teleseismic wave** that travels through the Earth's deep interior and can be detected thousands of miles away. In the decade following an August 1949 Soviet nuclear test, the United States and its allies deployed a network of seismic stations to monitor underground testing. Over the years, the Soviet Union conducted a number of tests at known testing areas, and these were relatively easy to identify. Tests at the USSR's Novaya Zemlya site in the far north could be detected at yields as small as a 0.01 kiloton (10 tons) explosion.[15] Today, underground explosions with yields of 0.1 kiloton (100 tons of TNT equivalent) in hard rock can be detected if conducted anywhere in the Northern Hemisphere. Monitoring is now conducted by the Vienna-based Comprehensive Test Ban Treaty Organization, which coordinates a worldwide network of seismic sensing stations. It identified, located, and characterized all of North Korea's underground nuclear tests, starting with the 2006 test and continuing to the 2017 testing.

The detection problem lies in distinguishing nuclear tests from earthquakes, but that turns out to be straightforward. An earthquake is created by rock plates deep in the Earth sliding against one another. That creates an up-and-down wave movement, much like an ocean wave. In contrast, a subsurface explosion creates a back-and-forth wave movement—a pressure wave. The two wave types travel at different speeds in the Earth, so their recorded signals look quite different.

The sensor used to detect underground explosions, called a seismometer, functions much like the movement sensor on today's smartphones and smart watches. That similarity has been used to create an app that, using crowdsourcing with other smartphone users who have downloaded the app, will give as much as a minute of warning before an earthquake strikes the area.[16] The developers have not indicated whether the app could also identify an underground nuclear explosion.

Cyber Geolocation

All electronic devices (computers, tablets, printers, smartphones, smart watches, etc.) connected to the internet have an internet protocol (IP) address, the unique identifier for each individual device. The protocol specifies that every message packet transmitted by a device must have a header that contains the device's IP address. A number of free and paid subscription geolocation databases exist that can locate the source device, with location accuracy ranging from the country of origin to state or city and down to the postal zip code level. They work by linking the IP address with the physical location. One technique used to geolocate a device is to transmit to the IP address from multiple locations, knowing the route each transmission takes, and measure the time delay of a return signal. Then geolocation follows a similar technique to that used in TDOA.

Most of the motherboards of computers, laptops, and mobile devices have inbuilt features for remote activation of an antitheft mechanism. These devices keep continuously gathering location information and consequently can be used to geolocate the host. The internet of things devices also have IP addresses and often include sensors that can provide audio or video intelligence. Of course, not everyone likes the idea of being located, and enterprising cyber companies have developed methods to thwart that. Concealing or even faking a device's location is discussed in detail in chapter 20.

IP addresses may be considered almost quaint in the field of geolocation, since GPS is now built in to all smart devices. Since the device knows its exact location, others can pinpoint that location with a much higher degree of accuracy than can be had with just an IP address. Of course, those with criminal intent find ways to avoid being found with GPS. Burner phones (an inexpensive phone, always paid for in cash, with prepaid minutes included) are typically used once and left in a trash bin. They are the bane of all drug enforcement and counterterrorism efforts.

Should *cyber* be a new INT? Many intelligence professionals think so. Social media, especially, is a cyber field replete with geolocation material. It typically contains geographic footprints, for example, in the form of locations from where tweets originate, or references in the media content to things that have a fixed location. These messages convey geospatial information, capturing, for example, people's references to locations that represent momentary social hotspots. Most of the popular social media applications such as Twitter, Facebook, and Instagram have integrated geolocation tagging for any images that are uploaded. Photographs taken by smartphones also are geolocation tagged.[17] Geospatial information derived from social media has turned out to be a boon for GEOINT in government, law enforcement, and, of course, commerce (a subject covered in detail in chapter 19). Social media also has enabled what may be the most significant advance in cartography since Eratosthenes: volunteered geographic information, also characterized as crowdsourcing. We'll come back to it in chapter 15.

Chapter Summary

Geospatial intelligence depends, obviously, on being able to geolocate an object, person, or activity of interest. There are many ways to do that. Like most other developments that have enabled GEOINT, they depend on applying evolving technology. Cyber technologies and wireless media have been at the forefront of new methods for geolocation in recent decades. Yet much geolocation that addresses national security issues continues to depend on imaging and radio-frequency sensors for observation from a distance. And in locating something on the Earth's surface, it also depends on the sensor being positioned well above the surface—the subject of chapters 8 and 9.

Notes

1. William Rosenau, "Special Operations Forces and Elusive Enemy Ground Targets," RAND Corporation, 2001, chap. 3, www.rand.org/content/dam/rand/pubs/monograph_reports/MR1408/MR1408.ch3.pdf.
2. Ken Connor, *Ghost Force: The Secret History of the SAS* (London: Cassell Military Paperbacks, 2002), 315.
3. Marconi Heritage Group, "The Wireless War at Sea," n.d., http://marconiheritage.org/ww1-sea.html.
4. Capt. Christopher Arnold, "Silencing the Arctic Mammoth," *Alaska Star*, May 25, 2016, www.alaskastar.com/2016-05-25/silencing-arctic-mammoth#.XJJHNvZFwV0.
5. Russian Space Web, "Tselina Electronic Intelligence Spacecraft," n.d., www.russianspaceweb.com/tselina.html.
6. Columbia Pictures, *Zero Dark Thirty*, 2012.
7. US Department of Energy, "RFID Technology Creating Jobs, Impacting Americans with Increasing Frequency," October 11, 2011, www.energy.gov/articles/rfid-technology-creating-jobs-impacting-americans-increasing-frequency.
8. US Department of Energy.
9. Mary Bowerman, "Wisconsin Company to Install Rice-Sized Microchips in Employees," *USA Today*, July 14, 2017, www.usatoday.com/story/tech/nation-now/2017/07/24/wisconsin-company-install-rice-sized-microchips-employees/503867001/.
10. Department of Homeland Security, "When Your Ship Comes In," July 2007, www.dhs.gov/when-your-ship-comes.

11. B. Kaushik, Don Nance, and K. K. Ahuja, "A Review of the Role of Acoustic Sensors in the Modern Battlefield," paper presented at 26th AIAA Aeroacoustics Conference, May 23–25, 2005, 2, https://pdfs.semanticscholar.org/916d/986918d1b5ce8c0986aa4c1de220182603e1.pdf.
12. ShotSpotter, "ShotSpotter Technology," n.d., www.shotspotter.com/technology/.
13. University of Rhode Island, "The Cold War: History of the SOund SUrveillance System (SOSUS), n.d., https://dosits.org/people-and-sound/history-of-underwater-acoustics/the-cold-war-history-of-the-sound-surveillance-system-sosus/.
14. US Geological Survey, "What Was the First Instrument That Actually Recorded an Earthquake?" n.d., www.usgs.gov/faqs/what-was-first-instrument-actually-recorded-earthquake?qt-news_science_products=0#qt-news_science_products.
15. National Research Council of the National Academies of Sciences, *The Comprehensive Nuclear Test Ban Treaty: Technical Issues for the United States* (Washington, DC: National Academy Press, 2012), 60, www.nap.edu/catalog.php?record_id=12849.
16. Timothy Revell, "Seismic Sensing App Detects 200 Earthquakes in the First Six Months," New Scientist, December 1, 2016, www.newscientist.com/article/2114851-seismic-sensing-app-detects-200-earthquakes-in-first-six-months/.
17. Peter Hannay and Greg Baatard, "Geointelligence: Data Mining Locational Social Media Content for Profiling and Information Gathering," in *Proceedings of the 2nd International Cyber Resilience Conference*, 2011, http://citeseerx.ist.psu.edu/viewdoc/download?doi=10.1.1.214.3970&rep=rep1&type=pdf.

Gaining the High Ground

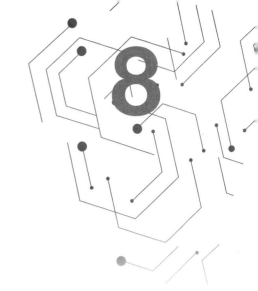

When cavemen began throwing rocks at each other, they quickly figured out that the side having higher ground had a definite advantage. Watching the enemy from above and throwing rocks downhill are both easier. In conflicts ever since, military commanders have attempted to get the high ground, both for observing their enemy's actions and for raining projectile weapons down on them.

Chapter 7 addressed the importance of geolocation. It's hard to assess an opponent until you know where they are. But once an entity (person, place, thing, or activity) of intelligence interest has been located, in almost all instances, you want to observe its movements or changes over time. That means conducting remote sensing—either by doing **reconnaissance** (periodic observation) or **surveillance** (continuous observation). And both those geospatial activities require observation platforms (which have been used for millennia) and remote sensors (which, if you include eyeballs, have an even longer history). While these two remote-sensing technologies go back to antiquity, developments have progressed remarkably over the last century or so. We'll focus on the platforms in chapters 8 and 9, then turn our attention to the sensors they carry.

Observation platforms (from towers to satellites) began their existence as tools for gaining geospatial intelligence about military opponents, and some of the original designs continue to be used today. From military beginnings, the platforms transitioned into use for other governmental and for commercial applications. Within military circles, though, the high ground continues to have an almost theological status. On few occasions has it had a historical impact matching that of the race to seize it at Gettysburg in 1863.

Gettysburg

The battle of Gettysburg on July 1–3, 1863, is well known to students of American Civil War history. Confederate general Robert E. Lee had moved his Army of Northern Virginia north into Maryland and then Pennsylvania, intending to draw the Union Army out and destroy it—hoping the result would be a peace settlement. When the Union Army of the Potomac indeed followed, Lee turned his scattered forces around and ordered them to converge on the town of Gettysburg.

But for Lee, there was just one problem. A Union cavalry division commanded by General John Buford had been close behind Lee, and it arrived in Gettysburg first, on June 30. Buford immediately recognized the importance of holding on to the high ground south of the town, and he deployed his two brigades to hold off the enemy. On July 1, the arriving Confederates attacked; Buford was able to hold the ground, taking heavy losses, until Union infantry arrived to reinforce him. During the day, the Union forces were pushed back, but were able to retreat and take up good defensive positions south of Gettysburg. Buford's holding action had deprived Lee of the high ground, and that proved to be key in the upcoming decisive battle.

By dawn on July 2, the Union Army had established strong positions in the shape of a fishhook starting at Culp's Hill in the north and running along Cemetery Ridge to the south, ending in a rocky promontory called Little Round Top.

Thousands of books have been written about the Battle of Gettysburg. At least one movie, aptly titled *Gettysburg*, dramatically memorialized Union colonel Joshua Chamberlain's tenacious defense of the Union flank on Little Round Top.[1] Much has been made of the terrain advantage that enabled the Union forces to hold off repeated Confederate attacks. But most history books, and the movie, touch only lightly on the most important geospatial advantage of holding Little Round Top that day. From its summit, Signal Corps observers could see most of the Confederate disposition and movements. Confederate general John Bell Hood's division was assigned to take Little Round Top. It was an impossible task. Hood could not move into position without being observed. He was forced to take a long trek through woods to avoid being seen. Meanwhile, more Union forces arrived to strengthen the line. By the time he finally got his tired and thirsty troops into position, he was surprised to find a reinforced corps waiting for him. Neither side prevailed in the resulting battle that day, but Union forces continued to hold. The next day, General George Pickett's charge against the Union center, again observed in detail from Little Round Top, resulted in a decisive Confederate defeat and Lee's withdrawal from Gettysburg.

If General Hood had been able to begin his attack as originally planned, the outcome on July 2—and at Gettysburg overall—might have been quite different; the outcome of the war might also have been different.

Observation Towers

If the terrain didn't afford the needed high ground, you built a tower for observation; and militaries did so, for millennia. Later on, those observation towers were used for surveying and triangulation in mapmaking, and for identifying and locating forest fires (again, using triangulation). After its invention in 1807, many ships were outfitted with a "crow's nest" for observation. It was simply a barrel or basket attached near the top of the tallest mast, but it meant that hazards, other ships, or land could be spotted at the greatest distance possible.

Lighter-Than-Air Craft

Observation towers had an obvious limit: they could extend only a short distance upward, limiting the area that could be observed. Military leaders wanted a larger area view, and that required better technology. The first step in that direction was the lighter-than-air craft: balloons, airships, and aerostats.

Balloons

The French brothers Jacques Etienne and Joseph-Michel Montgolfier made the first balloon flight on June 4, 1783; or rather, the three animals they placed in the gondola did. In October of that year, their balloon carried the first man aloft. European military leaders immediately saw the potential advantage of viewing a battlefield from above. The French were the first to use a balloon for battlefield reconnaissance. They formed the French Aerostatic Corps in 1794. Later that year, the corps launched the newly built balloon *l'Entreprenant* to observe Austrian army movements at the Battle of Fleurus. However, the overall report card on balloon observations was mixed. They were not easy to move around, especially over rough terrain. And attempting to fill them with hot air and launch them in windy conditions must have had results that were almost comedic. Misfortune seemed to follow the Aerostatic Corps. The Austrians captured an entire company along with its balloon in 1796 during the Battle of Würzburg. A second company was formed, but bad luck followed it also (though that was neither theirs nor the balloon's fault): During the French campaign in Egypt, the unit's equipment was lost when their transport ship was destroyed in the Battle of the Nile in 1799. What remained of the corps was then disbanded.[2]

During the mid-1800s, balloons had improved to the point that they could be used for aerial photography. The French photographer Félix Tournachon obtained the first such photograph, of a French valley, in 1858; unfortunately, the original has been lost. The oldest surviving aerial photograph was made in 1860 from Samuel Archer King's balloon, the *Queen of the Air*, and it shows the city of Boston.

The Civil War provided an opportunity for the US to test the effectiveness of balloons as platforms in battlefield reconnaissance. A number of civilian balloon operators offered their services to the Union Army. The most prominent of the group, Thaddeus Lowe, demonstrated balloon use to President Abraham Lincoln in June 1861. Shortly thereafter, Lowe was aloft in a balloon, providing reports on movements of Confederate forces in northern Virginia. Soon, the Union Army was sending observers up in balloons, typically relying on cavalry officers who were already experienced in reconnaissance.[3] One of these observers, George Armstrong Custer, who was to become famous after the war, recommended the use of night observation to monitor enemy encampments by observing their campfires during evening and early morning hours. This was the first known use of remote sensing from the air to observe nighttime activity—something that would later become common in a number of geospatial intelligence efforts, both for national security and civil applications.

Count Ferdinand von Zeppelin was assigned as Prussia's official observer with the Union Army during the American Civil War. In 1862 he visited Lowe's balloon camp and later took advantage of an opportunity to go up in one of Lowe's balloons.[4] The event apparently inspired him to create an improved platform, described next.

Airships

The balloon, being tethered, could observe only a line-of-sight region around its tether. A craft that could move about could cover far more area. That role was filled in 1874 by Count von Zeppelin's invention of a rigid airship. Also called a dirigible, it is a lighter-than-air craft powered by an engine-driven propeller that can maneuver and travel long distances.

Von Zeppelin's design was a rigid aluminum frame covered with a fabric envelope and containing several cells filled with hydrogen gas, with each cell having room to expand as the airship ascended. In 1895 he received a German patent for the invention, and the first model went aloft in 1900. By 1914 passenger travel in what had become known as the Zeppelin was a routine matter. During World War I, German Navy Zeppelins were often used in reconnaissance missions over the Baltic and North seas. Ten years after the war and the period of Germany rebuilding its capability, Zeppelins resumed passenger flights—until one named the *Hindenburg* burst into flames while landing at Lakehurst, New Jersey, in 1937, killing thirty-five passengers.[5]

The US began constructing military airships in 1917, in the form of the B-class blimp. Unlike the Zeppelin, a blimp is nonrigid, more like a balloon with no internal skeleton. Sixteen blimps were built during the war and used for antisubmarine patrols over the Atlantic. German U-boats would not attack a convoy having blimp protection. When US concerns about a possible war in Europe or Asia began to surface in 1933, the US Navy remembered that. It requested new modern antisubmarine airships. By December 1941, the US

Navy had ten technologically improved blimps that used nonflammable helium gas in place of hydrogen. During World War II, the Goodyear factory in Akron produced 154 airships—continuing the nonrigid blimp design.[6]

Blimps during the war continued the previous pattern of success; German U-boats sunk over 500 ships that were unescorted. But only one ship in a convoy having a blimp escort was sunk. While the blimps carried depth charges for attacking submarines, their primary benefit appears to have been in forcing submarines to submerge and avoid using periscopes—leaving the submarines effectively blind and negating their ability to attack convoys.[7]

Aerostats

The US blimp program continued through World War II because the blimps had a unique advantage over aircraft: persistence. A single aircraft could not perform long-term surveillance of a target area—an essential capability for what would later become known as activity-based intelligence (chapter 15). But a blimp could. The US Air Force rediscovered that advantage in the 1970s, though in a sense it went back in time to Thaddeus Lowe's idea. The newest version looked somewhat like a blimp but was tethered, like a balloon. It was called the **aerostat**. The first of this generation of lighter-than-air craft was deployed at Cudjoe Key, Florida, in 1980, with the mission of detecting low-flying aircraft. The official name for the vehicle was Tethered Aerostat Radar System (TARS), but the aerostat, shown in figure 8.1, was soon nicknamed "Fat Albert," in reference to a television and cartoon character of the time, and the name stuck.

The U.S. Customs Service then deployed a network of aerostats along the southwestern border to detect and track aircraft carrying illegal drugs into the US, and the US Coast Guard followed suit. They typically carry a mix of radar, optical, and SIGINT sensors. Their radar can track low-flying aircraft that are attempting to evade ground-based radars. Since 2013, the Fat Alberts in the US have been operated by the Department of Homeland Security.[8]

Military aerostats equipped with video cameras were used by the US in Iraq in 2004 and in Afghanistan in 2007. They soon became a part of everyday Afghan life, and many Afghans said that they hardly noticed them anymore. The Taliban insurgents definitely noticed them, and avoided areas under aerostat surveillance. In Kandahar Province, a Taliban stronghold, at the height of US presence it was difficult to find an area where no aerostat could be seen.[9]

Exotic Approaches to the High Ground

Military organizations are generally conservative in adopting untried technologies. But when the need is great, they can be highly creative. And as they began

FIGURE 8.1 A tethered aerostat radar system. *Source:* US Customs and Border Protection.

to recognize the advantages of aerial observation, they welcomed any means of obtaining this valuable information. At about the same time that balloons were being tested, some militaries experimented with two alternative means of getting remote sensors to the high ground. At least one of those turned out to be a winner.

Kites

In the late 1800s, several governments began experimenting with the use of kites for photography. The first known photograph from a kite was made by the French photographer Arthur Batut in 1889. In the next few years, British, French, American, and Australian enthusiasts made kite photography a popular pastime.

The next major advancement in kite reconnaissance was to send an observer aloft by kite. US Wild West showman Samuel Franklin Cody developed such a kite and patented it in 1901. It could carry an observer up to about 2,000 feet, which was a more than adequate height for battlefield reconnaissance. The British Army took note and adopted the Cody War Kite, using it during World War I. Balloons were still used as well. But the balloon could be sent up only in light winds, and the kite could operate only in strong winds. The combination was more effective than either one used alone.

Of course, after the airplane demonstrated its advantages for aerial photography, militaries abandoned the kite.

Pigeons

The homing pigeon was an interesting, if brief, experiment in aerial observation via camera. Homing pigeons will take the most direct route to their nest from the release point, so their flight path can be predicted. Obviously, the pigeon's carrying capacity is limited, so a camera had to be very light, and the timing of pictures had to be accurate. The Germans solved both problems in 1907. Their pigeon camera weighed 70 grams and automatically took a picture every 30 seconds. Believe it or not, during World War I, both Germany and the United States used pigeons to take pictures of enemy trench lines.[10]

Aircraft

Balloons and kites obviously had two major disadvantages as aerial platforms; the area that they could cover was confined to the general region around their tether point. And they were vulnerable to attack (as was the airship). Aircraft didn't have those problems. They were able to select and overfly areas to be photographed. And since aircraft moved far faster than airships, they were harder to pinpoint for attack. They found their first use for both observation and photography during World War I.

World War I

After the opening German offensive, World War I in the West became an almost static affair of trench warfare. Both sides had quickly recognized the advantages of aerial observation for obtaining geospatial intelligence, and airplanes were the preferred platform—first for visual observation, then for photography. Early in the war, both sides in fact thought that reconnaissance was the only useful purpose of the airplane. The trench patterns, and any forces mobilizing for an offensive, could be readily identified in photographs. And the aircraft provided a reliable and responsive platform for gathering photographic intelligence.

During the war, the major powers involved—French, British, Russian, Austro-Hungarian, Italian, and German—all developed reconnaissance aircraft, and improved them steadily as the war went on. The improvements initially were in making the aircraft a stable platform to get better image quality. Later, all sides developed fighter aircraft, and survivability—an issue to which we'll return in this chapter—became the driving force in designing reconnaissance planes. The belligerents quickly learned what became the reconnaissance pilot's creed: fly high and fast. Get the pictures and get out of the area quickly.

The Interwar Years

Up until the development of the aircraft as a photogrammetry platform, the high ground was important almost exclusively for military purposes. But in the aftermath of World War I, it revolutionized mapmaking for civil uses.

Recall from chapter 2 that Sherman Fairchild was a postwar leader in photogrammetry. His love of aerial photography led him into the business of making airborne platforms to carry his cameras. He bought a surplus World War I Fokker D.VII biplane to take aerial photographs, only to soon find out that the aircraft was less than ideal for the purpose. As a result, he founded Fairchild Aviation Corporation to build the FC-1, an aircraft specifically designed to provide accurate aerial mapping and surveying. The prototype FC-1 flew in 1926, and was later put into production as the FC-2. The company continued to build aircraft for military use throughout World War II and most of the Cold War. One of its products, the C-119, was not an imaging platform; it was a cargo aircraft. It nevertheless had an important part in the history of remote sensing, discussed in chapter 9.[11]

Talbert ("Ted") Abrams went through a transition similar to that of Fairchild. A Marine Corps aerial photographer in World War I, Abrams took any flying job he could find after his discharge. He flew as an air mail pilot. He organized barnstorming shows featuring aerial stunts, parachute jumping, and wingwalkers. Then he discovered that he could make more money from aerial photography than from stunt flying. He founded Abrams Aerial Survey Company in 1923 and quickly received numerous government contracts for aerial photographs. He replaced his old Curtiss JN-4 Jenny—it was a poor photography platform—and purchased several new aircraft and cameras for the work.

The new aircraft were better platforms, but not good enough for Abrams. Like Fairchild, he decided to build his own. In 1937 he founded Abrams Aircraft Corporation to build an aircraft specialized for aerial photography. The result was the revolutionary P-1 Explorer (fig. 8.2), which had its maiden flight in 1937. It was designed with a wide wing span for stability, and with a pusher engine to keep the front of the aircraft clear for the camera. The maps that Abrams's company produced from photographs were used for highway design and construction projects across the US.[12]

Abrams is regarded as the dominant figure in civil applications of aerial photography—so much so that the American Society for Photogrammetry and Remote Sensing each year recognizes the outstanding contribution to aerial photography and mapping with the Talbert Abrams Award.

FIGURE 8.2 The Abrams P-1 Explorer. *Source:* National Air and Space Museum, Gift of the Abrams Instrument Corporation.

World War II and Afterward

In World War II, mapping for civil applications took a back seat; the emphasis in aerial photography, along with the available resources, shifted back to obtaining intelligence from photo reconnaissance. The British Royal Air Force early in the war, remembering its World War I history, designed a specific aircraft for the task: a fast, small aircraft that would use high altitude and high speed to avoid being detected and attacked. It removed the armament and radios from some Spitfire aircraft and equipped them with aerial cameras. The reconnaissance version of the Spitfire was the first of the "fly high and fast" genre. The US followed suit with reconnaissance versions of its premier fighters. The best photo-reconnaissance aircraft of the war, however, probably was the British de Havilland Mosquito. Originally intended to be a bomber and one of the last aircraft made almost entirely of wood, the Mosquito could fly above most fighters at its operating altitudes above 40,000 feet.[13]

After World War II, the British began developing a jet-powered replacement for the Mosquito. The result was the English Electric Canberra, the best long-range aerial reconnaissance aircraft of its time. The Canberra first flew in 1950; it and its American version, the B-57, had service lives of more than fifty years. In 1957, a Canberra set an altitude record of more than 70,000 feet. In getting to the high ground, the aircraft had no peers.[14] But though the reconnaissance version, the RB-57D, could reach 70,000 feet, it could not stay there for a long mission. And staying above 60,000 feet was important for survivability. Soviet surface-to-air missiles were expected to be able to shoot down anything flying below that altitude by 1960. So in the early 1950s a US advisory panel recommended development of a replacement for the RB-57, and in 1954 President Eisenhower approved the project. He assigned the CIA the job of developing the aircraft, explaining that he did not want to see it entangled in the Defense Department's bureaucracy or immersed in interservice rivalries.[15]

The U-2 Program

Jake Kratt looked at his instrument panel and pondered his options. His reconnaissance mission training flight had been a success until now. But his lone engine had just flamed out. He couldn't make it to his home base in Nevada unpowered. He was going to have to make a dead-stick landing at some emergency field. But where? He was over the Mississippi River, headed West into Arkansas, and had his choice of US Air Force airfields that he could reach without power. Kirtland Air Force Base in New Mexico seemed a good choice. It was only 900 miles away. After a few message exchanges with his home base, the Pentagon notified the Kirtland base commander to open a sealed envelope that he had been provided for just such an event and to follow the procedures it contained. An aircraft would be making a dead-stick landing at his airfield in 30 minutes, and it would require special security protection. A half-hour later, the aircraft quietly glided in, and base security police watched in astonishment as Kratt, clad in a space suit, emerged from the cockpit of the strange-looking aircraft.[16]

In the spring of 1956, no jet aircraft in the world could glide for 900 miles—save one. The U-2 was built using the fuselage of Lockheed Aircraft's F-104 fighter, but with much longer wings that made it basically a jet-powered glider. Lockheed's choice of their F-104 fuselage was somewhat ironic when contrasted with the U-2's performance; with its short, stubby wings, the F-104 had a glide path approximating that of a brick.

The first of twenty original U-2s was delivered to the CIA in July 1955. The aircraft was capable of reaching 73,000 feet and traveling almost 3,000 nautical miles without refueling. Its first photoreconnaissance flight, on June 20, 1956, covered parts of Poland and East Germany. Overflights of the USSR began soon afterward, and continued until 1960, even though the Soviets' radar capability to track the aircraft was steadily improving, as was their surface-to-air missile coverage. Flights accordingly were planned to avoid known missile sites, but on May 1, 1960, a Soviet SA-2 missile brought down a U-2 piloted by Francis Gary Powers. Unfortunately, the site that fired the missile had not been previously identified for avoidance.[17]

No further flights of U-2s over the USSR were approved after the Powers shoot-down, but the aircraft continued to be used for photoreconnaissance over China (with Nationalist Chinese pilots), over Cuba during the 1962 Cuban missile crisis, and over Vietnam during the US involvement there.[18]

Survivability

By the 1970s, despite the improvements in reconnaissance aircraft, it was obvious that "high and fast" no longer worked over the territory of a sophisticated military opponent. The performance of fighter aircraft, air-to-air missiles, and surface-to-air missiles had outpaced the performance of reconnaissance aircraft. Since the need for reconnaissance never goes away, military forces developed two quite different alternatives.

The first solution was to move away from overflights, and instead do *standoff* reconnaissance. Airborne radar (addressed in chapter 12) could do that; it allowed imagery to be collected from aircraft at distances well removed from the area to be imaged. By 1990 the U-2 and the Joint STARS (JSTARS) aircraft, a modified Boeing 707, could both obtain radar imagery from ranges on the order of 150 miles.[19]

The other alternative to survivability turned out to be straightforward: Don't worry too much about it. Make the aircraft cheap, and don't put a person in it. If possible, make it stealthy. The result was the unmanned aerial vehicle (UAV).

UAVs

Like many "modern" innovations, UAVs have a lengthy history. During World War I, the US Army aircraft board asked the inventor Charles Kettering to build an unmanned aircraft—in effect, a flying bomb—that could hit a target at a range of about 120 kilometers. Kettering came up with a design, and the Dayton-Wright Airplane Company built it. It was called the Kettering Aerial Torpedo, later renamed the Kettering Bug. The Bug's gyroscope guidance system did not provide very good accuracy; and though forty-five Bugs were produced, it never was used in combat.[20] The British built a similar UAV, the Sopwith AT, in 1917, but it never flew.

In the interwar years, the British developed the first reusable UAV, called the Queen Bee, a target drone. During World War II, the German V-1—forerunner of the cruise missile—was produced in quantity, and more than 20,000 were launched at targets such as London and Antwerp.[21]

During and after World War II, many countries produced unpiloted drones for use as gunnery targets to train their pilots. One jet-powered US target drone, the Teledyne Ryan Firebee, looked like it might be useful as a reconnaissance platform. So in 1962, the air force began funding the development of reconnaissance versions of the Firebee, called the Ryan Model 147. Two years later, during the Vietnam War, Model 147Bs conducted photoreconnaissance missions over southern China and Vietnam. Versions of the UAV conducted reconnaissance at both high and very low altitudes. It was produced in SIGINT collection and photoreconnaissance versions. Its flights continued until 1975, a month after the Vietnam War ended.[22]

UAVs were relatively inexpensive, compared with conventional aircraft. But early models had two issues in conducting photoreconnaissance. They had to be programmed for a specific flight and specific targets, so they had no flexibility to take advantage of opportunities that developed inflight. And they had to return from their mission with their film, by which time the photographs they had taken were hours old.

Both problems were solved by the same innovation during the 1980s, and the solution had far-reaching consequences in both military and civilian matters. The solution was to put a video camera on the UAV and send the video stream directly to a ground site, with an uplink to send flight commands back to the UAV. Now, users could get a real-time picture of activity in an area, and could command the UAV to loiter over targets of interest. Suddenly UAVs didn't have to just conduct reconnaissance—periodic looks at an area. They could conduct surveillance—observing an area for an extended time. The first of the new generation was the Israeli Tadiran Mastiff, which saw action in the 1982–83 Israeli-Lebanon War. It was a model for future UAVs. The Mastiff could loiter and stream live video via a data downlink.[23] Other countries quickly followed suit with their own designs. These early models had a "tether," however; they had to remain within line-of-sight of their ground station in order to stream video and receive commands. Later on, UAVs began to uplink their video to a satellite, which relayed the video to a remote ground site. Flight commands then were sent back via satellite to the UAV. There was no risk to the pilot, who was sitting at a console a few hundred miles away or halfway around the world.

Since the 1990s, reconnaissance UAVs have proliferated worldwide. They are easy to buy or build. Today, they're so small and cheap that anyone can acquire them, and they come with a small camera onboard—so anyone can have a local espionage outfit. UAVs operate globally for many purposes unconnected with military matters. They help with law enforcement intelligence as they watch for poachers in African game preserves, for example. And they are used commercially to provide realtors with images and video for real estate listings, having nothing to do with intelligence.

Chapter Summary

Aircraft and later UAVs provided the essential platforms for photogrammetry during peacetime; in conflicts, they could be used for reconnaissance so long as the area to be overflown was not heavily defended. Where the air defense threat was too great, standoff reconnaissance and UAVs might operate (accepting the occasional loss of the UAV). But large areas of the world—Russia, China, India, and Pakistan, for example—were still off limits to US airborne reconnaissance platforms; standoff reconnaissance couldn't observe very far into either Russia or China. The solution to that strategic reconnaissance problem already existed, however, by the mid-1960s. The US, USSR, and other countries went to the ultimate high ground—space, the topic of chapter 9.

Notes

1. New Line Cinema, *Gettysburg*, 1993.
2. "The First Air Forces: A Century of Balloons at War," *Military History Now*, July 5, 2012, https://militaryhistorynow.com/2012/07/05/early-air-power-100-years-of-balloons-at-war/.
3. "First Air Forces."
4. "Ferdinand von Zeppelin: Biography and Facts," n.d., www.zeppelinhistory.com/zeppelin-inventor/ferdinand-von-zeppelin/.
5. "History of Zeppelins and Rigid Airships," n.d., www.zeppelinhistory.com/.
6. Naval Airship Association, Inc., n.d., www.naval-airships.org.
7. Naval Airship Association.
8. Dave Long, "CBP's Eyes in the Sky," US Customs and Border Protection, www.cbp.gov/frontline/frontline-november-aerostats.
9. Graham Bowley, "Spy Balloons Become Part of the Afghanistan Landscape, Stirring Unease," *New York Times*, May 12, 2012, www.nytimes.com/2012/05/13/world/asia/in-afghanistan-spy-balloons-now-part-of-landscape.html.
10. Andrea DenHoed, "The Turn-of-the-Century Pigeons That Photographed Earth from Above," *New Yorker*, April 14, 2018, www.newyorker.com/culture/photo-booth/the-turn-of-the-century-pigeons-that-photographed-earth-from-above.
11. Rebecca Maksel, "Cities from the Sky," *Air & Space Magazine*, January 12, 2009, www.airspacemag.com/history-of-flight/cities-from-the-sky-133299464/?page=1.
12. Wayland Mayo, "Talbert 'Ted' Abrams: Father of Aerial Photography," n.d., www.b-29s-over-korea.com/ted_abrams/ted_abrams02.html.
13. "De Havilland Mosquito," n.d., www.baesystems.com/en/heritage/de-havilland-mosquito.
14. Stephen Dowling, "Why NASA Still Flies an Old British Bomber Design," BBC, March 10, 2016, www.bbc.com/future/story/20160309-why-nasa-still-flies-an-old-british-bomber-design.
15. Gregory W. Pedlow and Donald E. Welzenbach, "The CIA and the U-2 Program, 1954–1974," CIA History Staff, 1998, www.cia.gov/library/center-for-the-study-of-intelligence/csi-publications/books-and-monographs/the-cia-and-the-u-2-program-1954-1974/u2.pdf.
16. Peter W. Merlin, *Unlimited Horizons* (Washington, DC: NASA, 2015), https://ntrs.nasa.gov/archive/nasa/casi.ntrs.nasa.gov/20150014963.pdf.
17. Pedlow and Welzenbach, "CIA."
18. Pedlow and Welzenbach.
19. US Air Force, "E-8C Joint Stars," September 23, 2015, www.af.mil/About-Us/Fact-Sheets/Display/Article/104507/e-8c-joint-stars/.
20. Jimmy Stamp, "Unmanned Drones Have Been around since World War I," *Smithsonian*, February 12, 2013, www.smithsonianmag.com/arts-culture/unmanned-drones-have-been-around-since-world-war-i-16055939/.
21. Smithsonian Air and Space Museum, "V-1 Cruise Missile," n.d., https://airandspace.si.edu/collection-objects/missile-cruise-v-1-fi-103-fzg-76.
22. "Ryan Model 147," n.d., https://enacademic.com/dic.nsf/enwiki/1675436.
23. "Tadiran Mastiff," *Military Factory*, n.d., www.militaryfactory.com/aircraft/detail.asp?aircraft_id=901.

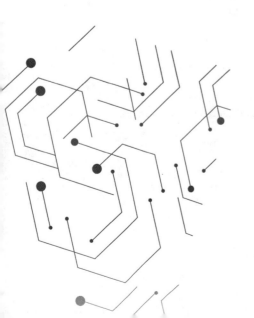

The Ultimate High Ground

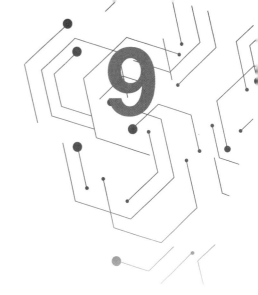

Soon after the launch of Sputnik I in 1957, governments recognized the potential value of satellites as a platform for sensors that could provide the raw material for geospatial intelligence. Driven by a need for intelligence about military developments within the USSR, the United States was the first into space with reconnaissance satellites. The Soviets soon followed. Other countries then developed space-faring technologies and launched their own remote-sensing satellites—some for military purposes, but increasingly for civil use. The ultimate high ground now is shared terrain for military, government, commercial, and research programs. The path to that ground had its beginnings in World War II–era hardware, and accelerated rapidly after the Sputnik launch.

The first move onto new "high ground" was not by a satellite, but by a rocket. After the end of World War II, the US Army acquired a number of German V-2 rockets and test-launched them from the White Sands Proving Grounds in New Mexico. Because the missile warhead compartment (the nose cone) was empty, it was available for scientific experiments. On October 24, 1946, the thirteenth V-2 test launch contained a motion picture camera, its film cassette protected by a steel capsule to allow it to survive the inevitable crash landing. And it did survive, providing the first picture of earth from space, 65 miles up (fig. 9.1). (A member of the Flat Earth Society made use of this image to point out that the Earth actually is flat, claiming that the slight curvature in the image is due to the angle of the picture.) More than 1,000 pictures of Earth were captured from V-2s between 1946 and 1950, some of them from 100 miles altitude.[1]

The next step, of course, was to put the camera into orbit so that more pictures could be taken, from a better angle. That began the era of remote sensing using Earth satellites.

Remote-Sensing Satellites

Remote sensing has become probably the most important tool supporting geospatial intelligence, largely due to emplacing sensors aboard satellites. During

FIGURE 9.1 The first picture of Earth from space. *Source:* US Army.

the 1950s, scientists recognized that satellites could be used to obtain details about the Earth for research. At about the same time, leaders of several countries became aware that satellite-based remote sensing could also be used to answer their intelligence questions. The answers that they would get depended on the sensors that the satellites carried (covered in chapters 10 through 12) and on the orbits selected.

The high ground has a somewhat different meaning when you go into space. Depending on what you want to do there, there are different altitudes and different orbits that a satellite can be put into. Planners have to choose between four commonly accepted orbital regimes, shown in figure 9.2. Each has advantages and disadvantages for geospatial intelligence:

- If you want to get detailed images of the Earth's surface, then you need to be as close as possible; image quality goes downhill fast as you move further out. But get too close, and the Earth's atmosphere slows the satellite and drags it out of orbit into a fiery death. So for imaging, you want a **low Earth orbit** (LEO), of between 200 and 1,000 miles above the Earth's surface. At those altitudes, though, only reconnaissance (not surveillance) is possible; a satellite can observe any given point on Earth for a short period (about 8 minutes).
- Go farther out, and you're in a **medium Earth orbit** (MEO). This orbit, at around 12,000 miles altitude, is too far out for good imaging. And, though satellites move more slowly at MEO, they still can't surveil an area for very long. MEO, though, turns out to be a perfect orbit for one of the essential GEOINT tools: the global navigation satellite.
- If you place a satellite in circular orbit at an altitude of 22,200 nautical miles, on a trajectory that takes it eastward along the equator, then the

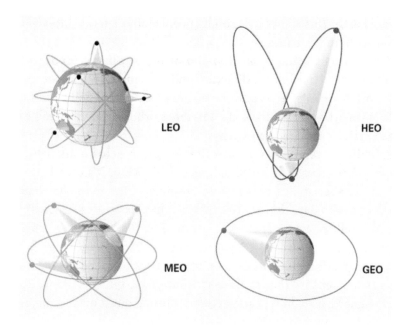

FIGURE 9.2 Satellite orbital patterns. *Source:* Robert M. Clark, *Intelligence Collection* (Washington, DC: CQ Press/Sage, 2014). Used by permission.

time it takes to circle the Earth is the same as that of the Earth's rotation (24 hours). This **geosynchronous equatorial orbit** (GEO) keeps the satellite over the same point on Earth at all times, so it can do surveillance. Yes, it's too far from Earth to be used for detailed imaging. But it is a useful orbit for monitoring global weather, that is, obtaining meteorological intelligence (e.g., cloud cover and storm paths).

- The **highly elliptical orbit** (HEO) is chosen to meet a geographic need: it allows a satellite to spend most of its time over a specific part of Earth. Areas near the North Pole are difficult to observe for satellites in GEO orbit. The higher up it is, the slower a satellite moves over the Earth's surface. So the HEO satellite spends much of its time near apogee (around 20,000 nautical miles), and very little time near perigee (around 300 nautical miles). This orbit was originally created to allow satellite communication with ground sites near the Arctic Circle. It turns out to be a good orbit for reconnaissance of that region, as well.

Once planners select a satellite's orbit, the next step is to establish its inclination, because that determines what areas of the Earth the satellite can observe. The **inclination** is the angle of the orbit measured from the equatorial plane, and some inclinations are critically important in geospatial intelligence applications. Different orbits use different inclinations.

A GEO satellite, for example, has zero inclination; it moves along the equator at the same speed as the Earth's rotation in order to maintain its fixed position over the Earth. This allows it to conduct continuous surveillance over about a third of the Earth's surface.

Some LEO satellites use what is called a **polar orbit**; they have a 90-degree inclination, moving in a north-south direction and crossing directly over the poles. The polar orbit provides access to the entire globe. Every point on the Earth is visible at one time or another. The satellite's orbit is roughly fixed in space while the Earth rotates underneath it.

Imaging satellites often use an orbit called a **sun-synchronous orbit**. At LEO altitudes, that orbit is about 98 degrees inclination. It's called sun-synchronous because the satellite passes over any given point on the Earth at about the same time each day (chapter 10 explains why this orbit is selected).

The first two satellites ever launched—Sputnik I by the USSR in October 1957, and Explorer 1 by the United States in January 1958—went into LEO and were for demonstration, the initial shots in what became called the "space race." Neither carried sensors that could perform remote sensing. The third such satellite, Vanguard, also didn't carry remote sensors, but it made a major contribution to understanding the Earth nevertheless.[2]

On March 17, 1958, the US Vanguard satellite was launched into an elliptical orbit with an apogee of 3,970 kilometers and a perigee of 650 kilometers. Today, it continues to be the oldest human-made satellite still in orbit. Its early tracking data resulted in our better understanding the size and shape of the Earth. Plots of Vanguard's orbit led to an interesting discovery: Earth is not a perfect sphere. It is somewhat pear-shaped, being elevated at the North Pole and flattened at the South Pole.[3] The result led cartographers to modify their world maps.

Since 1960, most satellites have been put into orbit for the purpose of Earth observation. In 2019, there were over 700 Earth observation satellites in orbit, the bulk of which are controlled by either the US or China.[4] Most such satellites operate in LEO and provide geospatial intelligence for other than military applications.

Government Nonmilitary Applications

Two of the early US launches placed satellites in LEO that could provide remote sensing for civil purposes—weather observation and monitoring the Earth's resources.

Weather Observation

The meteorological satellite named TIROS-1 (Television Infrared Observation Satellite) represented a series of firsts. Launched on April 1, 1960, it was the

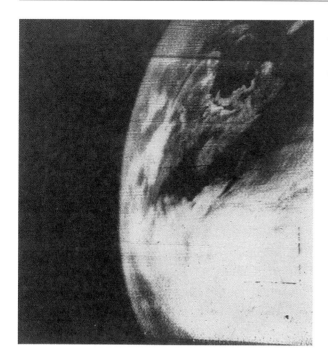

FIGURE 9.3 The first TIROS-1 image. *Source:* National Aeronautics and Space Administration.

first remote-sensing satellite put into orbit. It was the first satellite ever to carry a television camera. And it provided the first repeated images of Earth from space (the V-2 images could be obtained only once per launch). Figure 9.3 shows the first image returned, on May 9, 1960. TIROS-1's camera primarily imaged North America, but only for a limited time during each orbit. By 1972, a total of ten TIROS spacecraft had been launched.[5]

TIROS was the beginning of the global weather satellite program under what was to become the National Oceanic and Atmospheric Administration (NOAA). NASA developed the satellites, and NOAA managed the program, which later included the Nimbus satellite, NOAA Polar Operational Environmental Satellite, Geostationary Operational Environmental Satellite (GOES), and several others.

Earth Resources Monitoring

Landsat, introduced in chapter 3, continues to be the longest-running program for remote sensing in the world. The original satellite, named the Earth Resources Technology Satellite, was launched into an LEO sun-synchronous orbit on July 23, 1972. It was the first satellite designed to study and to monitor the Earth's surface, primarily the landmasses.

That Landsat was launched at all was remarkable, given the opposition in Washington agencies at the time. The US Bureau of the Budget argued that high-altitude aircraft could do the job with less risk and expense. The Defense Department was concerned about possible compromise of its ongoing satellite reconnaissance program. Foreign policy sectors of the US government feared a major blowback from foreign countries if images of those countries were taken without permission.[6]

But the US Geological Survey (USGS) wanted Landsat to support its survey mission. In what proved to be a brilliant political move, the USGS convinced the secretary of the interior to announce that the Interior Department would build its own Earth observation satellite. The threat drove NASA to push ahead with Landsat. Once Landsat 1 was in orbit, opposition quickly evaporated. The imagery made major contributions to research and environmental understanding in agriculture, forestry, geography, geology, mapping, oceanography, and water quality.[7] As the image quality improved, Landsat imagery was drawn into the web mapping services that proliferated in the early 2000s: Google Maps/Google Earth, MSN Maps, and Yahoo Maps. Landsat 8, the most recent in the program, went into orbit on February 11, 2013.

Military Applications

While scientists in the 1950s were excited about satellites as a source of remote sensing for research, military leaders in several countries were interested in their use for obtaining imagery intelligence. The US government had more than an interest; it had a pressing concern about national security, specifically a threat from the USSR. The Soviets had tested a hydrogen bomb in 1953. They had begun producing the long-range BISON bomber in 1955, and were believed to be developing intercontinental ballistic missiles. At the same time, the Soviets had become more secretive about their military progress. All of that amounted to the potential for a surprise nuclear attack on the US homeland. US leaders well remembered the Pearl Harbor debacle, and they needed intelligence that would explain the nature of the nuclear threat and provide advance warning of a sneak attack.

President Dwight D. Eisenhower had two possible solutions for obtaining the needed intelligence. The first was aerial photography of the USSR, which would have been a violation of national sovereignty—unless the Soviets agreed to it. So, at a Geneva summit meeting in 1955, Eisenhower presented the "Open Skies" proposal, under which the United States and the Soviet Union would exchange maps indicating the location of their military facilities. Each side would then be allowed to conduct aerial surveillance of those facilities in order to assure that the other side was in compliance with pending arms control agreements.

The Soviets rejected the proposal outright, as Eisenhower had expected. But it gave him the justification to begin U-2 reconnaissance overflights. The first flight was on July 4, 1956. And as we know from the previous chapter, the flights ended (over the USSR, at least) after the Francis Gary Powers shootdown on May 1, 1960.[8]

Eisenhower knew that U-2 reconnaissance would eventually have to end as Soviet air defenses improved. His second solution was to collect imagery from satellites, *if* that could be done without violating national sovereignty, as the U-2 had. It was well settled that a nation's sovereignty included the airspace over its territory. But how far up did that sovereignty extend? Did it have an upper limit? That question was not settled until 1957.

The Soviet launch of Sputnik 1 in 1957 established an important precedent in international law. Sputnik's low Earth orbit of necessity carried the satellite over almost every country in the world, including, of course, the US. The US government chose not to object to the overflight, recognizing the advantage of being able to overfly the USSR with reconnaissance satellites in the future. In fact, no country objected to the overflight. The right for satellites to orbit over any country was universally accepted after that.

And reconnaissance satellites were a game changer in strategic GEOINT. They have several advantages over airships, aircraft, balloons, and UAVs when it comes to collecting intelligence. But their main benefit is that they can gather information over denied areas (places where aircraft can't legally go)—for example, places like China, Russia, North Korea, and Iran.

Nevertheless, the operation of reconnaissance satellites was a sensitive subject that neither the US nor the USSR wanted to discuss openly. When the two countries were negotiating the Limited Test Ban Treaty of 1963, both agreed that neither would interfere with the other side's means of collecting intelligence to verify the treaty. But neither side wanted to use the term "spy satellites," and so they came up with the euphemism **national technical means**, or NTM. It continues to be used today, even though the rationale for the term is long gone. Another euphemism still in use is the term **overhead collection** assets, which is a puzzle. Don't aircraft collect imagery from overhead? But over the years, the term "overhead collection" has acquired an understood meaning: collection from satellites.

So, during the Cold War, as the space race was ongoing, both the US and the USSR moved into obtaining strategic GEOINT from space. The US program was managed by the National Reconnaissance Office (NRO). When it was established in 1961, the NRO was charged with the job of designing, building, launching, and maintaining America's intelligence satellites. It was headquartered in the Pentagon, and its existence remained classified for over thirty years, until 1992.[9]

During the NRO's history, it has launched and operated a series of imagery and SIGINT satellites that have provided both important national intelligence and extensive information about the Earth's features, both natural and human-made. While many NRO missions are classified, some of the most significant declassified missions are noted below.

Film Return Satellites

You could argue that imagery intelligence in the US began in earnest as a national asset to solve intelligence problems on August 18, 1960, when it launched the first imagery intelligence satellite that successfully returned a photograph from space. CORONA imaged denied territories—primarily the USSR and China—and returned the exposed film to Earth in capsules, which Air Force C-119 aircraft recovered in mid-air over the Pacific Ocean.[10] While the early mission targets primarily were Soviet strategic missile facilities, CORONA imagery also was used to produce maps and charts of all areas of the world for the Department of Defense and other US government mapping agencies.

A total of 145 CORONA missions between 1960 and 1972 produced over 800,000 images. It was a highly productive program, and imagery analysts in the intelligence and defense communities were kept exceedingly busy. To put that into perspective, according to the former director of the National Photographic Interpretation Center, Dino Brugioni, "The very first satellite mission that we flew captured a million square miles of Soviet territory. That was as much as twenty-four U-2 missions had captured in the Soviet Union over four years. So in one day, we got more film than all of the U-2 missions put together."[11]

The CORONA program was, in a sense, both the first and the last of its kind in space. US leaders were desperate for intelligence about the Soviet missile and bomber threats. It was a rare time when they were willing to spend heavily, and endure repeated failures, to obtain that intelligence. And CORONA had a special talent for failing. One mission misfired on the launch pad. Three failed to reach orbit, and two more went into eccentric and unusable orbits. Three experienced camera failures. One had its retro-rockets malfunction. The program managers had to repeatedly go back to Congress for more money to cover the escalating costs of the program. In fact, the August 18 success came after twelve unsuccessful attempts.[12] The patience of leaders in the face of repeated failures was remarkable, and an indication of how much they needed its intelligence. As former deputy director of the NRO Dennis Fitzgerald observed much later when discussing the program, "Do you know how many failures we'd be allowed today before being fired? At the most, two."[13]

Imaging satellites needed to fly as low as possible, in order to get good-quality imagery (i.e., high resolution[14]). CORONA had a perigee of 160 kilometers;

it was later lowered to 120 kilometers to provide better image quality. But this very low (for a satellite) altitude meant more drag from the upper atmosphere, and the satellite could not stay in orbit for long.

The compromise was to use an elliptical orbit. The satellites that followed CORONA did that. The last of the US film return satellites, called the KH-9 HEXAGON, had a perigee that was designed to occur around the regions of most intelligence interest (at the time, the USSR and People's Republic of China). The perigee was about 160 kilometers, with an apogee of 275 kilometers.

The KH-9 satellites provided imagery of Soviet and Chinese nuclear installations, missile sites, and other military activities. But they also provided material for cartography. Between March 1973 and October 1980, twelve KH-9 Mapping Camera System missions were flown. The USGS and the Defense Mapping Agency used this imagery for mapping and for creating terrain elevation data.[15]

Electro-Optical Satellites

The problem with film return satellites, of course, was the long delay between a photograph being taken and the imagery product being made available to decision-makers. The U-2 program could provide photographs a few hours after the aircraft landed. With film return, the delay was measured in days or weeks. Delay was not an issue for cartographers; Earth's features change very slowly, at least those of interest to mapmakers. But national leaders operate on a short time scale, and they demand current information. So an imagery platform that could produce quick results became a high priority. And in 1976, the NRO delivered. On December 19 of that year, it launched the KH-11 electro-optical reconnaissance satellite, which transmitted its images to earth via a relay satellite.[16] It allowed decision-makers near real-time access to developments around the globe—more quickly than even a U-2 could provide.[17]

Since KH-11, several countries have launched electro-optical satellites for intelligence purposes. One of the most recent is China's Gaofen-11, launched on July 31, 2018. Its orbit follows the same pattern pioneered by CORONA. It is in an elliptical near-polar orbit, reaching an apogee of 693 kilometers. Its perigee, which gives the best ground image resolution, is 247 kilometers.[18] (For comparison, the Landsat satellite never gets closer to Earth than 700 kilometers.) Because of some interesting features of its orbit, we'll revisit Gaofen-11 in chapter 10.

Current imagery satellites have another handicap in geolocating targets. They have to be relatively close to a target (i.e., LEO) in order to get the needed resolution for identifying it. That means (1) their area of coverage is limited (on the order of a few kilometers at any time) and (2) they can conduct periodic reconnaissance for only about 8 minutes during a pass; surveillance isn't possible. That appears to be changing. In a 2014 symposium, then–director of national intelligence James Clapper observed that "we will have systems that are capable of persistence: staring at a place for an extended period of time to detect activity; to understand patterns of life; to warn us when a pattern is broken, when the abnormal happens; and even to use ABI [activity-based intelligence] methodologies to predict future actions."[19]

Signals Intelligence Satellites

Imagery is the primary means for geolocating objects of military interest; but remember that it isn't the only means. Recall from chapter 7 that radiofrequency geolocation has a big advantage over imagery: it can be done day or night, in all weather. And communications emitters, which can be extremely small or concealed, may not be visible to imagers, but they can be detected and located when they emit radiofrequency signals.

Also, existing imaging satellites can't "stare" at a given area of the Earth continuously. In contrast, a SIGINT satellite (so called because it relies on receiving radiofrequency signals rather than optical energy) can conduct surveillance of about a third of the Earth's surface from GEO altitude. From there, it can detect and geolocate radiofrequency emitters throughout the region, 24 hours a day. Surveillance gives it another advantage: it can geolocate and track moving vehicles, ships, and aircraft (if they emit a signal)—something that current imaging satellites can't do for more than a short period of time.

Some SIGINT satellites operate at LEO, where being close to the target allows them to detect weak signals and to cover the entire Earth rather than a third of it. But like imaging satellites, they can only do reconnaissance—which is why they're often called electronic reconnaissance satellites.

First the US, then the USSR, and later several other countries developed SIGINT satellites that could geolocate radars for what is called **electronic intelligence** (ELINT) and communications systems for **communications intelligence** (COMINT). The NRO launched the first such satellite on June 22, 1960, two months before the first CORONA launch—making it the first reconnaissance satellite ever launched. It had a cover mission and story to conceal the actual mission from the government of its primary target: the USSR. The satellite was called the Galactic Radiation and Background (GRAB) experiment. A second GRAB went into orbit on June 29, 1961, and operated until August 1962, collecting Soviet radar signals. In 1962, the NRO launched a GRAB successor, called POPPY, with the same mission.[20]

The Soviets soon followed suit. In 1967, they launched the first in their series of ELINT satellites, called Tselina. Since then, both countries have steadily improved their capabilities to geolocate radar and communications emitters from space, and several others have done the same.

Commercial Imaging Satellites

The secret to a good intelligence service is *comparative advantage*: having intelligence sources that give the decision-maker an edge over opponents. During the Cold War, the US had that advantage in its reconnaissance satellites. It held a quasi-monopoly in imaging satellites (the USSR had them also, but their imagers were less capable than those the US put into orbit). And it wanted to protect its advantage.

But by 1980, it was apparent that the US edge was about to erode. Commercial imaging satellites were being developed around the world and would be in use by the end of the decade. Such developments posed a special problem for the US government, which had a substantial technological edge in the quality of its overhead imagery, and did not want to see a potentially lucrative industry develop overseas. At the same time, it definitely did not want high-quality imagery in the hands of unfriendly governments or groups who might use it to collect intelligence, conduct industrial espionage, plan terrorist attacks, or mount military operations. So the government made the decision to foster a commercial satellite industry within the US, where it might be able to maintain some control over distribution of the product.

Consequently, in 1984 Congress passed the Land Remote Sensing Commercialization Act, which allowed the development of a commercial imaging satellite industry. Under the terms of the act, NOAA transferred responsibility for Landsat data and resources to an industry consortium, the Earth Observation Satellite Company (EOSAT), and contracted with EOSAT to build, launch, and operate Landsat 6 and Landsat 7.

Four different government-commissioned studies at the time had warned that the commercial market for satellite imagery was not sufficiently developed to sustain a commercial remote-sensing industry.[21] The predictions proved to be correct. EOSAT was unable to create a successful commercial business from the Landsat program, even with NOAA funding support. In order to create a profit, the company raised imagery prices, driving away customers and resulting in even higher prices. As the customer set dwindled, EOSAT cut back on imaging operations and major gaps began to develop in Earth coverage. From 1989 to 1992, Congress had to provide emergency funding to NOAA to avoid having Landsats 4 and 5 shut down. The bid to commercialize Landsat imagery had failed. During 1992, EOSAT ceased to process Landsat data, and the government subsequently had to take back ownership of the program.[22]

Meanwhile, France had begun to put into orbit its SPOT (Système Pour l'Observation de la Terre, or System for Earth Observation) satellites, beginning with the launch of SPOT 1 on February 22, 1986. The SPOT program soon demonstrated commercial success where EOSAT had failed. The first two of the series, launched in 1986 and 1990, provided higher-resolution images than Landsat could offer, with a shorter revisit time, and at lower prices. By 1989, SPOT had passed ahead of EOSAT in the commercial imagery market.

By 1992, it was apparent that the US attempt to have it both ways—develop a US commercial satellite industry while protecting national security—had failed; in trying to get both, it had gotten neither. Other countries besides France were developing commercial imaging satellites that soon would offer better resolution, better prices, and global coverage. Congress repealed the 1984 act and replaced it with the Land Remote Sensing Policy Act of 1992. It eased restrictions on the US commercial imaging industry and provided a government market for the product.

A Procession of Commercial Imaging Companies

After the passage of the 1992 act, several commercial imaging companies were founded in the US, spurred by the promise that government would buy at least some of the imaging product. And the government did respond, in the form of both direct subsidies and guaranteed purchases of the product. The three US imaging companies operating in the mid-1990s—EarthWatch, Space Imaging, and OrbImage—were awarded contracts to upgrade their ground systems so that imagery could be transferred from their satellites to the National Imagery and Mapping Agency, and government purchases of commercial imagery steadily increased.[23]

It wasn't enough. The commercial demand, even with government subsidies, wasn't sufficient to keep three companies solvent in the face of foreign competition, and the industry went through a period of consolidation, illustrated in figure 9.4:

- EOSAT, founded in 1984, had struggled financially. It became Space Imaging in 1994 as a consortium of four investors: Lockheed Martin Space Systems, Raytheon Systems Company, Mitsubishi, and Hyundai. It was acquired by GeoEye in 2006.
- Orbital Imaging Corporation was founded 1992 as a division of Orbital Sciences Corporation. It changed its name to GeoEye in 2006 after acquiring Space Imaging for $58 million.
- WorldView Imaging Corporation was founded in January 1992. In 1995, it became EarthWatch Incorporated, and in September 2001, EarthWatch became DigitalGlobe and subsequently purchased GeoEye. As the single surviving US company in the business, it built an extensive business serving both governmental and commercial markets.[24]
- In October 2017, MacDonald, Dettwiler, and Associates acquired DigitalGlobe. The company then was renamed Maxar Technologies Ltd.

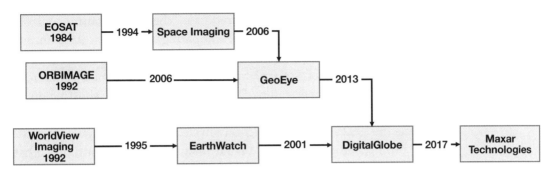

FIGURE 9.4 US commercial satellite imagery companies. *Source:* Author's collection.

While the US struggled with how to deal with commercial imaging satellites, other countries moved in to fill the demand. Since SPOT 1, there have been a series of SPOT launches by Spot Image, the company that distributes the imagery to customers. Spot Image now sells imagery from both optical satellites (France's Pleiades, Taiwan's Formosat-2, and South Korea's Kompsat-2) and radar satellites (Germany's TerraSAR-X, European Space Agency's ERS, and Canada's Radarsat).

The Challenge of Legislating a Commercial Satellite Industry

An important provision of the 1984 act had provided for the issuing of regulations to ensure that national security concerns were protected. So from the beginning, restrictions had been placed on commercial imaging from space by US companies. The 1992 act resulted in a surge of companies entering the industry. But both the 1984 and 1992 acts and subsequent regulations continued the policies that forced these companies to offer an inferior product—never a good business strategy. For example, currently,

- Regulations restrict the quality of imagery that can be sold (a number that changes regularly as higher-quality imagery becomes available from non-US providers). The US government for years banned companies from offering commercial satellite views with a pixel resolution better than 50 centimeters. Sharper images could be sold only to the government. In 1999, EarthWatch requested permission to sell 0.25-meter resolution satellite imagery. It took fifteen years for the government to grant permission, despite the ready availability of much sharper imagery from aircraft during that time.[25]
- A controversial regulation allows the government to exercise **shutter control**. That is, the US can prohibit the imaging of selected areas. The regulation is widely regarded as a violation of the Constitution's First Amendment. The US has never actually exercised that option, though it

has made use of a technique referred to as "checkbook shutter control," likely to avoid a court challenge. In preparing for the post-9/11 conflict in Afghanistan, the government bought exclusive rights to all high-resolution images of Afghanistan produced by the Ikonos commercial imagery satellite (operated by DigitalGlobe), effectively shutting down access to the imagery by anyone else.

- While the 1992 act eased restrictions on electro-optical imaging, an existing prohibition on commercial radar imaging satellites continues. That restriction effectively cedes the commercial radar imaging field to Canadian, German, and Italian firms.
- One section of the 1997 US National Defense Authorization Act bans US companies from releasing satellite images of Israel and Israeli-occupied territories of a quality higher than that available from non-American commercial sources. Under this act, US firms also must wait months for government approval to enter into foreign imagery sales agreements.

Figure 9.4 illustrates the result of thirty-five years of US pursuit of two basically contradictory objectives. The policy has not worked well, and regulations on image quality and use have constrained innovation within the US. And, the entire government effort to control commercial imaging worldwide has mainly served to encourage competitors.[26] In the last decade, however, another development in imaging satellites has emerged, one where US restrictions on image quality are not an issue—the move to what are called smallsats.

Small Commercial Imagery Satellites

Small satellites (known as smallsats) can now be built at relatively modest cost with automotive-grade electronics and can provide a resolution of about 3 to 5 meters, well above the US lower limit restriction on image quality. But that's good enough for a wide range of commercial and research missions, such as climate monitoring, crop yield prediction, urban planning, and disaster response. By 2019, one company, Planet Labs, had over 130 such satellites—each about the size of a loaf of bread—in orbit and obtaining massive amounts of imagery available for commercial sale.[27] But there's more. It seems that Planet Labs is capable of supporting geospatial intelligence as well. One of its satellites obtained the image of the al-Dawadmi missile test facility described in chapter 1.

Another commercial development appears imminent, and it would be an impressive leap forward. LEO satellites can observe a given point on Earth for only a few minutes, as noted above. But if you put enough of them in orbit, there will always be at least one satellite in position to view that point. In 2018 a start-up company, EarthNow, proposed to do exactly that. The company is backed by a multinational consortium including Microsoft's Bill Gates; the French aeronautical company, Airbus; the Japanese multinational conglomerate, SoftBank Group; and OneWeb founder, Greg Wyler. EarthNow proposes to build a constellation of small advanced imaging satellites that will provide continuous video of most areas of Earth. EarthNow's development schedule is ambitious; the company plans for an eventual network of 500 imaging satellites in LEO, each weighing about 200 kilograms (slightly too heavy to be called a smallsat). And the planned resolution does not run afoul of US resolution restrictions. Launches are proposed to begin in 2020, with an initial constellation in orbit by 2022.[28]

The EarthNow plan to provide video surveillance is likely to meet some resistance from governments and privacy advocates. Whether it will trigger a new round of restrictions is yet to be determined; the multinational nature of the consortium could make that difficult to do.

Chapter Summary

The primary reason for gaining the high ground was to use it for observations. Until the twentieth century, that almost always meant humans making visual observations and reporting them verbally or in writing, as Union observers on Little Round Top did at Gettysburg. With the advent of aircraft and satellites, sensors were needed that could be carried on these platforms to provide more detail than the unaided human eye could deliver, along with a permanent record. Cameras initially filled that need, a topic explored in chapter 10.

Governments (initially the US and USSR) took the lead in going to the ultimate high ground. The first Earth observation satellites in orbit monitored weather patterns, where good imagery resolution was not needed. Satellites to collect intelligence (imagery or radiofrequency signals) soon followed. The need for dramatic improvements in image quality and quicker access to an image led to a switch from film to electro-optical imaging. Monitoring Earth resources came next, and has become the most common application of imagery satellites.

Commercial imaging applications were slower to develop, because governments (the US in particular) were reluctant to yield the intelligence advantage that imaging satellite technology gave them. That has changed in the last three decades; companies are now developing inexpensive small satellites that can produce good-quality images, greatly expanding the market for commercial imagery products.

Notes

1. Tony Reichhardt, "First Photo from Space," *Air & Space Magazine*, October 24, 2006, www.airspacemag.com/space/the-first-photo-from-space-13721411/.
2. Constance McLaughlin Green and Milton Lomask, *Vanguard: A History* (Washington, DC: National Aeronautics and Space Administration, 1970), https://history.nasa.gov/SP-4202.pdf.

3. Green and Lomask.
4. "How Many Satellites Orbiting the Earth in 2019?" *Pixalytics*, www.pixalytics.com/satellites-orbiting-earth-2019/.
5. NASA, "TIROS," n.d., https://science.nasa.gov/missions/tiros.
6. NASA, "Landsat Science: History," August 8, 2018, https://landsat.gsfc.nasa.gov/about/history/.
7. NASA, "Landsat."
8. US Department of State, "U-2 Overflights and the Capture of Francis Gary Powers, 1960," https://history.state.gov/milestones/1953-1960/u2-incident.
9. NRO, "50 Years of Vigilance from Above," www.nro.gov/about/50thAnniv/50th-Flyer.pdf.
10. NRO.
11. NRO video clip, www.nro.gov/corona/videoclip.html.
12. Kevin C. Ruttner, ed., "CORONA: America's First Satellite Program," Central Intelligence Agency, 1995, www.cia.gov/library/center-for-the-study-of-intelligence/csi-publications/books-and-monographs/corona.pdf.
13. Dennis Fitzgerald, personal communication, November 2006.
14. Imagery "resolution" refers to spatial resolution, which describes the minimum separation distance between two objects that allows them to appear as separate in the image. The higher the resolution, the closer that the two objects can be and still be perceived as separate.
15. NRO, "NRO History and Heritage," n.d., www.nro.gov/Portals/65/documents/about/50thAnniv/50th-Flyer.pdf.
16. An electro-optical sensor is an electronic detector that converts light energy into an electronic signal—something that digital cameras do. A satellite is referred to as "electro-optical" when it then transmits the resulting image to earth using RF transmission.
17. NRO, "NRO History."
18. Andrew Tate, "China Closing the Satellite Imagery Capability Gap," *Jane's Defence Weekly*, August 14, 2018, www.janes.com/article/82366/china-closing-the-satellite-imagery-capability-gap.
19. Colin Clark, "DNI Clapper Teases 'Revolutionary' Intel Future; Big Cost Savings from Cutting Contractors," *Breaking Defense*, May 22, 2014, https://breakingdefense.com/2014/05/dni-clapper-teases-revolutionary-intel-future-big-cost-savings-from-cutting-contractors/.
20. NRO, "GRAB and POPPY: America's Early ELINT Satellites," n.d., www.nro.gov/history/csnr/programs/docs/prog-hist-03.pdf.
21. Ann M. Florini and Yahya A. Dehqanzada, "No More Secrets? Policy Implications of Commercial Remote Sensing Satellites," Carnegie Endowment for International Peace, July 1, 1999, https://carnegieendowment.org/1999/07/01/no-more-secrets-policy-implications-of-commercial-remote-sensing-satellites-pub-150.
22. Florini and Dehqanzada.
23. Ann M. Florini and Yahya Dehqanzada, "Commercial Satellite Imagery Comes of Age," *Issues.org*, n.d., http://issues.org/16-1/florini/.
24. Christopher Lavers, "The Origins of High-Resolution Civilian Satellite Imaging, Part 2: Civilian Imagery Programs and Providers," *Directions Magazine*, February 4, 2013, www.directionsmag.com/article/1646.
25. Walter Scott, "US Satellite Imaging Regulations Must Be Modernized," *SpaceNews*, August 29, 2016, http://spacenews.com/op-ed-u-s-satellite-imaging-regulations-must-be-modernized/.
26. Kevin M. O'Connell, "The Geospatial Revolution: What's a Government to Do?" Spacenews.com, July 3, 2017, www.innovative-analytics.com/newspress/geospatial-revolution-whats-government/.
27. Brit McCandless Farmer, "Can Commercial Satellites Be Used for Espionage?" CBS News, January 27, 2019, www.cbsnews.com/news/can-commercial-satellites-be-used-for-espionage-60-minutes/.
28. Alan Boyle, "EarthNow Fleshes Out Plan to Deliver Video That Shows Earth from Orbit in Real Time," *GeekWire*, February 4, 2019, www.geekwire.com/2019/earthnow-video-orbit/.

Visible Imaging

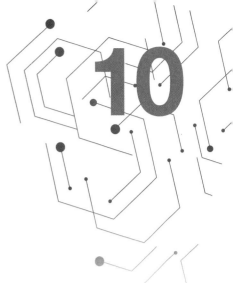

Platforms on the ultimate high ground, hundreds of miles above the Earth, are of little use without something more than 20/20 vision, the limits of our innate ability to see. The second piece of technology required to provide long-range remote sensing is, of course, a sensor. Over the last century or so, we've seen a procession of such sensors matching the progress of platforms. It is the synergy between the two that has paved the road to today's geospatial intelligence revolution. Let's turn our attention now to sensors; the first in the progression is still in use, albeit greatly improved: the camera.

Cameras, from their first use in 1816, produced black-and-white photographs in the visible part of the electromagnetic spectrum—that part of the spectrum that is visible to the human eye. They created the simplest type of optical image: a panchromatic image. Visible light represents only a very small part of the total spectrum, but it was the only part that mattered for more than a century. Later on, the idea of imaging outside the visible spectrum developed—the subject of chapter 11. This chapter is about understanding imaging in the visible band.

Cameras were important for their role in cartography, as we have seen. But cartographers were the incidental beneficiaries of camera development to support geospatial intelligence. During World War I and World War II, aerial camera improvements were driven by their value in military reconnaissance. After the wars, cameras then found use in mapping. During the Cold War, a similar need for intelligence drove the development of satellite-based cameras, and again cartographers benefited in better tools to use for mapping.

Aerial Film Cameras

During World War I, all the major powers used aerial cameras for reconnaissance. Germany had a scientific lead in camera design at the war's outbreak, and it adopted the first aerial camera, a Görz, in 1913. High-resolution photographs

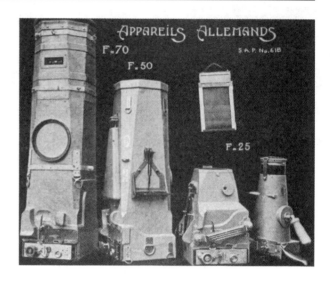

FIGURE 10.1 German World War I aerial film cameras. *Source:* Wikipedia, from Herbert E. Ives, Airplane Photography.

require long-focal-length cameras, and at the time, the Germans had better focal length cameras; several of them are shown in figure 10.1.[1]

Early film cameras had to overcome a number of challenges. And the operator or photographer did not have it easy. He had to lean over the side of an open-cockpit aircraft and manually take the photograph. Then he had to remove and store the exposed glass plate and insert an unexposed plate in order to take another image. And it all had to be done in what were typically subfreezing conditions. During World War I, British aircrews used the Watson Air Camera, and photography specialists invented a method of developing the glass plate negatives in flight so that they would be ready to create print copies when the aircraft landed. Still, handheld aerial cameras required a skilled photographer to operate them, and the picture quality depended on just how good the photographer was. If the camera could be affixed somewhere on the bottom of the aircraft, more photographs of predictable quality could be taken with far less stress on the photographer. As always, there was just one problem. Aircraft vibration and movement created image distortion. The solution was a camera with a fast shutter speed, and those became available in 1915.[2]

The next improvement in aerial cameras was to provide more coverage. The single-lens camera could cover only a relatively small area with a single photograph, and reconnaissance aircraft could not afford to loiter for an extended time over enemy territory to get more complete coverage by taking more photographs. After the United States entered the war in 1917, one of its aerial photography experts solved that issue. James W. Bagley was a topographic engineer who had worked for the US Geological Survey and transferred to the US Army Corps of Engineers after the war began. He and Captain Fred Moffit developed a camera that had three lenses. Their T-1 camera simultaneously

FIGURE 10.2 US Army Air Corps camera and aircraft. *Source:* US National Oceanic and Atmospheric Administration.

took one image vertically and two oblique images (looking at an angle sideways from the aircraft). It could photograph a much larger area in one pass than the single-lens camera. After the war, Major Bagley continued to improve aerial photography for military applications and to refine the art of photogrammetry. During this time, he developed a four-lens camera, the T-2, and later a five-lens camera, the T-3, that could give one vertical and three or four oblique images in one film exposure.[3]

The Fairchild Camera (introduced in chapter 2) arrived too late to be used in World War I. But Sherman Fairchild continued to create more sophisticated versions. Soon after the war, the US Army selected the most refined Fairchild model of the time as its standard aerial camera. Figure 10.2 shows Fairchild's 1928 T-2 camera along with the DH-4BP reconnaissance aircraft that carried it.[4] Fairchild's aerial cameras continued to improve during the interwar years, both for military reconnaissance and for mapping. By 1936, Fairchild was providing a nine-lens camera for the US Coast and Geodetic Survey's mapping operations.

Photo intelligence (later redefined as a contributor to geospatial intelligence) became an essential precursor to any operation during World War II. Long-range reconnaissance was necessary for the strategic planning that preceded combat operations. Fairchild responded to the military need, producing most of the aerial cameras used by Allied Forces during the war, including Bagley's T-2 and T-3 cameras. Aerial photographs were used both for making military maps and for creating battle damage assessments using the photography captured before and after bombing runs.

After World War II, emphasis in aerial photography returned to that of prewar days: camera manufacturers mostly were concerned with building better mapping cameras that could provide pictures with more detail and accuracy. But as the Cold War developed, the US once again needed aerial film cameras that could be used in reconnaissance: that is, cameras providing the detail that

FIGURE 10.3 Model A-2 cameras being loaded into a U-2. *Source:* US Air Force.

would allow analysts to identify threats, and do so quickly. So larger cameras were built that could keep pace with improved aircraft that had a higher flight ceiling, culminating in the U-2 camera system.

The U-2 carried three reconnaissance cameras that were specifically designed to take photographs from high altitudes. The cameras had to operate in extreme cold and under low atmospheric pressure; and because payload weight was limited in the U-2, the camera and film had to be lightweight. First produced in the late 1950s, the Model A-2 camera had much better resolution than previous aerial cameras—2 feet in resolution from an altitude of 60,000 feet. Each of the three cameras shown in figure 10.3 carried 1,800 feet of Eastman Kodak's lightweight Mylar-based film.[5] At the same time, engineers were working on the challenge of moving cameras into space.

Satellite Film Cameras

Making a camera work on an aircraft was not an easy task. But in comparison with making the camera work on a satellite, it was child's play. Putting one into space meant solving more than just one problem. In fact, there was a long list:

- First, it takes an incredibly tough camera not only to survive the vibration during a satellite launch but also the subsequent hostile space environment. And once it's in space, there are few options for repairing it, as the engineers who built the Hubble space telescope can attest. After its 1990 launch, engineers discovered that the $1.5 billion Hubble's main mirror was improperly ground; it couldn't deliver acceptable images.[6] Five space shuttle missions from 1993 to 2009 were needed to repair, upgrade, or replace Hubble components.
- Second, for both reconnaissance and mapping, it's desirable to get high-resolution images. That was a manageable problem for aircraft; the U-2 took its pictures from about 20 kilometers above the Earth. In comparison, a satellite takes its pictures from a distance of 160 kilometers or more. Images from a satellite using the U-2 camera would have had about eight times worse resolution than U-2 imagery, which was unacceptable. The solution was to use a long-focal-length camera. A satellite camera therefore has to be a high-resolution telescope with a film-handling device attached to it. But it's extremely difficult to fit a long telescope into a spacecraft bus; the early ones, at least, were not that big. The solution was to use mirrors to create what are called folded optics, where the focal length is much longer than the telescope itself. (Many binoculars and astronomical telescopes use this trick.) The telescope mirrors had their own set of challenges; they had to be much lighter than those used in large conventional telescopes because more weight translates to heavier (and more expensive) launch vehicles. At the same time, a lighter mirror is more likely to flex and ruin the image. A lightweight mirror that could handle the mechanical stresses and temperature changes in space required clever design and the use of exotic materials.
- Third, the film has to be specially designed to operate in a hostile space environment. It must be very-high-speed film, that is, film that is highly sensitive to light. And at the same time, it has to have a high resolution to match the quality that the telescope optics can provide. Kodak, drawing on its U-2 experience, was able to provide the required quality film for the KH-4 CORONA and its successors.
- Finally, some method had to be found to get the film back to Earth from the satellite. The solution was to run the exposed roll of film into a reentry capsule known as a "film bucket." The bucket then was jettisoned from the satellite and reentered the atmosphere, protected by a heat shield. At around 18 kilometers in altitude, the bucket slowed enough to deploy a deceleration parachute followed by a main parachute. The idea was to have an Air Force C-119 (ironically, built by Sherman Fairchild's company) capture the main parachute and bucket during descent (fig. 10.4). If the

FIGURE 10.4 A C-119 recovering a CORONA film bucket. *Source:* US Air Force.

aircraft missed its target, as sometimes happened, the US Navy would pick the bucket up out of the water. The bucket was designed to float for two days, then sink if it had not been recovered.

Those were just the major problems. There were others that had to be solved as they were discovered. For example, the camera could become out of focus due to temperature changes in the telescope mirror. Early CORONA missions encountered film fogging that eventually was discovered to result from static electricity discharges from camera components.

Despite these cumulative challenges, during the 1960s two US companies—Itek and Perkin-Elmer—claimed that they could build a space-qualified camera. Both companies got the chance, and both were successful.

Itek produced the first cameras used on CORONA. The cameras were initially 5 feet long, later extended to 9 feet to give a better focal length (and, therefore, better resolution). Initial camera resolution was around 40 feet, but by the end of the CORONA program the resolution had improved to about 6 to 10 feet.

The later KH-9 HEXAGON satellites carried a main camera built by Perkin-Elmer. It featured a larger lens than the CORONA camera, and an improved resolution of about 2 feet. Twelve of the HEXAGON missions carried a separate special camera that had a low resolution of 20 to 30 feet and was intended to provide photographs for mapmaking. These missions provided imagery of almost the entire Earth between 1973 and 1980. The images have since been declassified and transferred to the US Geological Survey.[7]

Digital Cameras

Film return from satellites was an expensive endeavor, and mishaps did occur. But the biggest issue was that intelligence customers often had to wait for days or weeks to see the photographs after they had been taken. And national-level intelligence customers can be understandably impatient; in a fast-developing crisis, several days is too long to wait for a decision advantage. Some method of capturing high-resolution images in electronic form was needed; in electronic form, they could be transmitted immediately to a ground station.

Video cameras had already demonstrated the ability to do that. On April 1, 1960, the US launched its first successful meteorological satellite, TIROS-1, introduced in chapter 9. It carried a television camera, called a vidicon, which could observe cloud cover and be used for weather prediction.[8] Vidicons are relatively rugged and cheap. And they can transmit the video images back to Earth; no need to deorbit a film return bucket. Still, they weren't the solution to the intelligence problem of avoiding film return. A vidicon camera's spatial resolution is very poor—an unimportant point, if all you wanted to see was cloud cover.

What was needed was a camera that could capture a high-quality image electronically. The solution to that problem was discovered almost by accident while the CORONA satellites were still in orbit, though the solution wouldn't go into space until seventeen years later.

In 1969, the physicists Willard S. Boyle and George E. Smith were in Murray Hill, New Jersey, working at Bell Labs, in the semiconductor division. The electronics department was divided into two sections: those who worked on semiconductor devices and everyone else. A group not in the semiconductor section was working on a new type of "bubble memory" technology for computers—one that used thin magnetic film, rather than semiconductors, to store data in tiny "bubbles" of magnetism. Upper management was excited about that idea and gave Boyle and Smith an ultimatum: come up with a better semiconductor memory device, or you'll likely lose your funding to the other group.[9]

The assignment greatly focused Boyle's and Smith's attention. In a one-hour brainstorming discussion, they sketched out the basics of a design. It was brilliant. But they didn't succeed in making a better memory device. What they developed, instead, was an integrated circuit array that was sensitive to light. When a light photon hit the array, it would create an electrical charge. And the charge pattern created by many photons hitting the array could be recovered and displayed as an image.[10] In layman's terms, Boyle and Smith had created a sensor for recording images that would eventually make film cameras obsolete.

This semiconductor circuit later received a name: the **charge-coupled device** (CCD). And forty years afterward, in 2009, it earned its inventors a Nobel Prize in physics. It also opened the floodgates to a steady stream of innovative devices, including at least one that most of us carry around and often use without thinking about how it came to be. It was Steven Sasson at Eastman Kodak who invented the first digital camera by placing the CCD array into the focal plane of a camera in 1975. Unfortunately for Kodak, its management didn't recognize the potential of Sasson's camera.

When Sasson showed Kodak management his new invention, their reaction was "that's cute—but don't tell anyone about it."[11] Also, the first digital images were inferior to what film could deliver. But Kodak, which derived most of its profits from selling film, made the mistake that its competitors did not: comparing a mature technology with a nascent one. Kodak's love affair with film was the proximate cause of its bankruptcy in 2012, ending a century of its dominance in the field of photography. CCDs from the beginning had 100 times more sensitivity to light than the best camera film, translating to far greater resolution in an image than film cameras could deliver *if* a large enough number of pixels could be placed in a CCD array. Sasson's first digital camera produced inferior images because it had only a 100 × 100 pixel array, or a total of 10,000 pixels. For comparison, in 2016 the iPhone 7's camera provided a resolution of 12 million pixels. In the same year, commercially available handheld cameras did even better; they delivered resolutions between 24 and 51 million pixels.[12]

CCD cameras were soon aboard reconnaissance satellites. The first of a new generation of satellites was the KH-11 Kennan, launched in December 1976, and it was also the first to carry a CCD camera.[13] Details about that camera have not been publicly released, but digital cameras soon thereafter made their way onto commercial satellites. The SPOT satellites have used CCD imagers to provide high-resolution images since the first launch.

What's the largest pixel camera to go into space? The number keeps growing. Russia's Elektro-L weather satellite, launched in January 2011, produces images of 121 million pixels. But that isn't the current known leader. The European Space Agency's Gaia satellite, launched in December 2013, features a camera of 1 billion pixels.[14] That's an array of about 33,000 × 33,000 pixels, a far cry from the original Eastman Kodak digital camera's 100 × 100 pixels.

Video Cameras

Video cameras predated the development of CCD imagers. But as smartphone users well know, digital cameras can create high-quality video as well as pictures.

So vidicon cameras gradually were replaced with CCD imagers much like those in current smartphone video cameras.

The CCD video camera came into its own as a reconnaissance and surveillance tool in the post-2001 conflicts in Iraq and Afghanistan, when mounted on UAVs and aerostats. The cameras provided **full-motion video** (FMV), a technology that can display a massive amount of data coming into the camera at such a fast rate that the images appear to be smooth and continuous. Full-motion video was used for a variety of missions in Iraq, Afghanistan, and Syria, ranging from time-sensitive activities such as targeting to detailed geospatial intelligence analysis. The cameras proved to be almost too addictive for battlefield commanders who could watch opposing forces in real time, and would tend to do so for hours. There is nothing comparable to observing an opponent's movements and, for example, watching them plant an improvised explosive device (IED).

UAVs provide excellent platforms for these steerable video cameras. The early UAVs had to be close to the target in order to get good resolution—needed for identification of the target and to establish a target signature. But digital imaging technology and improved UAV surveillance platforms today provide the high-resolution video needed to identify and track targets of interest at longer distances. An example is the camera/UAV combination that the US Department of Homeland Security uses for domestic and border surveillance. The MQ-9 Reaper UAV used by US Customs and Border Protection carries a video camera that reportedly can read a license plate number at a 2 mile distance.[15]

The trend in video observation is to use what is referred to as wide-area motion imagery. Video cameras that can do this produce a continuous stream of high-resolution images that allow the tracking and recording of vehicle and pedestrian movements throughout entire cities. They can be used to track moving targets over an extended period. For example, as people move around in the scene, they will sometimes be hidden by buildings or trees. Their image or signature has to be reacquired when they emerge. Increasingly, this requires automated tracking, followed by using signatures to automatically reacquire the targets when they emerge from cover.

The US Defense Advanced Research Projects Agency (DARPA) initiated a project called ARGUS in 2007 to do just that. ARGUS relies on an array of 368 smartphone cameras, each pointing at a different area, to create a video mosaic. Mounted on an airborne platform, ARGUS can reportedly automatically track every moving object—pedestrians and vehicles—in an area of 100 square kilometers. The US Air Force fielded ARGUS under the nickname Gorgon Stare (surely one of the most ominous possible nicknames for such a system) in 2014.[16]

It was only a question of time before that capability moved into space. Chapter 9 described the plans of the start-up company EarthNow to orbit a 500-satellite constellation of video imagers. Each satellite would carry four independently steerable telescopic video cameras, with an image resolution approaching

1 meter—good enough for most geospatial intelligence purposes, with the added benefit of conducting continuous surveillance. The EarthNow website suggests a number of possible uses of its real-time imagery, to include catching illegal fishermen, monitoring hurricanes and typhoons as they develop, knowing when volcanoes erupt, tracking whale migration patterns, measuring the health of crops, and observing conflict zones for quicker response times. The developers propose to offer users video imagery in real time, eventually from a smartphone or tablet, and the ability to zoom in on a particular location for a closer look.[17]

Getting the Image Right

Visible imaging is about choosing the right imaging system for the job, using it to get the right type and quality of image, and then analyzing the image to extract the information that you need—whether for mapping, intelligence, or other purposes. And the secret in imagery is much the same as the secret of a good map: to capture the things that you need to know, and to not include the things that contribute nothing to your knowledge base.

When you take a photograph with your digital camera or smartphone, you don't usually think much about getting the image right. The camera automatically focuses on the closest object near the center, and adjusts its sensitivity to the amount of incoming light to present a pleasing picture. Things are a bit more complicated when planning to acquire an image in remote sensing. There are many factors to consider, including the same sorts of things that a professional photographer thinks about before shooting: the right illumination; perspective and number of different perspectives; photo quality; and scene coverage—among others. In processing the image later on, it's also critical to get the right combination of contrast and brightness to allow an imagery analyst to identify the intelligence target.

A lot of the basics for getting the image right were sorted out with aerial camera use in World War I and World War II, and refined during the Korean War. By the time of the U-2, these were well understood for aircraft. Satellites added a layer of complexity and a new set of problems beyond getting the image back to Earth, as discussed above. Ironically, the imagery challenges today are the same that cartographers have long faced: timeliness, coverage, resolution, and accuracy. For intelligence purposes, image quality, measured in terms of spatial resolution, is perhaps the most highly prized.

Image Quality

For some purposes, image quality isn't paramount. For example, Landsat images have a resolution on the order of 30 meters—not useful for detailed images, but quite useful for things like monitoring the health of crops. Landsat was a big contributor years ago for determining the health of Soviet wheat crops and assessing how much wheat the Soviets might need to import. Traders in the worldwide crop futures market, and other futures markets that depend on crops (e.g., ethanol and sugar), rely on the imagery from Landsat and other low-resolution imagers in making their investments.

Image quality *is* paramount, though, in much of geospatial intelligence analysis. Analysts frequently depend on getting a detailed target signature, and that in turn depends on the spatial resolution of the image. A poor-quality image might tell you that the object on a runway is an airplane. With better quality, you could determine that it is a jet fighter, and an even better image would allow you to identify it as an F-35. Most useful of all, an exceptionally high-quality image would let you read the tail number and therefore identify a specific F-35.

National Photographic Interpretation Center director Art Lundahl had at one point told a review panel that there was no difference between the intelligence that could be derived from imagery with 10-foot resolution and that with 5-foot resolution.[18] By the 1980s, no one in the center would have agreed; the demand—which continues today—was for resolution measured in inches, not feet.

The problem, however, is that with the advent of digital satellite cameras, imagery analysts were suddenly flooded with imagery. It wasn't possible to look at every available image of a target. As the volume of imagery from satellites increased, it became important—in some cases, at least—to know how good the image was *before* looking at it.

So imagery analysts needed a measure of quality to tell them, in advance, how useful an image might be for identifying unique signatures and patterns. During the 1970s, they developed one that is still in use. It is called the National Imagery Intelligence Rating Scale (NIIRS). It uses numerical ratings from 0 to 9 (9 being the best quality) to serve as a shorthand description of the information that can be extracted from a given image. Table 10.1 illustrates the use of the scale in terms that are meaningful to imagery analysts. Analysts rely on the NIIRS scale to sift through a mass of incoming images and quickly select the ones that deserve a closer look. If you need to tell whether an aircraft is an F-35, you want at least NIIRS 4 imagery; no need to waste time looking through NIIRS 3 files.

Recall China's Gaofen-11 imaging satellite from chapter 9. At perigee (247 kilometers), it is believed capable of achieving an imaging resolution of 10 centimeters or less. That equates to a NIIRS of between 8 and 9—about the best that you can expect to achieve from a satellite with existing technology, and good enough for most geospatial intelligence purposes.[19] One way to get good spatial resolution is to get the imager as close as possible to the target, as Gaofen-11 does. So why don't we always go for higher resolution? Because it usually comes at a price.

TABLE 10.1 NIIRS Rating Scale

NIIRS	Description	Resolution
0	Not interpretable	
1	Detect a port or airport runway	> 9 m
2	Detect a large hangar or building	4.5–9 m
3	Identify a type of surface ship	2.5–4.5 m
4	Identify a type of large fighter aircraft	1.2–2.5 m
5	Identify refueling equipment on aircraft	.75–1.2 m
6	Detect a spare tire on a truck	.4–.75 m
7	Identify individual railroad ties	.2–.4 m
8	Identify auto windshield wipers	.1–.2 m
9	Detect spikes in railroad ties	< .1 m

If you want to look at the largest possible part of the Earth, put your satellite in a geosynchronous equatorial orbit. That will provide continuous coverage of about a third of the Earth. The GOES-17 meteorological satellite does that. If you're happy with a spatial resolution on the order of 1 kilometer, which is good enough for weather forecasting, then that's the right orbit. But if the goal is to see details on the Earth's surface, you have to move a lot closer, down to LEO, or better still, to the altitudes that UAVs use. When you do that, though, the area that is visible at any given instant shrinks. The camera on France's SPOT satellite has spatial resolution on the order of 1 meter, good enough to identify a tank on the battlefield. But the camera can see only a few kilometers of area, not enough to cover the whole battlefield. This can be a significant issue in geospatial intelligence, which often depends on being able to examine the relationship among objects that are widely separated.

Can you have both coverage and resolution? Yes, with clever design. The ARGUS/Gorgon Stare video camera accomplishes it by using a large number of cameras, each looking at a small part of the scene.

Flattening the Earth in an Image

We return to the inconvenient fact that Earth isn't flat. And a camera on an aircraft or satellite sees only one point on the ground from directly overhead. It views all others from an oblique angle, so that tall structures seem to lean away from the nadir.

Oblique images, however, are not a bad thing—in imagery analysis, at least. The World War I multilens cameras that Major Bagley developed provided oblique as well as vertical images. Analysts soon recognized that oblique imagery is the more useful of the two. It reveals features that cannot be seen clearly or may be obscured in pure vertical imagery. It turns out to be easier to interpret ground features from an oblique view. And multiple oblique views of the same

Orthographic view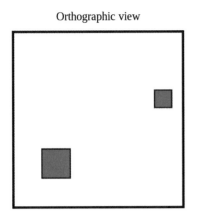

Perspective view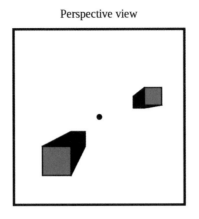

FIGURE 10.5 Orthographic and perspective views of two structures. *Source:* Pieter Kuiper, Wikimedia Commons.

entity give even more geospatial information. It's not possible to determine the height of a building, for example, from a vertical image—unless the building casts a shadow that can be measured.

But the additional information comes at a price. An oblique image has distortion; the same distance measured at two different points on the image seldom equates to the same distance on the ground. In figure 10.5, for example, the perspective view shows two tall structures, with the camera's nadir between them. Both structures are shown in an oblique view. As you get farther away from the nadir, distortion increases. Where 1 inch at nadir on a photograph might equate to 100 feet on the ground, as you get to the edge of the photograph, 1 inch could easily equate to 150 feet. Also, objects on the ground that are closer to the camera due to uneven terrain will appear to be larger.

So, for measuring objects and distances between them and for mapping, a flat Earth or **orthographic view** is needed. Digital imagery processors deal with that problem electronically. The process, called **orthorectification**, removes the effects of image perspective, and corrects for terrain elevation. The product is an orthorectified image; it has a constant scale, and features in the image are shown in their true positions with the correct dimensions.

Stereo Imaging

Soon after aerial photography began to be used in intelligence, it became apparent that an imagery analyst could obtain more useful detail from a three-dimensional image than from the two-dimensional image provided by conventional photography. There was a solution for that. To obtain the illusion of depth in a photograph, you created a **stereo image**. A stereo image requires at least two photographs of the same scene taken from different angles; a 30- to 40-degree angular difference is about right. And that required that the aircraft (and later, the satellite) had to precisely time the two exposures. The two images then formed a stereo pair or **stereograph**. They were placed side by side and viewed through a device called a **stereoscope**, which focused the right eye on the right image and the left eye on the left image. The imagery analyst then was able to view a three-dimensional stereo image.

Stereo imaging was pioneered in World War I. By the outbreak of World War II, the techniques for obtaining and interpreting it were well established. During that war, many of the combatants collected three-dimensional imagery using stereographs in aerial reconnaissance. Stereo imagery allowed you to extract cultural and geographic features such as buildings, roads, and other terrain features and human-made structures. It also enabled more accurate determination of the height of an object or the elevation of a point on the ground. Camouflaged weapons and structures could be difficult to spot in conventional aerial photographs, but aerial stereographs helped photo interpreters with identification.

Sun Angle and Shadow

Chapter 9 described the near-polar orbit called sun-synchronous. At LEO altitudes, the orbit is approximately 98 degrees inclination (8 degrees counterclockwise from true north). Imaging satellites often use this orbit, because the satellite passes over a given point on the Earth at about the same time every day.

Why would you want to pass a location at the same time each day? Because the target that you want to take a picture of is in about the same sunlight condition every day. That means that the shadow conditions—which are important for target identification and measurement, and for change detection—remain much the same from one day to the next. The best shadow conditions occur at around mid-morning and mid-afternoon. Let's take another look at Gaofen-11's orbit as an example. Its perigee occurs at 10:00 local time at 20° north latitude. Several areas of high intelligence interest to China are located around that latitude, including Hawaii (home base of the US Pacific Fleet), India (a major regional competitor), and the South China Sea (a region where China has contested territorial claims).[20] So Gaofen-11's orbit is optimized to obtain the best possible image of a sensitive region at the best possible time for imaging.

Polarimetry

Polarized sunglasses have been used for years by boaters to reduce reflected glare from the water. Skiers, cyclists, golfers, and joggers also use the sunglasses to reduce glare from the terrain around them. Most neither know nor care how the sunglasses work, so long as they do the job.

In analyzing imagery, however, we care a great deal. We can use the equivalent of polarized sunglasses to spot things of special interest in the image.

Natural light has random polarization, meaning that the electric field direction changes constantly and unpredictably. When such light is scattered from a rough surface, it remains unpolarized. In contrast, when reflected from a relatively smooth surface, such as a lake or road, the light has a strong polarization—which humans observe as glare.

Of course, human-made objects such as automobiles and buildings tend to have smooth surfaces. The light they reflect is more polarized than are natural features that tend to scatter light in an unpolarized way. So we use polarimetric imagery (where the camera is set to capture a specific light polarization) not to filter out glare but to detect it. In such imagery, human-made objects stand out from natural backgrounds such as grass, trees, dirt, and sand, making them much easier to detect. We also can use **polarimetry** to look at objects that can't be seen in conventional imagery because they are located in shadowed areas.

Analyzing the Image

Chapter 1 noted that an image, alone, isn't GEOINT—not even the highest-quality image of the right thing at the best time of day. Analyzing the image, and relating it to other intelligence, is required to produce GEOINT. And the first step in that is to extract useful information from the image. For many years, the professionals who did this were called photo interpreters, since they worked with photography. They examined the developed roll of film on an automated light table (film negatives give a better view than prints do, and it is easy to zoom in to look at features of interest). The table functioned much like the light tables that Warren Manning and other landscape architects used in their geodesigns (described in chapter 6). Figure 10.6 shows a light table that was

FIGURE 10.6 Film interpretation light table. *Source: Author's collection.*

used to look at U-2 and KH-4 film; the holder from which the film was unrolled is visible on the bottom left.

When digital imagery became common, the name "photo interpreter" was revised to "imagery interpreter." And later on, another change in nomenclature made it "imagery analyst"—which more accurately describes what these talented professionals do.

The switch to digital imagery allowed the use of computers to enhance the imagery analysis process. Features of no interest could be removed or grayed out, and features of interest highlighted. For example, the computer could in some cases remove thin cloud cover from a scene to allow viewing an otherwise hidden ground image. All these advances in geospatial technology reflect progress; nevertheless, imagery analysis continues to be what it has always been—a matter of searching for important signatures, analyzing them in a timely manner, and monitoring changes.

Visible Signatures

Chapter 1 touched on signatures as a concept, and chapter 7 described examples used in acoustic geolocation. We encounter visible signatures in everyday life without giving them much thought. On the road, we easily recognize at a distance the differing visual signatures of a sports car, a minivan, and a heavy truck. We know the differing leaf signatures on trees that indicate various seasons of the year. We can instantly identify the unique signatures that characterize a breed of dog or cat.

In the same way, imagery analysts depend on identifying and analyzing visible signatures in both civil and national intelligence applications. One type of signature tells a geologist that certain rock formations are a likely source of oil or tin deposits. For agriculture, analysts may look at signatures that indicate soil moisture and vegetation health. On a global scale, the formation of clouds having rotation and a hole in the center is the signature of a hurricane.

National intelligence imagery analysts might focus on signatures indicating enemy offensive and defensive installations, troop concentrations and movements, and a wide range of emerging threats. Consider again the example of the al-Dawadmi military base in Saudi Arabia, introduced in chapter 1. It had two launch pads, which have easily recognizable signatures. It was a bit trickier to recognize that the missile test stand had the same signature as one used by China. As another example, the manufacture, movement, and storage of nuclear weapons is of high intelligence interest. Their storage facilities have unique signatures such as heavily reinforced concrete bunkers with extensive fencing and guard shacks. Their transportation usually requires specially modified heavy trucks accompanied by a guard convoy.

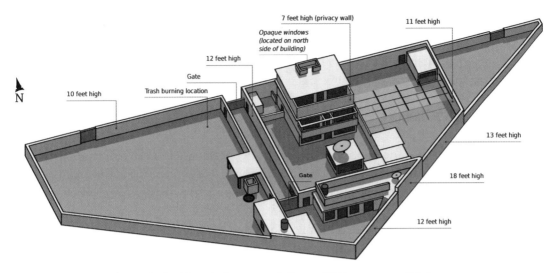

FIGURE 10.7 Osama bin Laden's Abbottabad compound. *Source:* US Department of Defense.

Unusual signatures will always draw an imagery analyst's attention. One such example is shown in figure 10.7. It had walls that were unusually high for its neighborhood, an unusual security layout, no phone or internet wires running into the area, and several satellite dishes. It later became famous when US SEALs raided the compound and killed Osama bin Laden in 2011.

There are entire specialties of signature analysis. An interesting one is cratology. It involves developing signatures of crates used for shipping material of intelligence interest. The first step, as with all signatures, is to catalogue or build a library of crate characteristics, such as dimensions, color, composition, and markings of various crates. A later observation of a unique signature allows you to identify what is inside. Imagery analysts do this all the time:

- Just before the Cuban missile crisis in 1962, imagery analysts spotted crates on the deck of a Soviet ship en route to Cuba. An astute analyst observed that the same crates were routinely used in shipping IL-28 medium bombers. The identification was one of the discoveries that prompted a closer imagery look at Cuba and the resulting discovery of ballistic missile complexes there.[21]
- In August 2006, a US reconnaissance satellite reportedly imaged crates being loaded onto a transport aircraft in Iran. The design matched that of crates used to ship Iran's C-802 Noor antiship cruise missiles. According to a press report, the identification set off a chain of diplomatic action that kept the missiles from being delivered to Hezbollah guerrillas who were fighting Israeli forces in Lebanon.[22]

Although signatures in the visible spectrum were (and are) essential for producing geospatial intelligence, the developments in spectral and radar imaging (the subjects of chapters 11 and 12) opened up entirely new classes of signatures for exploitation.

Timeliness

These days, decision-makers tend to want images in near real time. But they also want detailed analysis. That's a tough assignment. Still, for ongoing military operations or in crises, quick analysis of the imagery product becomes a requirement.

Even in commercial applications, speedy analysis can be in demand. Think about the crop futures trader mentioned earlier in this chapter; millions of dollars of profit (or loss) can depend on being the first to recognize a change in a crop estimate. Digital imagery made the analysis of imagery a more timely process. Today, pattern recognition algorithms allow rapid search of digital images to identify features that need a closer look. They also help in identifying changes that might have intelligence significance.

Change Detection

Imagery analysts want to observe how signatures change over time. **Change detection** is a pillar of GEOINT and is important for military, law enforcement, and intelligence applications—for example, movement of military forces, monitoring illicit arms shipments, and observing weapons test preparations, among others. Some changes are difficult to detect with the human eye; slight changes in the orientation of a dish antenna may be critical in an intelligence application, for example, and that requires digital processing. Change detection can be automated by identifying and highlighting individual pixels that have changed in two images of the same scene taken at two different times.

The bin Laden compound shown in figure 10.7 is an excellent example of change detection in GEOINT. The National Geospatial-Intelligence Agency had imagery of the compound dating back to the beginning of its construction. From that historical point of view, the agency's analysts were able to observe all of the construction details as the building came together, floor by floor—information that undoubtedly was of value to the SEAL team when planning the raid.[23]

Visible change detection is also widely used in specific industries such as waste disposal, mining, and energy. In waste disposal, it provides a visual representation of where there is additional landfill capacity. Change detection technology allows mining companies to track activity and manage their fleets. Pipeline and power line managers must monitor their rights of way for obstructions such as fallen trees or construction that would affect their lines. With change detection, they can easily identify these problems and quickly address them.[24]

Visible change detection has been a staple of GEOINT analysis since the beginnings of aerial reconnaissance in World War I. But what about changes that cannot be seen in the optical spectrum? That's where infrared, spectral, and radar imaging step in. Those capabilities allowed change detection to assume a new and greater dimension in GEOINT.

Chapter Summary

Visible imaging has long been a staple in creating geospatial intelligence. From its initial use in tactical intelligence during World War I and its application to photogrammetry between the wars, photography expanded into providing strategic intelligence during World War II. Aerial film cameras were followed by satellite film cameras and video cameras that provided steadily improving image quality. New technologies allowed imagery analysts to extract more types of information from visible images, especially after the digital camera replaced the film camera for imaging.

The digital camera also was the enabling technology for infrared imaging, opening new dimensions for geospatial intelligence—the subject of chapter 11.

Notes

1. The distance between a camera's lens and the film surface (where the image is in focus) is the camera's focal length.
2. Tim Slater, "British Aerial Photography and Photographic Interpretation on the Western Front," September 18, 2011, http://tim-slater.blogspot.com/2011/09/british-aerial-photography-and_18.html.
3. Revere G. Sanders, "Aerial Cameras and Photogrammetric Equipment: A Quarter Century of Progress," September 1944, www.asprs.org/wp-content/uploads/1944_sep_136-159.pdf.
4. NOAA, "Surveys from Above," n.d., www.ngs.noaa.gov/web/about_ngs/history/camera_timeline_web.pdf.
5. National Museum of the US Air Force, "Powerful New Cameras for the U-2," June 2, 2015, www.nationalmuseum.af.mil/Visit/Museum-Exhibits/Fact-Sheets/Display/Article/197566/powerful-new-cameras-for-the-u-2/.
6. NASA, "The Hubble Space Telescope Optical Systems Failure Report," November 1990, https://ntrs.nasa.gov/archive/nasa/casi.ntrs.nasa.gov/19910003124.pdf.
7. "America's Eyes: What We Were Seeing," NIMA Historical Imagery Declassification Conference, September 20, 2002, https://nsarchive2.gwu.edu/NSAEBB/NSAEBB392/docs/50.pdf.
8. "TIROS," NASA, n.d., https://science.nasa.gov/missions/tiros.
9. Nokia Bell Labs, "2009 Nobel Prize in Physics: The Charge Coupled Device (CCD)," n.d., www.bell-labs.com/about/recognition/2009-charge-coupled-device-ccd/.
10. Nokia Bell Labs.
11. Chunka Mul, "How Kodak Failed," *Forbes*, January 1, 2012, www.forbes.com/sites/chunkamui/2012/01/18/how-kodak-failed/#377c546a6f27.
12. Bing Putney, "How Much Resolution Do You Really Need?" *SLR Lounge*, n.d., www.slrlounge.com/how-much-resolution-do-you-really-need/.

13. Jeffrey Richelson, *The Wizards of Langley* (New York: Basic Books, 2001), 200.
14. ESA, "Gaia Telescope," n.d., https://m.esa.int/Our_Activities/Space_Science/Gaia/Telescopes.
15. Shirin Ghaffary, "The 'Smarter' Wall: How Drones, Sensors, and AI Are Patrolling the Border," Vox, May 16, 2019, www.vox.com/recode/2019/5/16/18511583/smart-border-wall-drones-sensors-ai.
16. "Sierra Nevada Fields ARGUS-IS Upgrade to Gorgon Stare Pod," *FlightGlobal*, July 2, 2014, www.flightglobal.com/news/articles/sierra-nevada-fields-argus-is-upgrade-to-gorgon-stare-400978/.
17. Alan Boyle, "EarthNow Fleshes Out Plan to Deliver Video That Shows Earth from Orbit in Real Time," *GeekWire*, February 4, 2019, www.geekwire.com/2019/earthnow-video-orbit/.
18. Jeffrey T. Richelson, "Civilians, Spies, and Blue Suits: The Bureaucratic War for Control of Overhead Reconnaissance, 1961–1965," January 2003, https://nsarchive2.gwu.edu//monograph/nro/nromono.pdf.
19. Andrew Tate, "China Closing the Satellite Imagery Capability Gap," *Jane's Defence Weekly*, August 14, 2018, www.janes.com/article/82366/china-closing-the-satellite-imagery-capability-gap.
20. Tate.
21. Dino Brugioni, *Eyeball to Eyeball: The Inside Story of the Cuban Missile Crisis* (New York: Random House, 1990), 73.
22. John Diamond, "Trained Eye Can See Right through Box of Weapons," *USA Today*, August 17, 2006, www.usatoday.com/news/world/2006-08-17-missiles-iran_x.htm.
23. NGA Director Robert Cardillo, interview, *60 Minutes*, CBS-TV, January 27, 2019.
24. Identified Technologies, "Identified Technologies Unleashes New Change Detection Technology," March 9, 2017, www.identifiedtech.com/press-release/identified-technologies-unleashes-new-change-detection-technology/.

Spectral Imaging

There is much more to the optical spectrum than our eyes or the "eye" of a camera can see. A device with which most of us are familiar—the prism—led to the discovery of "invisible" parts of the spectrum. And those unseen parts have turned out to be valuable bands to image for both military and civilian geospatial intelligence applications.

The British amateur astronomer William Herschel made the initial discovery of these unseen parts of the spectrum in 1800. Herschel knew the different colors that constitute sunlight could be separated by passing them through a glass prism. He suspected that each color might produce a different level of heat, so he created an experiment to test his theory. The experiment involved a prism that refracted sunlight onto a flat surface to create a color spectrum display. He then placed thermometers with blackened bulbs in each part of the spectrum and measured the temperatures of the different colors.

Herschel was a meticulous scientist, and that turned out to be the key to his revelation. He placed two additional bulbs on the surface on either side of the visible spectrum to measure the ambient temperature for comparison. His measurements contained a surprise. As he had suspected, the temperatures varied in different parts of the spectrum; they increased steadily as the thermometers moved from violet through blue, green, yellow, orange, and red light. What he didn't expect was what he found when he checked the temperature of the control thermometer just outside the red part of the spectrum, where there was no visible light. It had the highest temperature of all.[1]

Herschel had uncovered an invisible form of light energy that extended beyond red light. His experiments led to the discovery of **infrared** (IR) light.

Herschel's results came to the attention of Johann Ritter, a researcher at the University of Jena in Germany. Ritter thought that if invisible light could exist beyond the red part of the spectrum, it might also exist on the other side—beyond the violet end. One year after Herschel's discovery, in 1801, Ritter found that it did. He repeated Herschel's prism experiment, but instead of a thermometer he used a silver chloride compound—knowing that silver chloride darkens more in violet light than in red light. When he placed samples of silver chloride in the dark area beyond the violet end of the visible band, he found that those samples darkened even more than those in the violet part. Ritter had discovered **ultraviolet** (UV) radiation.[2]

Herschel's and Ritter's breakthroughs would in time open up new remote-sensing vistas. Suddenly, previously unknown signatures became available for GEOINT analysis. But it would take over a century for that to happen. First, there had to be some way to obtain imagery in the IR and UV bands.

The Infrared Bands

Herschel's experiments concerned one type of infrared radiation, called **emissive IR** because it depends on energy being emitted (in Herschel's case, by the sun). The warmer an object is, the more IR energy it emits, so that it can be distinguished in an image. But all objects around us emit some IR energy, which means that it's possible to obtain IR images at night as well as in daytime.

But there is more. Like visible light, IR energy can be reflected from objects. We refer to it as **reflective IR**. And the amount of reflection tells us something about the object reflecting—much like tree leaves reflect green light, helping us to identify a tree. So IR energy could be observed as a result of reflection, or by emission. Either way, this offered a new set of imagery signatures for analysis.

Over time, researchers found that the IR spectrum could be characterized in distinct bands, each having different reflective or emissive characteristics. The development of imagery in these nonvisible optical bands opened a new dimension for civil and military applications. There are four major IR imaging bands, as shown in figure 11.1. Each band identifies something unique about the area it is imaging, so each serves a unique purpose in remote sensing.

The Near Infrared Band

The near infrared (NIR) band is the closest to visible light, and shares many of the same characteristics—except that our eyes can't see it. But it was fairly straightforward to build night-vision devices to work in this band. A German company, Allgemeine Elektricitäts-Gesellschaft AG, developed the first night-vision equipment in 1935, and the German Army was using it by 1939. Later in the war, NIR searchlights were being mounted on German tanks and NIR sniper scopes were being used in combat.[3] (Sensing in the NIR band normally depends on reflected energy—hence, the use of NIR searchlights.) These devices gave German soldiers a definite advantage in night fighting, particularly because their opponents couldn't tell that they were being illuminated. NIR sensors are also frequently used by the military to defeat camouflage, which doesn't reflect NIR like natural vegetation

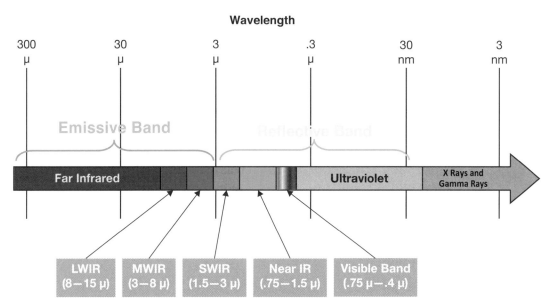

FIGURE 11.1 The optical spectrum. *Source:* Author's collection.

does. For civil applications, the band is useful for vegetation mapping. The longer wavelengths in the NIR band reflect strongly from minerals, crops and vegetation, and moist surfaces; different tree types can be distinguished, for example.

The Short-Wave Infrared Band

Next to the NIR band is the short-wave infrared (SWIR) band—which also depends on reflected energy. Sensors in this band can detect the presence of water and allow vegetation analysis. They can discriminate oil on water and allow scientists to determine the moisture content of soil and vegetation.

The Mid-Wave Infrared Band

The mid-wave infrared (MWIR) band can be used to sense both reflective and emissive energy. This is the "heat seeking" band; antiaircraft missiles use it to steer the missile toward the hot jet engine exhaust. Of course, large explosions and ballistic missiles in flight both emit massive amounts of heat energy, and satellite-based sensors are used to detect them.

The Long-Wave Infrared Band

The long-wave infrared (LWIR) band is referred to as the "thermal imaging" region. The Earth and objects on it emit electromagnetic energy strongly in this band. You can get an image of the Earth based on thermal emissions only, requiring no external energy source such as the sun.

For example, using an LWIR sensor, you can tell if the engines are running on an aircraft that is sitting on the ground. You might also see a shadow on the tarmac, indicating a cooler surface, telling you that an aircraft had been there but just recently left. The aircraft prevented the sun from warming that area of the tarmac.

There is also a band below LWIR in frequency, called far infrared (fig. 11.1 shows it). This band extends to the bottom of the optical spectrum, where the microwave bands are. However, much of it is not useful for imaging purposes; the atmosphere is opaque across most of the band.

The Ultraviolet Spectrum

The UV part of the optical spectrum (shown in fig. 11.1) lies on the opposite side of the visible spectrum from infrared—that is, at higher frequencies or shorter wavelengths than the visible light band. We don't notice it, because our eyes can't see UV light. But reindeer can. A reindeer's eyes see much the same visible band as we do, but they can see UV as well. It turns out that lichen (on which reindeer feed) absorbs UV energy and appears black against a snow background to reindeer. So does wolf fur, even if it is white. And urine (which indicates the presence of predators or competitors) also appears darker on snow. So the reindeer's eyes evolved to allow it to both find food and to avoid becoming food.

While not as widely used in geospatial intelligence as IR imaging, UV imagery shows some features that are not apparent in the more common visible and IR varieties. Some gases are observed only in UV light, and UV imagery of an archaeological site may reveal artifacts or ancient traffic patterns that are not otherwise visible.

Imaging outside the Visible Band

What remained was to develop cameras that could obtain IR and UV images. In 1909 the physicist Robert Williams Wood developed the first known IR photographs. He did it by creating an experimental film that required very long exposures to create the photographs. It took time, however, for Wood's experiment to make it into the commercial world. Kodak's IR film first became commercially available in the 1930s.[4]

After the charge-coupled device camera was invented, digital cameras were built that could also image outside the visible band—first in near IR, then extending through all the IR bands as new sensing technologies were developed to capture IR energy. Other emerging technologies allowed digital cameras to image in the UV band. So cameras could obtain a series of images of a scene, with each image in a different band of the optical spectrum. This enabled a powerful new type of remote sensing: spectral imaging.

Spectral Imagers

Beginning in chapter 10 and up to now, we've been discussing a single image in the visible or IR bands. In the years before the launch of Landsat 1, it became obvious to researchers that, because images highlight different features in each spectral band, you might be able to learn more about objects in a scene by in some way combining the images from different bands. Their efforts to do that led to the development of spectral imaging, and Landsat 1 became the first spectral imager in orbit.

We see the world around us in color—at least, most of us do. But any color that we perceive can be created by combining just three colors—red, green, and blue—in the right combination of intensities. We can view a scene through a filter—a red lens, for example—in which case, we'll see only objects that reflect red energy, and the brightest objects will be ones that appear to be red in normal vision. That filtered image is monochromatic; it contains only one narrow part of the spectrum. The IR images that Kodak's film captured were also monochromatic; they simply showed the intensity of IR energy as a black-and-white picture.

However, we normally view our surroundings as a spectral image—that is, one that uses three bands across the visible spectrum, a combination of red, green, and blue. The different colors that our eyes perceive are a blend of different intensities of each of these three bands.

It is possible to create spectral images that include bands outside the visible ones. The problem is that we see only some combination of red, green, and blue. If you add another band to the image, one that we can't normally see, something has to give. But there is a solution. It just requires that we apply a bit of processing trickery, called false color, to create the image.

"Seeing" Spectral Images

Images that include nonvisible bands need to use **false color**. That means shifting some parts of the visible spectrum to become invisible and moving parts of the nonvisible spectrum into the visible region. The traditional technique is to display NIR as red, red as green, and green as blue. Blue then disappears (becomes black). The image this produces doesn't look anything like what we expect to see, so explaining it to an intelligence customer can be challenging.

FIGURE 11.2 A false color image of Las Vegas. *Source:* National Aeronautics and Space Administration.

For example, figure 11.2 is a traditional false color image of Las Vegas; an imagery analyst will explain that the "red" areas actually are green vegetation (many of which are golf courses). NIR reflects strongly off healthy vegetation.

Spectral Imaging Applications

Spectral imaging in the IR band proved invaluable for a wide range of Earth resources applications. One of the major benefits of Landsat was that it provided multiple images in different parts of both the visible and IR spectra that could be combined to obtain imagery that was useful for environmental monitoring, agricultural predictions, geology, and hydrology, among others. All of that depended on developing an understanding of the full optical spectrum and how material on the Earth's surface either emitted or reflected energy in different bands.

Spectral imaging is one of the new frontiers in geospatial intelligence. We use it to gain unique insights from a scene. For instance, we'd like to be able to detect and classify gases that don't appear in visible imagery. We'd like

FIGURE 11.3 German Marder 1A5 tank with infrared camouflage. *Source:* Sonaz, Wikimedia Commons.

to identify and classify the objects and areas in a scene. Spectral imaging has demonstrated the ability to do all these things, and more. With it we can, for example, characterize underground facilities, identify buried roadside bombs, and find tunnel entrances.

IR images have long been used to identify older forms of military camouflage. In response, countries have developed countermeasures in the form of camouflage that will reflect NIR like normal vegetation does. So modern camouflage better mimics vegetation, and it blends into the background vegetation in conventional NIR imagery. Figure 11.3 illustrates an IR-camouflaged German tank that can be concealed from most IR imagers. But newer spectral imaging systems cover a wide spectral range and so are able to separate even the best camouflage from surrounding vegetation, especially in the SWIR portion of the spectrum.

Law enforcement agencies also make use of spectral imaging. The most common use probably is in monitoring illicit crops. Law enforcement organizations use spectral imagery to identify marijuana crops cultivated under forested canopies; marijuana leaves, it turns out, have a different spectral image than the surrounding vegetation. And coca or opium poppy crops that have been damaged by herbicides appear different from undamaged crops in spectral imagery. Such imagery is routinely used in drug eradication efforts.

Spectral imaging has many commercial and environmental applications as well. It is used in agriculture, ecology, oil and natural gas exploration, oceanography, and atmospheric studies to gain knowledge that single-band IR images can't provide. It is frequently used to identify minerals for mining. It was used in northern Chile to identify the prospects that are now major copper deposits.[5] Certain land features are obvious to a trained analyst. Oil-bearing rock appears different from other rock types. Specific types of iron and clay minerals show up in spectral images.

In 2016 a World Bank–sponsored study used spectral imaging to estimate oil production by Daesh (ISIS) in the areas that it controlled at the time. The study was of intelligence interest because oil was a major source of Daesh's revenue. Imagery analysts combined spectral imagery from the Visible Infrared Imaging Radiometer Suite on a NASA/NOAA satellite with prewar output data to measure oil production in Daesh-controlled areas. The imagery allowed measurements of methane gas flaring (which correlates to oil production) at the extraction sites. The study found that production levels decreased from a peak of 33,000 barrels per day in July 2014 to an average of 19,000 throughout 2015—indicating that the coalition-led attacks on Daesh's oil production during that time were having some success.[6]

Spectral imaging is also used extensively in agriculture and environmental studies. It is especially powerful for mapping specific invasive species based on their spectral characteristics. In agriculture, healthy vegetation can be distinguished from stressed, or dying, vegetation. Healthy vegetation appears green because chlorophyll absorbs most of the other two primary colors—blue and red. Stressed vegetation (due to inadequate water) absorbs less of the blue and red light waves, so it appears different to spectral sensors.

If you want to start a controversy among wine enthusiasts, ask a question about vineyard irrigation. It was long banned in Europe by the European Union's wine laws because it was believed that natural rainfall was the only acceptable way to maintain the proper wine quality for different varietals. That law was never applied in the wine-growing regions of California and Australia. Irrigation was essential to producing wines in those dry climates, and the law may be changing in Europe.

But how much water to use? That depends on the stage of grape development. The basic rule was called controlled water stress: keep irrigation to a minimum, allowing more water during budding and flowering, and then less during the ripening stage. That requires careful management of vineyards. And spectral imaging recently has demonstrated the ability to provide vineyard water stress management.[7]

How Spectral Imaging Works

So spectral imaging can do great things. But how does it work?

It makes use of an optical device called a **spectrometer** that takes in light and, much like a prism, breaks it into different spectral bands. Each separate band then can be used to create an image of a scene at a specific wavelength. An imaging spectrometer does exactly that.

Think of spectral imaging as using this "electronic prism"—the spectrometer—to acquire many images of the same scene, at different wavelengths, all simultaneously. You thereby create a "stack" of images, each looking somewhat

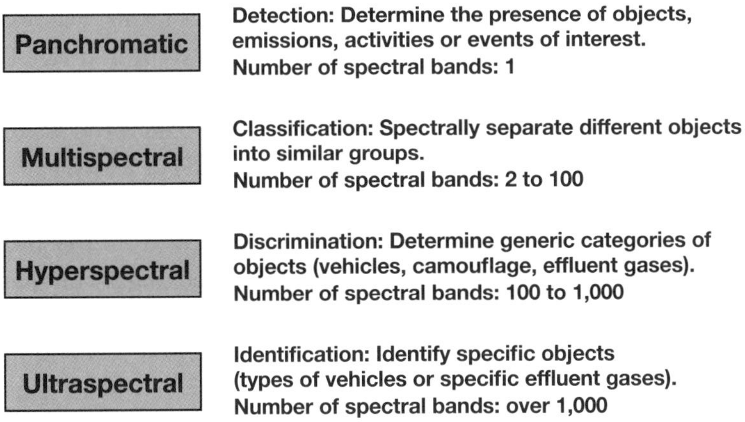

Panchromatic	Detection: Determine the presence of objects, emissions, activities or events of interest. Number of spectral bands: 1
Multispectral	Classification: Spectrally separate different objects into similar groups. Number of spectral bands: 2 to 100
Hyperspectral	Discrimination: Determine generic categories of objects (vehicles, camouflage, effluent gases). Number of spectral bands: 100 to 1,000
Ultraspectral	Identification: Identify specific objects (types of vehicles or specific effluent gases). Number of spectral bands: over 1,000

FIGURE 11.4 Classes of spectral imagers. *Source:* Robert M. Clark, *Intelligence Collection* (Washington, DC: CQ Press/Sage, 2014). Used by permission.

different from the others. Put another way, you're creating *layers* of information about a scene—the sorts of layers that geographical information systems (discussed in chapter 6) use.

The layers of information allow identifying objects in the image, because each object reflects (or emits) energy differently in different parts of the total spectrum. The more wavelength bands used to create separate images, the more information you get about objects in the images and the more accurately you can determine what the object is. As figure 11.4 indicates, increasing the number of spectral bands allows a progression from detecting a type of object, to classifying it, to identifying a specific object or gas.

The most common spectral images today are in the form of **multispectral imagery** (MSI), which is collected from aircraft and from satellites. But the goal is always to acquire as much detail as possible over as much of the spectrum as possible. **Hyperspectral imagery** (HSI) and **ultraspectral imagery** (USI) provide that: they contain far more spectral information, so they're information-rich. They can provide details about an intelligence target that you simply cannot get with conventional or MSI imagery. For example, thin layers of oil are absorbed differently (in different parts of the spectrum) compared with thicker layers of oil. Multispectral imaging can locate oil slicks by their distinctive color on ocean water. Hyperspectral imaging goes one step further: it will indicate the amount of oil present. Those images can be useful for identifying polluting drilling platforms or ships that are leaking oil.

Spectral Signatures

Spectral images, rendered in false color, have many civilian and military applications, as we've just discussed. But even more insights are possible by looking at the images in a different way.

Let's return for a moment to the stack of images—in GIS terms, a set of layers that spectral imaging produces. What do you do with a stack of images? How do you use them? You can look at only one image at a time, after all, even if it is an image that combines a few bands using false color. And we're increasingly talking about hundreds or thousands of bands, all of which can't be meaningfully represented in a single false color image.

The secret? This is important: what you really want to know from spectral imaging doesn't come from looking at the total scene. You're looking for unique signatures of objects in the scene, or signatures of uniform areas, such as crops, forests, and polluted areas. The trick is to use the stack to create something that isn't an image; it's another example of a signature, usually represented as a graph of a single pixel in the scene, as a function of frequency or wavelength. The objective is usually to determine the composition of materials present in the target pixel.

Suppose we want to know more about a single pixel of special importance. In the most basic terms, the first step is to measure the intensity of the energy received by the pixel in each spectral band. The second step is to use those measurements to create a graph of intensity versus wavelength for that pixel. And the final step is to compare signatures in each pixel with available signatures contained in a spectral library—a set of spectral signatures of different solids, liquids, and gases. The US Geological Survey maintains such a library.

Figure 11.5 illustrates the result. It also makes an important point about spectral imaging: even two quite similar gases can be distinguished because they have different spectral signatures. The distinction can be critical in looking at gases containing phosphorus that are escaping from a chemical plant; they could be coming from pesticide production or nerve agent production, and spectral signatures are needed to tell the difference. Note that the LWIR band in the figure (around 10 microns) is where the biggest differences are. This is the primary band used for signatures to characterize gaseous effluents—such as those from a polluting smokestack or a chemical weapons production plant.

Spectral Imaging Platforms

The first spectral imagers flew on aircraft. By 1970 multispectral imagers were being built for airborne use, and technology quickly advanced into hyperspectral imagers. By 1985 hyperspectral imagers were scoring major successes in agricultural research and in investigating wetland vegetation canopies. In 2018 airborne spectral imagers were being used in many countries for resource management: monitoring the development and health of crops, and providing early warning of plant disease outbreaks.

Manned aircraft, however, are costly to operate. As UAVs have become less expensive, with longer ranges, and the capability to operate autonomously, they

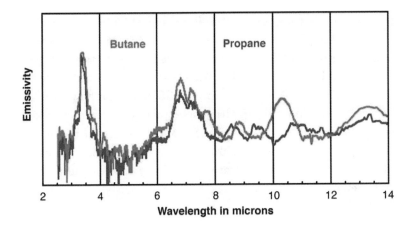

FIGURE 11.5 Comparison of spectral signatures. *Source:* Robert M. Clark, *Intelligence Collection* (Washington, DC: CQ Press/Sage, 2014). Used by permission.

present an attractive alternative platform. Manned aircraft spectral sensors have been redesigned to be lighter and smaller, and thus more suitable to be carried by UAVs for hyperspectral mapping. In 2018 a large number of such sensors were commercially available.[8]

But airborne sensors can't easily cover the entire globe. That requires satellites. And since 1972, a number of spectral imagers have been launched by several countries. For over four decades, Landsat spectral imagery has been widely used in regional planning, education, emergency response, disaster relief, and global change research. The spectral sensor on the most recent of the series, Landsat 8, includes eleven bands covering the spectrum from visible light to LWIR.

Of course, in spectral imaging, more is better. So satellites operating in more spectral bands than Landsat are in use today; they are able to locate valuable metal ore deposits such as copper, aluminum, tin, and gold globally—something that otherwise would have to be done on the ground or by airplane only at enormous cost. Spectral imaging technology has spread internationally, and many countries have spectral imagers in orbit for geology, water resources, vegetation, and land cover missions. Russia's Resurs-P series of commercial Earth observation satellites acquires high resolution (up to 1.0 meter) multispectral imagery. The trend, however, is to hyperspectral imaging from space. In 2018 China launched its Gaofen-5 remote-sensing satellite carrying a hyperspectral imager, and at least five HSI imagers in other countries were being planned or nearing launch.[9]

Chapter Summary

Despite its technological advances over the last two decades, spectral imaging is still in its early growth stages for geospatial intelligence. It depends on the availability of ever more complex spectral libraries that catalog the signatures from different materials. And the older libraries of MSI signatures aren't of much use in interpreting an HSI signature, just as HSI libraries don't help much in characterizing a USI signature. Spectral libraries are constantly being updated. These, in turn, will enable applications in GEOINT that haven't even been thought of yet. Processing and exploiting spectral imagery for geospatial intelligence remains a challenge, in part because of the costs and time delays involved. Processing and analyzing MSI data are expensive; HSI is even more demanding, often requiring custom software and very expensive expert labor. A similar set of challenges characterizes radar imaging, discussed in chapter 12.

Notes

1. NASA, "Herschel Discovers Infrared Light," n.d., https://spaceplace.nasa.gov/review/posters/herschel/Herschel-ir-activity.pdf.
2. Michael W. Davidson, "Johann Wilhelm Ritter," Florida State University, November 13, 2015, https://micro.magnet.fsu.edu/optics/timeline/people/ritter.html.
3. Night Optics, "The History of Night Vision," n.d., www.nightoptics.com/history-of-night-vision.htm.
4. LifePixel, "History of Infrared Photography," n.d., www.lifepixel.com/infrared-photography-primer/ch1-development-of-infrared-film.
5. Floyd F. Sabins, "Remote Sensing for Mineral Exploration," *Ore Geology Reviews* 14 (1999): 157–83.
6. Quy-Toan Do, Jacob N. Shapiro, Christopher D. Elvidge, Mohamed Abdel-Jelil, Kimberly Baugh, Jamie Hansen-Lewis, and Mikhail Zhizhin, "How Much Oil Is Daesh Producing? Evidence from Remote Sensing," World Bank, December 19, 2016, http://documents.worldbank.org/curated/en/239611509455488520/How-much-oil-is-the-Islamic-state-group-producing-evidence-from-remote-sensing.
7. Kyle Loggenberg, Albert Strever, Berno Greyling, and Nitesh Poona, "Modelling Water Stress in a Shiraz Vineyard Using Hyperspectral Imaging and Machine Learning," *Remote Sensing* 10 (2018): 202.
8. Telmo Adão, Jonáš Hruška, Luís Pádua, José Bessa, Emanuel Peres, Raul Morais, and Joaquim João Sousa, "Hyperspectral Imaging: A Review on UAV-Based Sensors, Data Processing and Applications for Agriculture and Forestry," *Remote Sensing* 9 (2017): 9, 10.
9. Deyana Goh, "China Launches Gaofen-5 Hyperspectral Imaging Satellite for Atmospheric Research," *Spacetech Asia*, May 9, 2018, www.spacetechasia.com/china-launches-gaofen-5-hyperspectral-imaging-satellite-for-atmospheric-research/.

Radar Imaging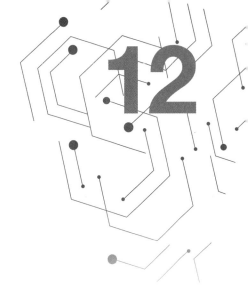

On September 3, 2017, North Korea conducted its largest underground nuclear test yet observed. The explosion had an estimated yield equivalent to about 250 kilotons of TNT. The test took place inside the Mount Mantap test site near Punggye-ri, creating a seismic disturbance equivalent to a magnitude 6.3 earthquake.[1]

A multinational research team from Singapore, China, Germany, and the United States later investigated the changes in Mount Mantap as a result of the detonation. It found that the explosion had pushed the mountain's surface outward by up to 11 feet and had shortened the mountain by 20 inches. Eight minutes after the initial explosion, an aftershock occurred about 2,300 feet from the center of the blast. The aftershock occurred about halfway between the blast's center and an access tunnel, likely as a result of the tunnel collapsing. The team also concluded that a large volume of fractured rock had later compacted over an area about a mile across, causing the mountain to further subside.[2]

What happened next was a bit of theater. On May 24, 2018, North Korea closed the test site permanently, apparently demolishing it with explosives in front of a group of foreign reporters brought in to witness the event. The North Korean government proudly proclaimed the demolition as a step toward peaceful resolution of demands that the North denuclearize. Many scientists took a more logical view: that the demolition was likely an effort to make the best of an embarrassing miscalculation. The test engineers may have failed to take into account the consequences of the detonation. When the aftershock and mountain subsidence were caused by tunnel collapse, the site was rendered unusable for further tests.[3]

That leaves a question: How did the scientists measure the changes in Mount Mantap so precisely? The short answer is GEOINT. But, more specifically, they were able to do so thanks to a remote sensor that can measure changes to the Earth's surface to within a few millimeters—imaging radar. Two such radar systems—one German, one Japanese—had imaged the mountain

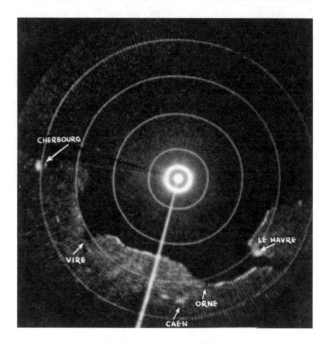

FIGURE 12.1 Radar image of the Normandy coast, 1944. *Source:* US National Archives.

before and, of course, after the test, and the image comparison allowed accurate calculations of the changes caused by the detonation.[4]

Radar is the third class of imaging sensor on which geospatial intelligence relies—because it provides insights that neither visible imagery nor infrared and spectral imagery can deliver. Like infrared imagery, it has an advantage over visible imaging: it can be done any time, day or night. And it has an added benefit over both other imaging types: it can be done under almost all weather conditions. But radar imaging's biggest advantage has come from the new class of signatures that it provides. Radar images can show features that aren't apparent in any type of optical imagery. Nevertheless, for a long time radar was hampered by its chief disadvantage: the poor quality of the image, compared with the best available visible imagery. That changed beginning in the 1970s.

Conventional Radar

In 1935, the British physicist Robert Watson-Watt demonstrated the first radar—to search the sky and provide warning of approaching thunderstorms. By the end of the decade, several countries—the major participants in World War II—independently and secretly had developed radars for military use.

Soon after the outbreak of war, both Great Britain and Germany were placing radars in aircraft. In 1943 the British developed the first airborne radar that could create an image of the Earth's surface. Named the H2S, its purpose was to identify targets for night and all-weather bombing. The US improved the radar by going to a higher frequency band with the H2X radar in 1944, which gave better images.[5] Figure 12.1 shows an H2X radar display of the Normandy coast before D-Day. It illustrates the problem of a conventional radar for imagery use: cities show up as bright blobs. Such radars can spot ships at sea, but cannot provide the detail desired for imaging. So conventional radar gives a geospatial picture that can be useful, but it's not a true image. To do that, engineers had to develop more complex radars, such as **side-looking airborne radar** (SLAR).

Side-Looking Airborne Radar

The first actual imaging radar system was side-looking airborne radar. It uses a long phased array antenna mounted on the side of an aircraft fuselage.[6] The antenna transmits a very narrow beam and, as the aircraft flies along, the radar returns are recorded on photographic film (later, digitally) to produce an image.

The first SLARs were developed for battlefield reconnaissance in the 1950s. The US Army OV-1B Mohawk carried one; its most prominent feature was the radar antenna, which was carried on an 18-foot-long boom below the aircraft (fig. 12.2). The Mohawk's radar also could detect moving targets, relying on the Doppler effect. This combination of the ability to obtain imagery and detect moving targets gave the Mohawk a long service life. Its mission lifetime of over thirty-five years spanned both the Vietnam War and the First Gulf War. In between, Mohawks flew missions along the dividing line between East and West Germany and near the Demilitarized Zone in Korea to provide warnings of hostile intent.[7]

SLARs proved at least as useful for civil geospatial applications. Early civilian research using SLAR was concerned with mapping and charting. By 1962 it became apparent that SLAR imagery would be useful for terrain and geological

FIGURE 12.2 A US Army OV-1B Mohawk with a side-looking airborne radar system. *Source:* AVIA BavARia.

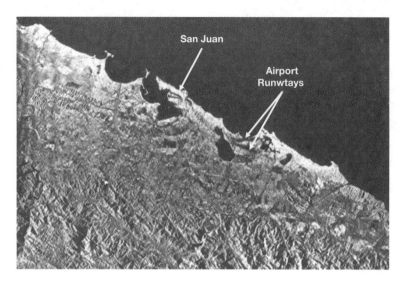

FIGURE 12.3 US Geological Survey side-looking airborne radar image of San Juan. *Source:* US Geological Survey.

studies as well. And SLAR imagery was being used for petroleum exploration by the early 1970s. The US Geological Survey began to collect SLAR imagery of the US in 1980. Since then, the imagery has been used for a number of geologic studies as well as for petroleum exploration. The quality of SLAR imaging became better over time; the 1999 image of the region around San Juan (fig. 12.3) has a resolution of 10 to 15 meters.[8]

Synthetic Aperture Radar

SLAR was a definite step forward in imaging radar, but it still lacked the resolution needed for military reconnaissance. Higher-resolution images mean more detailed signatures, which are always what we want. In any type of image (visible, spectral, or radar), given enough resolution, an imagery analyst can distinguish a military truck from a bus of civilians, which makes a big difference when making targeting decisions. Current SLAR image technology, as illustrated in figure 12.3, is good enough to identify airfields. The Luis Muñoz Marín International Airport runways near San Juan show up as dark streaks because, like the ocean in the image, they reflect energy away from the radar. But it's not nearly good enough to identify an aircraft at the airfield.

In the late 1940s, the US Army had a requirement for an airborne imager that would work over clouded areas, day or night. Visible and infrared imagers couldn't fulfill that job. Radar just might possibly work; it could see through clouds or fog, and it wasn't affected by darkness. It seemed to be the logical choice for an imager, *if* it could get images at a good resolution. Getting that resolution, however, would require some electronic trickery.

To get a good radar image, a small pixel is needed—one that corresponds to a ground area of 1 foot × 1 foot would be especially nice to have. That means getting *both* range resolution (along the radar beam) and azimuth resolution (crosswise to the radar beam) of 1 foot simultaneously. There was just one problem. The two require completely different techniques.

Range resolution was relatively straightforward. The secret was to transmit a very short pulse, lasting about a billionth of a second, but one with a lot of energy in it. Through a bit of engineering magic called pulse compression, radar engineers did exactly that. The result was a radar having a range resolution of a foot or even less, but still transmitting enough energy to get a return signal from distant targets.

Azimuth resolution was much harder to obtain, and required a completely different approach. The issue is that a radar beam spreads out as you move farther from the radar. Figure 12.1 illustrates the phenomenon; the radar beam (the bright streak from the center, aimed toward the Normandy coast) broadens as it gets further from the aircraft. Think about a flashlight beam; as you shine it on a very close object, it will be a few inches wide. Shine it on a distant object, and the light will spread over several feet. Continuing the analogy, a flashlight needs to have a brighter beam (more energy) to show distant targets. Radar beams work exactly the same way.

The Mohawk's SLAR was a good example of the problem. It had a beamwidth less than 1 degree, which is exceptionally good for conventional radar.[9] And it produced good images for its time. But at a distance of 60 miles, the Mohawk's radar beam was about 500 feet wide, an azimuth resolution far worse than the army needed. A pixel size of 1 foot × 500 feet simply wouldn't do the job for high-resolution imagery.

Radar engineers could calculate the needed antenna size, and they did. The answer was that, in order to obtain a useful level of azimuth resolution, the radar antenna would need to be at least as long as several football fields, far too large for a reconnaissance aircraft to carry. At least, that was the consensus of radar engineers until Carl Wiley tackled the problem.

Building an Antenna That Is 100 Kilometers Long

In 1950 Carl Wiley was hired by Goodyear Aircraft Company (later Goodyear Aerospace) to work on airborne radar technology. He has been described as a "brilliant if eccentric engineer."[10] He liked to work on difficult problems, and the challenge of building high-resolution imaging radar was one that especially appealed to him.

There are moments of discovery in history that occur when someone cuts through to the heart of a problem, and identifies a basic principle that changes everything. Such a moment came to Wiley in 1951. He asserted that you *could* put a phased array antenna on an aircraft that was indeed as long as several football fields, and even much longer. Here is the secret: *the individual elements in the array do not all have to transmit at the same time.* Wiley pointed out that each object in a radar beam has a slightly different speed relative to a moving antenna; so each object will have its own Doppler shift, and that shift will change for each pulse transmitted by the antenna as it moves along the flight path. Because the Doppler shift history is unique for each object in the scene, if you can capture and record all the radar returns and then process them, you can create a detailed image of a scene. A radar antenna about 1 meter in length can be made to acquire an image that otherwise would have required a much longer one—even one 100 kilometers in length—simply by illuminating an area with continuous pulses of radar energy while flying a 100-kilometer distance.[11]

Wiley went on to build the first **synthetic aperture radar** (SAR), which opened a new era in remote sensing and became one of the most important technologies to be applied in creating geospatial intelligence.

Evolution of Synthetic Aperture Radar

Wiley's first SAR to fly didn't achieve impressive results—it was worse, in fact, than a conventional radar of the time. It was called *Douser* and was mounted on a C-47 cargo aircraft for testing. The imagery product had an azimuth resolution of about 500 feet, about the same as the Mohawk's SLAR, but it was a start. Any new technology will follow an S curve in its performance over time. Performance will start slowly, then accelerate rapidly, until it approaches some sort of limit—the top of the S—before leveling off. And, during the 1950s and 1960s, SARs climbed that performance curve with a vengeance. By the late 1950s, engineers had improved the resolution; their radar could show objects that were only 100 feet across, then 50 feet. By the latter part of the 1960s, airborne SARs were operating with a resolution of just 1 foot in range and azimuth, producing high-quality imagery.[12] Later on, SARs were designed to provide a three-dimensional or stereoscopic image, just as visible imagers do, with exquisite terrain detail.

From its beginning and continuing today, a major constraint on the value of SAR imagery is that customers of GEOINT don't understand the unique contributions that it can make. For many decision-makers, seeing is believing. And SAR imagery can be intimidating because, like infrared imagery, it is hard for the untrained observer to understand what he is looking at. SAR images are affected by artifacts (discussed below). Military customers appreciate the advantages of SAR's ability to image day and night and in all weather. They like SAR's ability to "see through" materials such as wood and nonmetal roofing, to peer inside buildings and occasionally into tunnels. But it's fair to say that even they prefer visible imagery when they can get it. That places the GEOINT analyst in the sometimes-uncomfortable role of advocate for, and interpreter of, SAR imagery. Both roles have assumed importance over the years as a number of new applications have been introduced that have made SAR attractive for both military and civil use. Here are the most important ones.

Polarimetry

Chapter 10 introduced polarimetry, which has several uses in optical imaging. In optical imaging, filters are used to select the polarization in an image, but you can't control the polarization of the illuminator (the sun). Using SARs you can control both what you transmit and what you receive. SAR images are generated by sending and receiving microwave pulses that can be either horizontally or vertically polarized—or both. Most newer SARs transmit *and* receive both polarizations, and are called polarimetric SARs. They provide additional details about targets that are valuable for monitoring crops, soil moisture, forests, snow cover, sea ice and sea conditions, and for geological mapping. Some of these applications are relevant in geospatial intelligence. A polarimetric SAR is better at distinguishing targets of military interest such as tanks and warships, for example. By selecting the polarization combination, you can often see targets that would otherwise be hidden in clutter.

Foliage Penetration

Most SARs operate in the microwave bands. Radar systems in these frequency bands can penetrate many surfaces and materials that are often used to hide equipment or facilities. Canvas tents and most wooden sheds, for example, are transparent to radar. But higher-frequency SARs such as TerraSAR-X don't penetrate foliage well; the moisture in leaves reflects some microwave energy. At lower frequencies, however, radar can penetrate even dense foliage to detect objects located underneath. SARs in these lower frequency bands can also penetrate into dry earth for short distances. Known as foliage-penetrating SAR, these radars are mostly used on aircraft or unmanned aerial vehicles. They have demonstrated success in identifying military and terrorist activities concealed under jungle canopy.[13]

Artifacts

Conventional radar such as SLAR gets all the information it needs to form a small part of an image from the returned signal from a single pulse. Then, over time, as the radar beam moves, the image gradually takes form, pulse by pulse.

A SAR doesn't work that way. It must wait until it has transmitted thousands or millions of pulses, and collected and stored details from the returned energy of each pulse, to form even part of the image. But subsequent SAR processing assumes that nothing changes in the scene while an image is being formed, and that is seldom the case. Things often change on the ground that the SAR is imaging, and that produces consequences in the image, known as *artifacts*. For example, soon after SARs came into operational use, imagery analysts found that if an object in the image was moving while the image was being created, it would appear to be displaced from its actual position and also often blurred. And targets of intelligence interest—such as ships at sea, battlefield vehicles, trains, and automobiles—tended to be in motion. It was disconcerting to see a freight train in an image and note that it appeared to be traveling along the landscape parallel to the train track but some distance away. On one hand, those artifacts were just annoying.

But SAR developers and imagery analysts soon recognized that the "annoyance" could be useful, especially in monitoring activity on a battlefield or in countering terrorist activity. A SAR can be operated so as to create imagery that *highlights* moving targets in an area. This feature is commonly called **moving-target indicator** (MTI) or **ground-moving-target indicator** (GMTI). The term "MTI" is typically used because it is a more accurate description. MTI can be used to detect aircraft, helicopter, or ship movements, as well as ground vehicular movement.[14] Of course, their movement causes them to be displaced in the image, and they have to be relocated (by a GEOINT analyst) to their correct position.

The JSTARS aircraft, introduced in chapter 8, is an example; it routinely tracks the movement of ground vehicles and helicopters and relays tactical imagery to ground and air theater commanders. JSTARS evolved from separate US Army and Air Force programs to develop technology to detect, locate, and attack enemy armor at ranges beyond the front lines of combat.

Change Detection

Changes in a scene over time are detectable in visible imagery, as we noted in chapter 10. But that is also the case for any type of imagery: visible, spectral, or radar. You obtain an image of a scene, then get a second one the next day, and see what is different. Are there objects or entities that have entered the scene? Left the scene? Moved or changed within the scene? Over time, computational techniques have allowed automation of imagery searches to look for such changes and to color code those for easy identification. You can, for example, have successive images of a major port facility over time, and process the images so that ships that have arrived in the port are colored blue, and the area that a ship departed from is colored red. The technique has a special name: it's called **incoherent change detection**, only to differentiate it from what we discuss next.

SARs can do that type of detection, of course. But they can also measure very small changes in a surface through a technique called **coherent change detection**. It requires taking an image of an area, and then—days or weeks later—taking another image using an identical flight path and altitude. Electronically processing the two images then highlights even minute changes that have occurred—changes that would not be visible in optical imagery. Examples of surface changes that can be observed include vehicle tracks, crops growing or being harvested, and soil excavation. Changes in the terrain surface due to underground construction also can be observed using change detection. The underground excavation causes both vertical settlement and a horizontal straining that SARs can detect.[15] TerraSAR-X is an example of a SAR that can make such accurate observations; it was one of two spaceborne SARs used to measure the changes at Mount Mantap due to North Korea's nuclear test.

Civil Applications

From the beginning, spaceborne SARs were sent into orbit for intelligence purposes, primarily because they offered both night and all-weather imaging. But civil uses were not far behind.

The first obvious civil application was ocean surveillance. Early satellite optical imagery was taken only over land, with rare exceptions. Imaging the ocean had limited value because the spaceborne optical imagers could not cover large ocean areas with sufficient resolution to identify ships. But numerous maritime issues require broad area ocean imaging, for example, spotting ice floes and other hazards to navigation, and monitoring ship traffic to combat piracy, illegal fishing, and several forms of illicit goods traffic. That capability—broad area ocean imaging and detailed resolution, day or night—could be provided by radar imaging satellites.

So in 1974, NASA's Jet Propulsion Laboratory and NOAA teamed up to create a spaceborne SAR that could make oceanic observations. The joint effort resulted in the launch of the Seasat satellite in 1978—the first civilian application of synthetic aperture radar and a major step forward in remote sensing from space.

Seasat had several instruments that were designed to collect data on many ocean and atmospheric features. Its SAR had specific missions that relied heavily on polarimetry. It measured ocean phenomena such as surface waves, waves beneath the surface (known as internal waves), currents, upwelling, shoals, sea ice, wind, and rainfall.

The SAR also *allegedly* was able to detect the wakes of submerged submarines, a discovery not anticipated before launch. Seasat was abruptly shut down in October 1978, reportedly because it was malfunctioning. Soon thereafter, a conspiracy theory developed, claiming that once the alleged antisubmarine capability was discovered, the military deliberately shut the satellite down, using a cover story of a short circuit in the power supply.[16]

Seasat was the beginning of an era in the use of SAR for civil remote sensing. Since then, from space and from airborne platforms, SARs have monitored sea ice, measured glacier changes, observed oceanic wind patterns, warned of storm surges, identified areas vulnerable to landslides, and enabled drought prediction. A SAR's ability to do change detection is now used widely for monitoring volcanic activity. The pressure of magma building up inside volcanoes typically will cause slight distortions on the surface. SARs can detect these small variations via coherent change detection, so they have become valuable tools for monitoring volcanoes and their lava flows and warning of impending eruptions. The same capability has been shown to allow damage assessment after natural disasters such as earthquakes.[17]

Laser Radar

The laser was invented in the early 1960s. During the next fifty years, it found use in a wide range of fields: medicine, surveying, precision-guided bombs, and laser weapons, for example. It even found use in everyday items such as a carpenter's level and a laser pointer for presentations.

Early on, it was discovered that a laser could be used as a radar transmitter. Laser radars, nicknamed LIDARs, operate much like microwave radars. But they have some definite advantages. A laser beam is very narrow, so a LIDAR can illuminate an extremely small surface (less than a meter at aircraft-to-ground distances). It can measure distance much more accurately than can microwave radar. Unlike microwave radars, though, it cannot penetrate clouds.

The narrow beam and short pulses allow laser radars to produce three-dimensional images without having to use the SAR techniques discussed above. They can measure the dimensions of features (e.g., forest canopy or building height relative to the ground surface).

One of the laser radar's most important GEOINT uses is for seeing beneath camouflage or forest canopy. The laser takes many measurements of the same target area, so that it in effect finds holes in the covering material. It therefore is able to penetrate through the camouflage mesh or foliage to obtain a return from the object beneath. Using this technique, laser radars have demonstrated a capability to provide three-dimensional imagery of military vehicles concealed under foliage or camouflage. The images are of sufficient quality for analysts to perform object classification and identification.[18] Figure 12.4 illustrates an example of the ability of LIDAR to image under forest cover. The image is of the Effigy Mounds National Monument in Iowa, which has more than 200 prehistoric Native American mounds. Most have forest cover around them, making

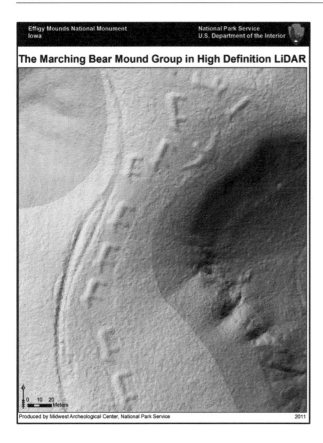

FIGURE 12.4 Laser radar image of Effigy Mounds National Monument. *Source:* US National Park Service.

the mounds difficult to identify from above. It's not a problem for LIDAR, as the example of the Marching Bear Mounds indicates; the tree cover has been neatly removed from the scene, showing the bare earth beneath.[19]

Laser radars also are used to remotely identify surface materials. Many chemical and biological agents, and spoil from excavations, fluoresce when exposed to ultraviolet (UV) light, so UV lasers are used for fluorescence sensing. Rare earth elements and some heavy atom elements such as uranium fluoresce, for example. For intelligence use, UV light has an obvious advantage in such roles: it is not visible to the human eye, so UV illuminators are less likely to be detected (at least, by humans; we're generally not concerned about the reindeer).

Chapter Summary

Imaging radar is the third sensor in the triumvirate of remote-sensing tools that fueled the evolution of geospatial intelligence. It provides insights that neither visible imagery nor infrared and spectral imagery can deliver. For example,

- SAR permits detailed ocean surveillance over a large area.
- SAR's ability to detect slight variations in the Earth's surface over time allows change detection for both military and civil uses.
- SAR can be optimized to detect and track vehicle movement.
- Both SAR and LIDAR can "see through" forest cover and obtain details of the ground surface and targets beneath the canopy.

As critical as these tools are, we have said repeatedly that GEOINT is much more than imagery. The production of geospatial intelligence requires the integration of nonimagery sources, other emerging technologies, and, of course, analysis expertise. Before diving back into the tools, let's first examine the explosion of drivers or needs that led to the rapid rise of the discipline.

Notes

1. Liza Lester, "2017 North Korean Nuclear Test Was Order of Magnitude Larger than Previous Tests," *Newscenter*, n.d., https://news.ucsc.edu/2019/06/nuclear-test.html.
2. University of California, Berkeley, "Radar Reveals Details of Mountain Collapse after North Korea's Most Recent Nuclear Test," *Phys.Org*, May 10, 2018, https://phys.org/news/2018-05-radar-reveals-mountain-collapse-north.html.
3. "North Korea Nuclear Test Tunnels at Punggye-ri 'Destroyed,'" BBC News, May 24, 2018, www.bbc.com/news/world-asia-44240047.
4. "North Korea Nuclear Test Tunnels."
5. Bernard Lovell, "The Cavity Magnetron in World War II: Was the Secrecy Justified?" *Notes and Records of the Royal Society London* 58, no. 3 (2004): 283–94, https://royalsocietypublishing.org/doi/pdf/10.1098/rsnr.2004.0058.
6. A phased array is a line or grid of individual antennas, configured so that they act as a single larger antenna. They can be individually fed radiofrequency energy so as to create a beam of energy that is electronically steerable, much as one would move a flashlight beam to illuminate objects.
7. John Sotham, "The Last of the Mohawks," *Air & Space Magazine*, March 1997, www.airspacemag.com/military-aviation/the-last-of-the-mohawks-1649/#yfiMGA2AiqLiFCPH.99.
8. US Geological Survey, "Side-Looking Airborne Radar Image of Puerto Rico," 1999, https://pubs.usgs.gov/of/2000/of00-006/htm/slar.htm.
9. "Evaluation of Two AN/APS-94 Side-Looking Airborne Radar Systems," US Coast Guard Research & Development Center, n.d., https://apps.dtic.mil/dtic/tr/fulltext/u2/a108404.pdf.
10. Lockheed Martin, "Synthetic Aperture Radar: 'Round the Clock Reconnaissance,'" n.d., www.lockheedmartin.com/en-us/news/features/history/sar.html.
11. Carl A. Wiley, "Synthetic Aperture Radar," *IEEE Transactions on Aerospace and Electronic Systems*, vol. AES-21, no. 3 (May 1985): 440.
12. Wiley.
13. Wiley.
14. Zhou Hong, Huang Xiaotao, Chang Yulin, and Zhou Zhimin, "Ground Moving Target Detection in Single-Channel UWB SAR Using Change Detection Based on Sub-Aperture Images," *Heifei Leida Kexue Yu Jishu*, February 1, 2008, 23.
15. J. Happer, "Characterization of Underground Facilities," JASON Report JSR-97-155, April 1999, www.gwu.edu/~nsarchiv/NSAEBB/NSAEBB372/docs/Underground-JASON.pdf.
16. Dwayne A. Day, "Radar Love: The Tortured History of American Space Radar Programs," *Space Review*, January 22, 2007, www.thespacereview.com/article/790/1.

17. Luca Guida, P. Boccardo, I. Donevski, and L. Lo Schiavo, "Coherent Change Detection on SAR Imagery," ResearchGate, April 2018, www.researchgate.net/publication/324851159_POST-DISASTER_DAMAGE_ASSESSMENT_THROUGH_COHERENT_CHANGE_DETECTION_ON_SAR_IMAGERY.
18. Richard M. Marino and William R. Davis Jr., "Jigsaw: A Foliage-Penetrating 3D Imaging Laser Radar System," *Lincoln Laboratory Journal* 15, no. 1 (2005): 23.
19. US National Park Service, "The Marching Bear Mound Group in High Definition LiDAR," n.d., https://upload.wikimedia.org/wikipedia/commons/4/4b/Effigy_mounds_lidar.jpg.

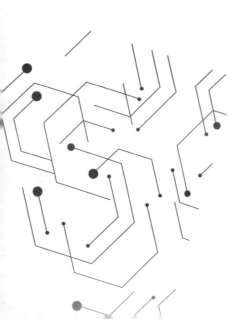

The Drivers of Geospatial Intelligence

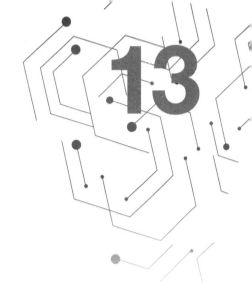

In 1961, COMINT analysts at the US National Security Agency began to see a troubling trend. They were intercepting messages about a number of Soviet ships headed for Cuba. The cargo manifests were blank, suggesting something suspicious, such as that the shipments could be carrying cargo of military significance. Other ships appeared to be heading for Cuba after having reported another destination, and declaring less cargo than the ships could carry. Meanwhile, separate National Security Agency intercepts indicated that the Cubans were unloading ships at night and taking unusual precautions to keep the deliveries secret. In one communication, Cuban port authorities mentioned the arrival of tanks; in others, Cubans discussed "highly unusual aircraft" and radars.[1]

During 1962, the pattern of suspect deliveries continued, and US leadership became concerned. If the military deliveries were purely for defensive purposes, that was a problem, but not one that called for military action; Cuba had the right to defend itself. If offensive weapons were being delivered, that was another matter entirely.

Which turned out to be the case. Soviet premier Nikita Khrushchev had decided to emplace nuclear-equipped SS-4 and SS-5 ballistic missiles in Cuba to counter the increasing US edge in ballistic missiles aimed at the Soviet Union. His objective was to present the US with a deployed strategic missile threat from Cuba when it would be too late for a US response. That required concealing the deployment until the missiles were combat-ready, and the Soviets had constructed an elaborate denial and deception (D&D) program that combined HUMINT, IMINT, OSINT, and diplomatic deception to do just that. It almost succeeded.

But the Soviets made several mistakes in underestimating US imagery analysts. They shipped their medium bombers in recognizable shipping crates (mentioned in chapter 10). In Cuba, they deployed missile units using their standard operating procedures—creating characteristic imagery signatures that analysts immediately spotted. SIGINT analysts identified the presence of

SA-2 surface-to-air missile sites, and imagery analysts noticed that these Cuban sites were arranged in patterns similar to those used by the Soviet Union to protect its ballistic missile bases. The US resumed reconnaissance flights that had been suspended, and on October 14, 1962, a U-2 flight located Soviet ballistic missile units at San Cristobal, Cuba, leading to what has been referred to ever since as the Cuban missile crisis.[2] Confronted with a US ultimatum, the Soviets ultimately agreed to pull the missiles out and end the crisis.

The unfolding of the Cuban missile crisis illustrates a defining point about geospatial intelligence: it is more than imagery. The critical pieces of information and intelligence often come from sources other than imagery—as happened in 1961 and 1962. A complete picture, not just an image, of the situation is required. So **fusion**—the integration of material from all intelligence sources, and nonintelligence sources as well—is essential to the production of GEOINT. Following are a few of the driving functions for fusion of imagery with other sources—many of which have been present for a long time—that fueled the rise of GEOINT.

Denial and Deception

Defeating Soviet D&D was a major success story of the Cuban missile crisis. But D&D has a long history, and sometimes the successes are on the other side.

During World War II, D&D often took the form of decoy tanks, military vehicles, missiles, even decoy submarines to confuse or mislead an opponent in military operations. Eighty years hence, decoys are still popular among military forces who must fight superior opponents. They are so popular that commercial firms now produce them. Choose your decoy: jet fighters, tanks, rocket launchers, antiaircraft missiles, and more, created from metal and fiberglass. They're readily available, and Middle Eastern countries have been big buyers for years.[3] During the 1991 Gulf War, Iraq deployed decoy SCUD missile launchers to draw fire from coalition aircraft. Decoys over the years have ranged from crude to very sophisticated. With each evolution in imagery quality, analysts have built success in identifying features indicating that a target is a decoy. Vendors, however, always respond by building better decoys. Fusion of imagery with other intelligence and nonintelligence sources has become essential in dealing with the threat, as it was in the Cuban missile crisis.

D&D is widely used in tactical military operations. But it is also used to gain a strategic advantage. By the 1990s, most countries were aware of what overhead imagery could do, and they were getting smarter about how to defeat IMINT. Since then, D&D has become more difficult to counter by visible imagery alone, and some countries have scored successes at deception based on their knowledge of imagery sources and methods. India did so in its nuclear test at Pokhran.

Pokhran

During 1995, the government of India began preparations to conduct an underground test of a nuclear warhead at its Pokhran test site in its northwestern desert province of Rajasthan. In December of that year, US intelligence analysts observed the test preparations in satellite imagery and provided a warning of the impending test.

In the US view, such a test was unacceptable; it would likely provoke a nuclear arms race between India and Pakistan, with a destabilizing effect on the Indian subcontinent. The US ambassador to India consequently presented a démarche demanding that the Indian government stop the test preparations.[4] The Indians responded by, in effect, saying, "Why do you think we're going to conduct a nuclear test?" In response, the ambassador displayed US imagery, showing the movement of vehicles and test equipment at Pokhran to top Indian officials and forcing them to back down.[5]

The Indian government had learned a lesson in the démarche, but not one that the US intended to teach. In 1998, a Hindu nationalist government led by Prime Minister Atal Bihari Vajpayee took power with a platform that included building a nuclear weapons capability. It secretly began preparations for a nuclear test, aided by an elaborate deception plan.

The deception was carefully designed to conceal preparations from US imagery. Indian test personnel did their work at night, always returning heavy equipment to the same parking spot at dawn and removing evidence of any activity. When they needed to dig shafts, they did so under a camouflage net. Sensor cabling was carefully covered with sand, and native vegetation was replaced to hide evidence of digging. Excavated sand was placed to look like wind-shaped sand dunes.[6]

Just before the test, Indian leaders added an imagery deception feature. They knew that the US closely monitored ballistic missile tests at their Chandipur missile test range in the east. Consequently, they began preparations for what appeared to be a ballistic missile test there.[7] As a result, reconnaissance satellites reportedly were focused on Chandipur, more than a thousand miles from Pokhran, with little or no coverage of the nuclear test site.[8] Consequently, when the Indians detonated three nuclear devices at Pokhran on May 11, 1998, it was a complete surprise for the US government and its allies.

The Pokhran deception succeeded, in part, because the US apparently relied primarily on imagery for a warning of likely test preparations, and the Indians were very successful at denying other intelligence sources that might have provided warning. US intelligence, aided by the Israelis, did much better in countering deception a few years later in al Kibar, Syria.

Al Kibar

Syrian intelligence undoubtedly learned from the Indian experience with Pokhran, and in 2003 they attempted to conduct their own deception. It had a less satisfactory outcome—for them, at least.

The Syrians were planning to build a nuclear reactor to produce weapons-grade uranium, located near the desert town of al Kibar. It was to be built with North Korean assistance, its design modeled on an existing one at Yongbyon, North Korea. Syrian intelligence realized that the US had a detailed profile of the Yongbyon reactor; US imagery analysts would undoubtedly recognize the similarity in a Syrian design. They consequently developed a sophisticated deception designed to mislead imagery analysts. First, they avoided adding signatures that might provoke notice: no special site security; no air defenses nearby; no transportation signatures such as roads, rail tracks, or airfields. Then the reactor itself was partially buried under a false roof to hide its characteristic shape from observation above. Finally, they built an outer shell over the reactor that appeared in imagery to be a Byzantine fortress similar to others found in Syria.

The Syrians had succeeded in minimizing any visible signatures that would tip off the purpose of the facility. Reportedly, in 2005 US imagery analysts found the building but could not figure out its purpose; it was "odd and in the middle of nowhere."[9]

The deception was exposed, as imagery deceptions often are, with help from a combination of intelligence from other sources. In early March 2007, special agents of Israel's Mossad broke into the Vienna residence of Ibrahim Othman, head of the Syrian Atomic Energy Agency. They hacked into Othman's computer and copied top secret material from it. The material included photographs of the interior at al Kibar, confirming that it was a copy of the Yongbyon reactor. Following up on that discovery, in June an Israeli special operations unit infiltrated Syria and approached al Kibar. They took photos of the site and brought back soil samples for analysis.[10]

The Israelis were convinced that the site was intended to produce material for nuclear weapons, and they shared their evidence with the US. But US leaders were not inclined to conduct a military strike on the site. As Secretary of Defense Robert Gates reportedly observed, half in jest, "Every Administration gets one preemptive war against a Muslim country, and this Administration has already done one."[11] The Israelis chose to act instead. On the night of September 5–6, 2007, Israeli F-15s, escorted by F-16s, took off from Israeli airbases and headed for al Kibar. The seventeen tons of precision-guided bombs that the F-15s dropped on the reactor building completely destroyed it.

Dealing with Imagery Deception

Even the best imagery alone can be deceived. The way to deal with D&D is by fusion: drawing on as many other intelligence sources as possible. It was SIGINT analysts who provided the first indications of a deception operation in the Cuban missile crisis. HUMINT made the difference at al Kibar. And, at Pokhran, more sophisticated imagery such as spectral or radar imaging might have uncovered the visible imagery deception. We know from chapter 12 that nighttime SAR or optical imagery could have identified the test preparatory activity at Pokharan. COMINT possibly could have picked up indications that something was happening there. A determined opponent may be able to defeat imagery alone, but in combination with other sources, effective countermeasures to the deception can usually be found. Depending on a single source has never worked well in a D&D environment.

Another prominent need that led to increasing GEOINT enabled operations emerged in the late 1950s—tracking mobile targets.

Fleeting Targets

By 1959, militaries had identified the challenge of dealing with **fleeting targets**, comprising entities and objects that move such as "concentrations of troops, vehicles of all kinds, watercraft, and aircraft."[12] It was a term well known in law enforcement, which had always had to deal with crime suspects fleeing on foot or in automobiles.

The fleeting target problem intensified for the military during the 1960s and 1970s as more battlefield units became mobile in order to move on a rapidly changing front line or simply to survive. Radar systems, communications units, artillery, and all types of missiles were mounted on wheeled and tracked vehicles. The problem was not confined to land, either. Governments increasingly became concerned about fleeting targets on the seas—pirates, narcotraffickers, and clandestine arms traffickers being three of the most problematic.

Large military forces traditionally had predictable movement patterns, which made tracking them comparatively easy. But unconventional forces such as terrorists, insurgents, and weapons traffickers seldom follow predictable patterns. They can make good use of terrain (mountains, forests, and cities) to mask their movements from imagery. In fact, irregular forces seldom exhibit the characteristics that aid the tracking of conventional forces. They are not just mobile, but also fleeting and elusive. They can immerse themselves in the local culture and simply disappear.

Against these targets, periodic reconnaissance, the best that a single LEO imagery satellite can offer, isn't enough. But if the target emits a radar or communications signal, SIGINT has a critical role to play in this type of fast reaction geospatial intelligence. It can geolocate radar and communications units and monitor their movements; and it is especially effective against targets carrying cell phones. Furthermore, SIGINT is invaluable in giving indications on where to focus imagery collection—known in the intelligence business as "tip-off."

Video and radar surveillance, however, have become the most effective remote sensing methods against fleeting targets. Recall from earlier chapters that in the 1980s and 1990s, video-equipped UAVs and radar surveillance aircraft such as JSTARS filled the need—often in conjunction with SIGINT. But video in particular has the greatest effect against fleeting targets when it is mounted on a UAV. Why? Because it dramatically reduces the response time of both military and law enforcement units. In the Mexican city of Ensenada, two hours south of Tijuana, a single video-equipped Quadcopter drone has been responsible for a 10 percent drop in crime rates, with a 30 percent drop in home robberies. It is effective because, as the drone provider noted, "thieves have a clock in their head, from when they break a window to when the police arrive. What the drone has done is dramatically decrease that time window."[13]

Video and radar surveillance also helped fill another requirement that has been threaded throughout the history of GEOINT's development: the need for precision and accuracy.

Precision and Accuracy

Precision and accuracy are two different things, though easily confused. In a geospatial context, precision is concerned with the numbers you use to specify a location. If you say that the Washington Monument is located at 38°53'N, 77°02'W, your locational precision is on the order of a mile; you've specified the location to within one minute, and one minute of latitude is a distance of a nautical mile. If you specify a location of 38°53'22"N, 77°02'07"W, you're being much more precise—down to less than 100 yards. Accuracy, in contrast, measures how well the location you specify matches the actual location. Both the above locations are accurate, though the more precise one might be preferred by a tourist.

For the US Defense Mapping Agency during the Cold War, the challenge of supporting new weapons systems meant characterizing geography and geodesy to a level of precision *and* accuracy previously unimaginable. Ballistic and cruise missiles require a precise model of the shape and size of the Earth to an accuracy of a few feet to support their navigation and guidance. They need to be programmed with the exact locations of both their launch sites and their target locations.[14]

That requires a detailed model of the Earth—its size and shape, along with a reference point for the coordinate system used—whether the standard geographic coordinate system or the UTM system (described in chapter 2). That model is called a **geodetic datum**. Unfortunately, there are several such models in existence today, and it's important to know which geodetic datum you're working with. If you use a map with one datum for your location, and a map with a different datum for a target or destination location, you could be precise to within a meter, and inaccurate by hundreds of meters.[15]

Again, we'd like to have both precision and accuracy, though accuracy is more important in most applications. If you specify a location to within 1 meter (precision), but you're off the actual location by 300 meters (accuracy), you could cause an international incident—as actually happened during the NATO bombing offensive in Yugoslavia in 1999.

One of the targets US analysts selected for the offensive was a Yugoslav military depot headquarters located in the city of Belgrade. They drew on several sources, including imagery, maps, and databases. Unfortunately, both the maps and the databases contained inaccurate information. Specifically, they incorrectly pinpointed the Chinese embassy, which had relocated in 1996, as the target to attack instead of the actual target, which was located 300 meters away. On May 7, 1999, US bombers struck the embassy with five GPS-guided bombs, severely damaging the building and killing three embassy personnel. The US later apologized and paid reparations for the incident, but many Chinese today still believe that the attack was deliberate.[16]

Even when you can precisely and accurately locate a target, sometimes more is needed. On occasion, you have to draw on outside expertise to understand what you are observing.

Outside Expertise

Dealing with familiar imagery targets is not too taxing, after you deal with issues such as deception and fleeting targets. Answering questions about something you haven't seen before is definitely taxing. Drawing conclusions about relationships among items of interest in an image or series of images can be downright difficult. Those are occasions when imagery analysts routinely make use of other intelligence sources. Analysis of installations and equipment discovered in imagery, for example, often requires drawing on outside expertise as well. Since the beginning of World War II, many issues have required specialized expertise from academia or industry. Three interesting examples concern advanced military equipment, underground facilities, and intelligence enigmas.

Advanced Military Equipment

When looking at military equipment that they've seen before, imagery analysts have a relatively easy job: report the presence, equipment name, and location of what they have observed. If they haven't seen an item before, they can turn to what once were called "photo interpretation keys"—photographs kept in a library, later available in digital form—to help identify the item.

During World War II, however, photo interpreters were encountering new types of military hardware, for which no interpretation keys were available. They needed experts to help with analysis. The first internal sources they turned to were the all-source analysts who, during World War II, were stationed in the same group and worked closely together. In the UK, the best known of these analysts was R. V. Jones, an Oxford professor who became a scientific intelligence officer in the Air Ministry. He had a long record of success in assisting British photo interpreters during the war:

- Early in the war, when reconnaissance photos revealed mysterious German antennas located along the French coast, Jones was able to identify them as two new radar types designed for air defense, subsequently named Freya and Würzburg.
- Later on, in June 1943, when a Royal Air Force photo reconnaissance aircraft obtained evidence of the German V-2 rockets at Peenemünde, Jones was able to identify the purpose and characteristics of the V-2 well before the first one was launched against London in September 1944.[17]

After the war, the need for outside expertise grew during the Cold War as overhead imagery of the USSR and China became available. As the Soviet Union and China secretly developed and tested new weaponry—intercontinental ballistic missiles, air and missile defense systems, surface craft and submarines, and tanks—NATO imagery analysts turned to their resident experts in these fields to help explain what they were observing in imagery.

Underground Facilities

Over the past five decades, many facilities of strategic importance moved underground. The US built its Cheyenne Mountain complex in Colorado, originally to house the North American Air Defense Command (NORAD). The Soviets built a complex of underground facilities (UGFs), the largest being Yamantau Mountain (in the Bashkir tongue, "evil" or "wicked" mountain) in the Ural mountain range. Spanning about 400 square miles underground, Yamantau Mountain has been considered by US analysts as a possible strategic command-and-control center, weapons production center, or weapons storage area designed to survive a nuclear attack.[18] Russian officials have made several conflicting claims about the purpose of the facility. Some obviously have been intended for deception: that it is a mining and ore-processing complex or an underground warehouse for food and clothing. More likely, Yamantau Mountain is intended as a shelter for the Russian national leadership in case of nuclear war.[19]

But Mount Yamantau is only one of many underground facilities that doesn't yield its secrets to conventional imagery analysis. By the end of the twentieth century, UGFs had become a serious problem. Countries such as Iran, Iraq, Libya, Cuba, and China were building UGFs to protect critical assets such as nuclear facilities, house leadership in the event of a war, or even to store foodstuffs. In 1999, a Defense Intelligence Agency report assessed the threat for the future, concluding that "the proliferation of underground facilities (UGFs) in recent years has emerged as one of the most difficult challenges facing the US Intelligence Community and is projected to become even more of a problem over the next two decades."[20]

The good news is that imagery can locate them and provide valuable insights. But again, it cannot deal with UGFs unaided. Geophysical details are essential in the analysis. In-depth analysis requires drawing on expertise from the SIGINT and MASINT communities for assessments of such topics as communications from the facilities, effluents or gaseous emissions, and seismic information. The assessments sometimes also draw in organizations from outside the community, such as the US Geological Survey on geological topics.

Intelligence Enigmas

Geospatial analysis frequently must confront unidentified facilities, objects, and activities, and they represent perhaps the toughest GEOINT problems. These are often referred to as intelligence enigmas. For such targets, the key often is change detection. We observe the facility, object, or activity over time. And along with that, it's a good idea to accept the help of outside expertise. The key there, however, is to guard against a problem that geospatial intelligence analysts often encounter: a tendency of some experts to mirror image.

Imagery analysts encountered many intelligence enigmas during the Cold War. One of the most interesting had a colorful and evocative nickname. It was called the "Caspian Sea Monster."

During the 1960s, a massive aircraft suddenly showed up in US satellite imagery. Located in the Caspian Sea near the port of Kaspiysk, the craft was almost 100 meters long, making it the largest aircraft in the world at the time. Since it was sitting in the water, it had to be some type of seaplane. Except for one problem: there was no way that it could fly. Its wings were far too short to support the craft in conventional flight. US intelligence analysts drew on the

FIGURE 13.1 The Caspian Sea Monster. *Source:* US Defense Intelligence Agency.

expertise of the aviation industry for help. The aeronautical engineers who examined the imagery concluded that what was now named the Caspian Sea Monster *could* fly—after a fashion. So long as it stayed just above the water surface, it could take advantage of a phenomenon called the "wing in ground" effect. Being close to the surface, with the engines mounted ahead of the wings, as illustrated in figure 13.1, gave it enough extra lift.

The engineers nailed it on the monster's flight profile. They weren't quite so successful in identifying its intended mission. Most of their estimates were the result of mirror imaging: a cognitive bias that assumes that others would do things as you would do them. After the plane was destroyed in a 1980 accident, the US discovered that it had been intended as a high-speed military transport for the rapid deployment of troops across ocean areas—not a mission that US experts had envisioned for the craft.[21]

Characterizing Oceans and Ocean Traffic

As noted in chapter 12, early SARs could not offer many insights about land areas that visible imagers couldn't obtain. But they could garner information about sea conditions and sea ice that would have been a difficult prospect for a visible imager.

Analysts also could use radar imagery as a starting point for monitoring ships of intelligence interest. But today at least 50,000 merchant ships are at sea at any given time, and the number of small craft (pleasure and fishing craft) afloat worldwide is several times that. Sorting out the ships of intelligence interest from the vast morass requires more than just radar imagery, not to mention there are not enough imagery analysts on the planet. Nonimagery sources provide the help.

A GEOINT analyst might begin by removing from a ship traffic display all ships that are readily identified. Chapter 7 discussed the automated identification system (AIS)—a radio frequency identification tag required to be carried by most international voyaging ships. Those can be identified and tracked as illustrated in figure 13.2. But most vessels are not required to carry AIS units—commercial fishing vessels, for example. To deal with the growing problem of illegal fishing, many states having fisheries require that commercial fishing vessels carry a vessel monitoring system that transmits the vessel's position and speed, among other things.

Vessels that are required to carry AIS or the vessel monitoring system occasionally do not activate it in order to conceal their presence. Such vessels are of intelligence interest; they may be discharging waste, leaking or discharging oil, on piracy missions, or conducting illegal fishing, among other nefarious acts.

The next check on such ships would come from SIGINT. Most oceangoing vessels carry radar for navigation and collision avoidance, and a radio communications unit as well. Countries with a spaceborne SIGINT capability can use

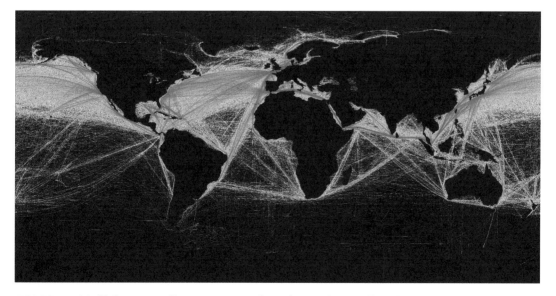

FIGURE 13.2 World shipping traffic. *Source:* B. S. Halpern (T. Hengl; D. Groll), Wikimedia Commons.

it not only to detect, locate, and identify a ship's radar, but even to associate it with a specific ship by a technique called "radar fingerprinting." And any communications that the ship uses can also be tracked by SIGINT. But a ship transmitting no AIS signal, no vessel monitoring system signal, and emitting no radar or communications signal *immediately* becomes a subject of high intelligence interest. Then closer surveillance is required: high-resolution optical or radar imagery, either from satellites or aircraft, would be in order.

The need for such geospatial intelligence is not limited to military and law enforcement applications. In 2007, the European Maritime Safety Agency began operating CleanSeaNet, an oil spill and vessel monitoring service that processes SAR images to detect possible oil spills on the sea surface, to monitor accidental pollution during emergencies, and to identify acts of deliberate or accidental pollution.[22] By 2012 several services were using CleanSeaNet to routinely provide reports from surveillance of ship traffic correlated with the AIS (or with a lack of an AIS) to commercial customers. The reports rely on combining the AIS with imagery from the SPOT and Pléiades optical satellites and the TerraSAR-X and TanDEM-X radar satellites. They provide detailed information on a ship's characteristics (position, dimensions, heading, and speed) to customers about 15 minutes after an image is obtained.[23]

Characterizing oceans and ocean traffic was only one of many new issues driving the development of GEOINT. The evolution of geospatial intelligence developed over several decades, well before the name was coined. A major factor was the identification of emerging issues that GEOINT could address as new sensors, platforms, and analytical techniques came into being.

New Issues

Consider the progress. By the 1970s, satellites and aircraft could do more than obtain photographs of small areas, to use for looking at specific targets. They could cover large areas (albeit at reduced resolution), showing what was happening over major parts of the globe. At about the same time, spectral imagery began providing a different view of the terrain and allowing new types of analysis to be done. Finally, radar imagery could provide a quite different look at the landscape and measure elevation to a high degree of accuracy. As the combination of surveillance and high-resolution capabilities progressed, providing ever more detail, it makes sense that GEOINT has become an intelligence centerpiece for national leaders and defense establishments. Of course, commercial imagery has brought the playing field to a new level, both inside and outside governments and the military. As mentioned in chapter 9, commercial imagery quickly found applications in, among other things, agriculture, cartography, hydrology, forestry, oceanography, and geology. Issues such as agricultural production, environmental pollution, and mass population movements now can be observed and analyzed.

The British author Michael Herman once correctly observed that "the appetite for intelligence is inexhaustible."[24] That turns out to be especially true for geospatial intelligence, and the list of customers and the topics that they are interested in is a long one. Following is a partial account of needs, just from within governments, from past decades to today:

- **Agriculture and food security.** After the Soviets succeeded in what became known as the Great Grain Robbery (discussed in chapter 19), governments took an interest in GEOINT for agricultural forecasting. That interest increased after a series of famines in various regions of Africa began during the 1980s. Food and water shortages are currently predicted to become more severe in Africa and Asia during the first half of this century, with water supplies being increasingly inadequate to support agriculture.
- **Arms control and treaty monitoring.** Starting during the Cold War and continuing thereafter, a series of treaties was negotiated—at first bilateral, between the US and the USSR, and later multilateral, under UN auspices. They first provided restrictions on nuclear weapons and nuclear testing. Later agreements addressed biological, chemical, and missile nonproliferation and disarmament. Government and international organizations concerned with monitoring treaty compliance have turned to GEOINT as the primary means of verification.
- **Prisoners of war.** Like infectious disease, this periodically becomes a high-interest topic. It became one during the Vietnam War, when the US was attempting to identify the locations where US prisoners of war were being held, in planning for a rescue. Decades later, it became an international issue, as Daesh began capturing military and civilian prisoners and executing them in Syria and Iraq.
- **Energy security.** This became a priority issue for oil-consuming countries and the energy industry in 1973, when members of the Organization of Arab Petroleum Exporting Countries declared an oil embargo aimed at eroding US support for Israel. While the embargo ended in 1974, the issue continues to be of interest to policymakers in a larger sense: it includes warning of major energy supply disruptions and intelligence about activities that could affect the price or availability of energy resources.
- **Demographics, migration, and population movements.** A mass emigration of Cubans to Florida in 1980, known as the Mariel Boatlift because of its point of origin, caught US authorities by surprise. A 2010 earthquake in Haiti triggered a similar flight of Haitians to the US and other countries. In 2011, the first Libyan Civil War erupted, and over 200,000 refugees fled the country to neighboring Tunisia and Egypt. Warnings of such events have

become a priority for the US (and Europe, which had to deal with population migration even before the 2015 mass migration from Syria).

- **Illicit drugs.** Illicit drug use, a relatively minor problem in the 1950s, began to increase during the 1960s, and cocaine use exploded during the 1980s. As demand for drugs increased worldwide, organized crime stepped up to deliver in quantity. Governments have responded with programs to trace the cultivation and production areas, identify trafficking routes, and monitor foreign and domestic counter-drug programs.
- **Environmental changes and natural resources.** Issues such as desertification, industrial pollution, and climate change can be addressed with the broad area coverage and spectral sensing capabilities offered by Landsat and its successors. This is an issue of wide interest within all sectors: government, industry, and academia.
- **Weapons of mass destruction development and proliferation.** Closely tied to arms control and treaty monitoring, this remains a high-priority issue for governments and international bodies such as the UN.
- **Human rights and war crimes.** A topic of interest to many governments, international and nongovernmental organizations, it came to the fore particularly during the breakup of Yugoslavia in the 1990s. The term "ethnic cleansing" was introduced then, though the acts themselves have a much longer history. It remains a significant issue in countries such as Myanmar, South Sudan, and Syria.
- **Humanitarian disaster and relief operations.** This includes responses to natural or human-made destructive events having attendant large-scale human suffering that necessitates individual government or international assistance.
- **Homeland security.** It involves border security and monitoring illegal immigration, of course; but monitoring ocean traffic near national borders to identify potential seaborne threats is a major concern for many countries. In the US, the 9/11 attacks prompted the creation of the Department of Homeland Security.
- **Piracy.** The Somali pirates get a lot of current attention, but piracy is also a continuing problem in the Indian Ocean and the Strait of Malacca near Singapore. Pirates are even a problem in parts of major rivers such as the Amazon and Danube.
- **Trade.** Governments want warning of efforts to circumvent or violate multilateral sanctions, including subissues such as identifying violators and their trade routes.
- **Infectious disease and health.** A topic of continuing concern for the UN's World Health Organization and Doctors Without Borders, it periodically also becomes a high-priority topic for specific governments—the severe acute

respiratory syndrome outbreak in China in 2002–3 and the 2018 Ebola outbreak in central Africa being two recent examples.
- **Terrorism.** The big concern of this century to date, terrorism has a long history of use by both states and nonstate groups. Terrorist groups have been characterized as the opposite of the targets that both sides had during the Cold War. Those targets (e.g., missile silos and underground bunkers) were described as "easy to find; hard to kill." Terrorists fit the opposite mold: "hard to find; easy to kill."

The topics on this partial list require more than imagery. Geographic knowledge and especially terrain knowledge are important in answering questions about agriculture, demographics, energy, environmental changes, humanitarian disasters, infectious disease spread, and prisoners of war, for example. Imagery is a primary factor in almost all GEOINT, but geospatial analysts go well beyond the exploitation of images, including taking inputs from sources from both inside and outside their fields of expertise. Consequently, there arose a demand from intelligence customers for a synthesis of all these sources into a complete picture.

A Complete Picture

This chapter has listed some of the needs or driving factors that led to the explosion of GEOINT. In addressing all of them, however, there was an overarching driver: the need for a complete picture of an area or situation, tailored to the needs of the intelligence customer. An imagery analyst might be able to deal with D&D and fleeting targets. He or she might be able to precisely and accurately specify a target location. And all of that might still not be enough to meet the customer's needs.

Consider, for example, the challenges of providing GEOINT to forces operating in mountain terrain. This terrain—of the sort encountered in Afghanistan and the Balkans, for example—has always been difficult for military operations. The remoteness of the area, the vertical challenge, limited avenues of approach, degraded communications, and adverse weather conditions all have had to be dealt with for many centuries. By 2001, remote sensing could provide a wealth of GEOINT unavailable to those in previous decades. Military commanders had much of the imagery and SIGINT needed to operate in mountainous regions: they had accurate terrain knowledge, knew the trafficability of roads, and could be provided with the types of forest cover and locations of possible ambush sites. They could even get the location of enemy units. Still, there was just one problem. All these pieces of information came from different sources. But the commanders don't want to get their knowledge in pieces—a map here, an image there, SIGINT and HUMINT reports from somewhere else. They need the complete picture that results from fusion.

Or consider a completely different type of difficult terrain—that encountered in urban operations. Fighting there is tough enough to make a military commander ask for a mountain assignment. But terrorists and insurgents in Iraq, Syria, and the Gaza Strip find it advantageous to fight there, especially against a superior force. Why? The physical uniqueness of urban terrain: it provides barriers that are difficult to assault, a human-made form of high ground for defenders, and often also a subterranean level.[25]

Also, in both types of terrain, there inevitably is a noncombatant population—a large population in urban areas. And there are facilities that must not be attacked by mistake: shrines, mosques, hospitals, and, as noted above, embassies. So militaries in many countries now operate under rules of engagement designed to protect civilians and certain facilities. Consequently, in the types of combat that became more common late in the twentieth century, militaries needed detailed information about inhabitants and structures in their operational areas. They needed a detailed understanding of the human geography (discussed in chapter 15), which led to a need for fine-grained intelligence about the locations and identities of people. That initially was aided by outside experts such as anthropologists.[26]

Getting a complete picture, then, required looking beyond individual targets and becoming more than an imagery analyst. Analysts had to not only draw on other sources but also treat the targets as part of a *system*. For example, in supporting Iraqi army efforts to retake the city of Mosul from ISIS in 2016–17, before calling in an air strike, coalition forces had to assure that (1) they knew exactly where an ISIS target was, and (2) no civilians were in the immediate area. That required high-resolution, real-time imagery or video of the sort referred to in chapter 10, of course; but it also required current HUMINT.

The same challenges apply in nonmilitary intelligence. Think about a building in a compound that is rail-served. The purpose of the building needs to be established, but so does its function in relation to other buildings around it and the processes that are going on in each—or that might be tied to a region far away. Assessing a country's oil production capability might require examining exploration, drilling, production, transportation, storage, refining, and distribution. Anticipatory intelligence (defined in chapter 1), another defining function of GEOINT, especially requires taking a systems perspective.

Pity the poor imagery analysts. As more capable satellites have been put into orbit, one after another, and UAVs have proliferated around the world, more detailed imagery has become abundant. The existence of high-quality imagery and nonimagery sources has spawned not only more requests from the traditional customers but also a large new set of customers wanting answers about ever more complex problems. This demand requires reaching well beyond traditional imagery and mapping practices; it reinforces the need for a multidisciplined approach and the integration of different sources, and taking a systems view.

Chapter Summary

Historically, the job of the photographic interpreter (later to become known as the imagery analyst, and now known as a geospatial analyst) was to annotate an aerial photograph, describing what was in it. Imagery interpreters could deal with facts, and an image provided those: they could answer the who, what, where, and when questions. That product served most military and governmental needs. And because it tended to focus on a single installation or target, the analysis was relatively straightforward.

But as requirements for intelligence expanded and remote-sensing technology became more sophisticated, GEOINT analysts could do in-depth analysis to answer the other two important questions: how and why (though even answering the first four questions began to require looking beyond what was in the image). Analysts who formerly had focused on a single target found themselves examining large regions with interrelated targets and taking a more strategic view. They also often had to answer "so what?" questions. It isn't always obvious why a customer should be concerned. All these issues require looking at targets over a period of time to observe changes in facilities and the movements of people and mobile objects, a capability realized when imagery moved from periodic reconnaissance to surveillance. Perhaps above all, imagery analysts learned to draw on all available intelligence sources, currently referred to as fusion, and to view targets as part of a larger system.

On many issues, customers need anticipatory intelligence. Telling European governments that a large influx of Syrian refugees had happened wasn't helpful. What they need is some warning about the next wave of refugees, preferably with an indication of where, when, and how many. Informing governments that a North Korean nuclear test has occurred, and the explosive power of the device, is of some use; even better is advance warning that a test is about to occur and where it will happen. To do that, GEOINT analysts needed another technological advance: a set of sophisticated tools that could support the needed in-depth analysis.

Notes

1. NSA Center for Cryptologic History, "NSA and the Cuban Missile Crisis," n.d., www.nsa.gov/about/cryptologic-heritage/historical-figures-publications/publications/coldwar/assets/files/cuban_missile_crisis.pdf.
2. James H. Hansen, "Soviet Deception in the Cuban Missile Crisis," *CIA: Studies in Intelligence* 46, no. 1 (2002).

3. Marlise Simons, "War in the Gulf: Decoys: A Firm's Fake Weapons Have Real Use—Deception," *New York Times*, January 27, 1991, www.nytimes.com/1991/01/27/world/war-in-the-gulf-decoys-a-firm-s-fake-weapons-have-real-use-deception.html.
4. A démarche is a political or diplomatic step, such as a protest or diplomatic representation made to a foreign government.
5. Tim Weiner and James Risen, "Policy Makers, Diplomats, Intelligence Officers All Missed India's Intentions," *New York Times*, May 25, 1998.
6. Weiner and Risen.
7. "Strategic Deception at Pokhran Reported," *Delhi Indian Express*, May 15, 1998, in English.
8. Weiner and Risen, "Policy Makers."
9. Andreas Persbo, "Verification, Implementation, and Compliance," May 13, 2008, www.armscontrolverification.org/2008/05/syrian-deception.html.
10. David Makovsky, "The Silent Strike," *New Yorker*, September 17, 2012, www.newyorker.com/magazine/2012/09/17/the-silent-strike.
11. Makovsky.
12. "Visual Search Techniques: Proceedings of a Symposium, Held in the Smithsonian Auditorium, Washington, DC, April 7 and 8, 1959," National Academies, 1960, https://archive.org/details/DTIC_AD0234502.
13. Jack Stewart, "A Single Drone Helped Mexican Police Drop Crime 10 Percent," *Wired*, June 11, 2018, www.wired.com/story/ensenada-mexico-police-drone/.
14. Al Anderson and Larry Ayers, "Geospatial-Intelligence Impact on the Cold War," n.d., www.ngaaeast.org/10.html.
15. Amanda Briney, "Geodetic Datums," *ThoughtCo*, February 17, 2019, www.thoughtco.com/geodetic-datums-overview-1434909.
16. US Central Intelligence Agency, "DCI Statement on the Belgrade Chinese Embassy Bombing," July 22, 1999, www.cia.gov/news-information/speeches-testimony/1999/dci_speech_072299.html.
17. R. V. Jones, *Most Secret War* (Ware, UK: Wordsworth, 1998).
18. Michael R. Gordon, "Despite Cold War's End, Russia Keeps Building a Secret Complex," *New York Times*, April 16, 1996, www.nytimes.com/1996/04/16/world/despite-cold-war-s-end-russia-keeps-building-a-secret-complex.html.
19. "A Rare Look Inside a Russian ICBM Base," *Congressional Record*, June 19, 1997, H3943, http://fas.org/spp/starwars/congress/1997/h970619_a.htm.
20. Defense Intelligence Agency, "A Primer on the Future Threat, the Decades Ahead: 1999–2020," July 1999, 139.
21. Lester Haines, "In Search of the Caspian Sea Monster," *The Register*, September 22, 2006, www.theregister.co.uk/2006/09/22/caspian_sea_monster/.
22. "CleanSeaNet: Ten Years Protecting Our Seas," *Copernicus Observer*, November 24, 2017, http://copernicus.eu/news/cleanseanet-ten-years-protecting-our-seas.
23. Airbus Defence and Space, "Ship Detection and Tracking," 2018, www.intelligence-airbusds.com/en/4445-ship-detection-tracking.
24. Michael Herman, *Intelligence Services in the Information Age* (London: Routledge, 2001), 140.
25. Scott Gerwehr and Russell W. Glenn, *The Art of Darkness: Deception and Urban Operations*, RAND Corporation, 2000, www.rand.org/pubs/monograph_reports/MR1132.html.
26. Gerwehr and Glenn.

14

The Tools of Geospatial Intelligence

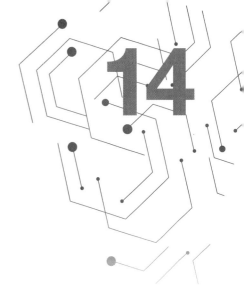

Charlie Allen, according to a *US News & World Report* article, was "more of a legend than a man around CIA. . . . A workaholic, Allen had served as an intelligence officer for 40 years and earned a reputation as a plain-spoken professional who regularly bucked the bureaucracy."[1]

In the summer of 1990, Allen was an extremely worried workaholic. As the national intelligence officer for warning, he had the responsibility of providing warning of major conflict outbreaks worldwide. And he was seeing signs of an outbreak that could seriously affect all NATO members.

During the final months of the Iran-Iraq War in 1988, Saddam Hussein had ordered the use of chemical weapons against Iran. After the war's end, Saddam had kept his massive standing army intact. What, Allen wondered, did he plan to do with it? It was large enough to overrun Israel. What was Saddam's next target?

He soon found out. Hussein had long claimed that oil-rich Kuwait was historically a province of Iraq, and on July 17, 1990, he reasserted that claim, along with a denunciation of Kuwait's leaders and a list of demands. In the next few days, imagery showed the movement of Iraq's Republican Guard units toward Kuwait and a logistics buildup just north of the border. On August 1, imagery showed that armored and infantry brigades were arrayed in attack formations 2 kilometers from the Kuwait border.

It all could have been a bluff to wring concessions from the Kuwaiti government, and many in the US intelligence community believed that. But for Allen, the critical piece of evidence was the disruption to Iraqi civilian life; resources were being diverted from the economy to support the buildup. From long experience, Allen knew that happened only when an attack was planned. Accordingly, on August 1, he issued a "warning of attack" to the White House and Defense Department, adding that there would be no further warning. On August 2, Iraqi forces crossed the border and occupied Kuwait within a few days.[2]

The astonishing part of this story is that US policymakers, for the most part, ignored Allen's warning. So the August 2 invasion caught them by surprise. Why? Because US officials had discussed the warning with leaders in the

Soviet Union and some Middle Eastern countries—all of whom denied that Saddam intended to attack. The officials accepted the opinions of those leaders.[3]

That government decision-makers would give more weight to foreign leaders' opinions than to the conclusions of their own intelligence service seems absurd. But it does happen. When intelligence must provide bad news to decision-makers, it has to overcome a high hurdle, like the one Charlie Allen faced. US leaders in 1990 did not want to confront an Iraqi invasion of Kuwait, so they accepted any indication—even one from foreign leaders whose motives were suspect—that it would not happen.

In a situation like that, intelligence has an obligation to do more than report; it must *convince*. Geospatial intelligence, which Allen used in preparing his warning of attack, can be quite persuasive. But the pieces of evidence need to be assembled, organized, and presented in a fashion that can sway a skeptical audience. In 1990 all the tools to accomplish that weren't readily accessible to Allen. Since then, a large suite of them has become available, including many that help to meet the emergent needs listed at the end of chapter 13.

There are plenty of challenges in GEOINT, but a primary one is that you have many potential layers of georeferenced information available to analyze for any given customer set. A single hyperspectral image, for instance, can have up to a thousand layers. And, each layer can have massive amounts of associated information constantly being updated from various sources. The United Nations Global Geospatial Information Management project estimates that 2.5 quintillion bytes of data are created daily, and that a significant amount of this quantity includes location.[4] How do you process and present volumes of information visually so that it can be analyzed not only to draw conclusions about present reality but also to estimate the likelihood of future realities? You do it by using robust suites of quantitative and visualization tools—in short, with information technology. An analyst has to select the relevant layers of data, extract geospatial intelligence from them, add the time element (looking into the future), and, most important for intelligence officers such as Charlie Allen, present the results in a convincing fashion.

One of the most powerful tools today was available for dealing with layers of data by 1990: GIS, introduced in chapter 6. It has made significant strides since its debut in the 1960s, and now has become the essential technology for producing geospatial intelligence. It is one of several tools that fall within the realm of geomatics.

Geomatics

The term **geomatics** has its origin in the disciplines of geodesy and photogrammetry. Michel Paradis, a photogrammetrist working in the Ministry of Natural

Resources in the Quebec Provincial Government, is credited with originating the term *géomatique* (English: geomatics) in a 1981 paper. He wasn't the first to use the term, but he gave it a broad scope. Paradis attached to it the vision of a unifying construct for the many tools, methods, and technologies associated with geospatial knowledge. Those include methods of Earth mapping, land surveying, remote sensing, cartography, GIS, global navigation satellite systems, photogrammetry, geophysics, and geography. And most importantly, it includes the synergy among them.[5] The most general definition of geomatics, according to *Science* magazine, is "the science and technology of collecting, manipulating, presenting, and using spatial and geographic data (notably data pertaining to Earth) in digital form."[6]

Since its introduction by Paradis, geomatics has become a unifying construct that incorporates the collection systems that have been covered in chapters 10 through 12, the geographical information systems discussed in chapter 6, and much more. It is now a distinct educational discipline taught in many universities worldwide. The term is clearly connected to geospatial intelligence, but with an emphasis on the tools and techniques for acquiring and handling geospatial data; GEOINT, of course, emphasizes more of the analysis side.

Two important geomatics tools are the ones for dealing with big data and visual analytics, and both are tied to GIS.

Geographic Information System

GIS, introduced in chapter 6, is a ubiquitous acronym today. Yet when asked exactly what it means, blank stares can be the result. At its most basic, GIS is a software and hardware combination capable of capturing, storing, manipulating, analyzing, and displaying geographic information (the GI). But that omits an essential component of the "system"—the S. The human analyst completes the definition when he or she properly defines the problem to be solved, crafts the relevant queries, and extracts the critical information (or intelligence, in the case of GEOINT) from GIS results.

The GIS introductory discussion in chapter 6 was about layering visible images and maps. By the late 1980s, you could also layer different types of images—visible, spectral, and radar—or any combination. That development allowed analytic conclusions to be drawn that no one image type alone could offer. Furthermore, because spectral imagers provide many possible choices of spectral bands, it became possible to choose the most useful band in combination with a visible or radar image.

The question was how to best display what GIS could deliver. Imagery had been used to produce maps (photogrammetry) since World War I. Maps were used to plan imagery missions. And imagery was being used to obtain terrain elevation for maps. If you're going to move your tank battalions across the Iraqi desert, you need to know the terrain as well as the locations or possible locations of enemy forces. You don't want to have to look at both maps and imagery to get that information. Visualization technology would allow you to have both in one electronic display with which you could interact. And this technology, known as **geovisualization**, has advanced remarkably in the last three decades.

Geovisualization

There are many versions of the saying that "one picture is worth a thousand words." In the original Chinese proverb, the number is 10,000 words. Maps have a power that everyday language doesn't possess. They can convey knowledge that cannot be easily passed on using just words. But the paper map can only provide knowledge that its creator put there; you can't ask it questions and expect to get answers. Even if the map is in the digital form represented by a raster display, what you see is what you get. And that's all you get. Video displays represent a step forward; they have even greater power to send a message because they display actions, but they still can't answer questions.

In contrast, georeferenced databases contain volumes of information that can potentially answer questions. But pulling the relevant information from a database for analysis isn't easy. The data can be visualized in a spreadsheet, but you can't analyze spatial relationships or make comparisons easily from a spreadsheet.

Geovisualization has changed all that. It is the term used to describe a set of tools and techniques that support the analysis of geospatial data. Enabled by the advances in technologies supporting GIS, it allows users to interact with a map display: to explore different layers that GIS provides, to zoom in or out, and to change the visual appearance of the display.

During the early 1960s, three digital display technologies were developed, each of which is still with us today—in greatly improved form. Light-emitting diodes (LEDs), liquid crystal displays (LCDs), and plasma displays all became available. Each has its advantages, and all have been used for geovisualization.

The integration of imagery and mapping in a visual display was an obvious step once GIS became widely available. And in 2001, a system for doing that was being used in geospatial intelligence. It would later gain prominence in the commercial world.

In 1999, the CIA's Directorate of Science and Technology established a private nonprofit venture called In-Q-Tel to act as a gateway for bringing advances in information technology into the CIA and other US intelligence community components. In-Q-Tel invested in several Silicon Valley startups, including a company called Keyhole that in 2001 had built EarthViewer, a software platform that enabled users to combine maps and satellite imagery, and to display changes in imagery over time. NIMA at the time had a research arm named the InnoVision directorate, and it joined with In-Q-Tel to fund Keyhole. Soon thereafter, EarthViewer was the tool of choice to support coalition troops deploying for Operation Iraqi Freedom.[7]

The product soon came to the attention of Google, which at the time was struggling with the demand for geospatial information in user searches. In 2004, Google acquired Keyhole, and in 2005 it placed its own version of EarthViewer online. Renamed Google Earth, it completely changed the nature of online mapping.[8]

Other tools for displaying geospatial intelligence soon followed, and existing ones became more widely used. Thematic cartography, introduced in chapter 5, was already well known. But geovisualization technology allowed it to be a tool consistently used in intelligence reporting. A recent example is the display of ivory trade worldwide shown in figure 14.1, which effectively conveys to policymakers the magnitude of the problem as well as some possible indicators of where to deal with it. The CIA cartographer who produced it noted that "most maps I make are thematic maps, or maps that tell a story."[9]

Augmented reality displays, introduced in chapter 6, began to be used commercially in 1992. Since then, they have made their way into geospatial intelligence and appear to be under development for use by nongovernmental organizations in dealing with global humanitarian problems.[10] Another advancement

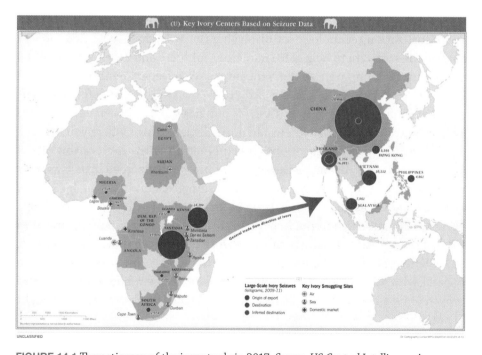

FIGURE 14.1 Thematic map of the ivory trade in 2017. *Source: US Central Intelligence Agency.*

is to take augmented reality to its next logical step—display of geospatial events over time. Four-dimensional projection mapping, also sometimes referred to as video mapping, uses overlays over time to both show changes up to the present and to show likely futures.

If Charlie Allen had been able to deploy geovisualization tools in 1990, he surely could have convinced US policymakers that the Iraqis weren't bluffing. A detailed pattern of the Iraqi troop dispositions and movements, shown along with the other implications indicating a pending attack, would have been very persuasive.

Visualization and GIS together form an effective combination in many GEOINT applications besides military ones. Law enforcement has used it repeatedly to solve crimes, as the following case illustrates.

Clive Barwell was a truck driver during the day around the city of Leeds. At night, he pursued his other interest: criminal assault. But in 1995, the police didn't know that. They only knew that a man had abducted and raped five women over a fifteen-year period. A police task force from Leicestershire, West Yorkshire, and Nottinghamshire counties had been searching for the perpetrator for years. They had the DNA from a blood sample, but at the time no DNA national database was available to search. They also had a partial fingerprint of the rapist, but it was insufficient for use in an automated fingerprint search. And they had the locations of purchases that the rapist had made in the Leeds area with a credit card stolen from one of the victims.[11]

So they called in the criminologist Kim Rossmo, a pioneer in geographical profiling and the developer of crime analysis software that relies on GIS and visualization technology for displaying the analysis results. Rossmo had applied the software to help solve numerous murders and rapes as head of the geographic profiling unit at the Vancouver Police Department, beginning in 1995. In 1996 the UK task force asked him to help in the search for their perpetrator. He first looked at the pattern of purchases. As he observed, they were all "routine purchases that you would normally make near where you live."[12]

Using the purchase locations in combination with the location of the abductions, Rossmo developed a geographic profile that zeroed in on two areas where the rapist most likely resided: the Killingbeck and Millgarth districts of Leeds, shown as the two red areas in figure 14.2. Local police accordingly began a manual search of the fingerprint records at the two police stations in those districts. After a painstaking examination of more than 7,000 records, in 1998 the police found a match to the partial fingerprint. That led police to Barwell, who had previously been imprisoned for armed robbery. The DNA also proved to be a match. Barwell subsequently pleaded guilty and received eight life terms.[13]

The Leeds case had a relatively modest amount of data that Rossmo needed to arrive at his answer. Today, the handling of different layers continues to be

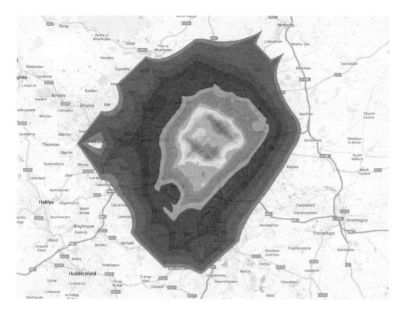

FIGURE 14.2 The likely residence of the Leeds crime perpetrator, according to Kim Rossmo. *Source:* D. Kim Rossmo, Center for Geospatial Intelligence and Investigation, School of Criminal Justice, Texas State University.

a challenge. Many of them are huge; data sets may have multiple thousands of points. But only the information relevant to the question or issue at hand needs to be extracted—meaning that the data must include a geospatial reference and (especially if four-dimensional mapping is planned) a time reference as well. Naturally, the next technology to enable GEOINT analysis would be for handling massive volumes of fast-appearing, multifaceted, complex items of information.

Big Data

Unless you happen to be in the software business, data can be a boring subject. But, along with GIS and visualization tools, it is an essential component of geospatial intelligence. The internet was one of the major new sources driving GEOINT; it was also the source of data on a scale not previously encountered, and it engendered a new expression: **big data**.

The term has been around for about twenty years. It does not simply mean a large volume of data, though that is one of its features. It describes data that also have complex features, so that normal data-processing applications are inadequate for making any valuable use of the information. All the tasks associated with data—capturing and storing, searching, sharing, analyzing, transferring, visualizing, querying, updating, sourcing, verifying—required a different approach. An extensive set of tools consequently has been developed to capture, curate, manage, and draw relevant knowledge from such data.

Big data is characterized by what are called the "five Vs": volume, velocity, variety, value, and veracity:

- *Volume* refers to the amount of data produced every second across all available channels. This includes internet channels such as social media platforms, mobile devices, and online transactions. For government intelligence, this also includes OSINT translations and the products of HUMINT, SIGINT, MASINT, and IMINT collection. All of these data must be gathered as they become available.
- *Velocity* characterizes the speed at which new data are being created, collected, and analyzed at any moment. Both online sources and government intelligence sources generate many billions of items each day. The proliferation of tablets, mobile devices, and the internet of things (discussed in chapter 20) is enabling a steady increase in velocity as well as volume. New data need to be analyzed as they become available; once the collection and analysis effort starts to fall behind, it loses the ability to catch up.
- *Variety* is defined as the different types of data that have to be examined. Long ago, when you talked about "data," that meant handling text, numbers, graphics, and photographs. Today data cannot be so neatly categorized. The types of information that must be dealt with continue to expand, largely driven by social media, such as updates, profiles, blogs, pictures, videos, audio files, and podcasts.
- *Value* describes the usefulness of the data. Just because they are referenced to a specific location does not mean that data are relevant in answering a question in which you are interested. But, inevitably, the most valuable data for answering your question are likely buried in a mass of irrelevant material, and extraction can be difficult.
- *Veracity* characterizes the quality or trustworthiness of the data you collect. All intelligence analysts are aware of the importance of verifying both the sources and the data. (It's a lesson that internet users eventually learn after a number of painful missteps.) We generally value quality above quantity, and that means examining the credibility of the types of data collected and their sources.[14]

These characteristics aren't unique to geospatial data, of course. SIGINT and OSINT have much the same challenges of dealing with big data. That turns out to be a benefit, because the tools for dealing with big data in one intelligence discipline can often be applied, with some modification, to other disciplines.

The term "big data" is also applied to predictive analytics, user behavior analytics, and other advanced analytics methods, in intelligence and many other disciplines, which extract value from data. But these really are a separate class of tools, many of which are optimized for handling and displaying big data in usable form. Let's consider them next.

Data Analytics and Visual Analytics

Though imagery does not give us insight into human thought processes, the patterns of activity in a region do allow us to infer what future activity and actions might occur. Unit deployments, facility construction, and road and bridge repair all can indicate certain types of future actions by a government. The patterns of Iraqi force movements and disruption to civil life were key in Charlie Allen's decision to provide warning of attack. To be able to address those things, however, imagery analysts had to turn to something called data analytics.

Data analytics (synonymous with "big data analytics") is the process of examining data sets in order to derive insights about the information they contain. It typically involves searching through several data sets to look for significant patterns or correlations. Data analytics technologies and techniques are widely used in government and industry to enable leaders to make better-informed decisions. They are used by scientists and researchers to verify or disprove existing scientific hypotheses, models, and theories.

Data analytics has proved to be valuable also in law enforcement GEOINT. A tool called Memex, designed to help officers hunt down human traffickers, is a typical application. DARPA announced the existence of Memex in 2017. It searches for and analyzes data collected from the internet—both the open web and the dark web.[15] Then it displays the geographical patterns of human trafficking. As a DARPA spokesman explains it, "What we're looking for is online behavioral signals in the ads that occur in these spaces that help us detect whether or not a person is being trafficked.... Victims of sex trafficking are often sold as prostitutes online, and a number of websites are the advertising point where people who want to buy and people who are selling can exchange information, or make deals."[16]

Data analytics has been developing gradually since the beginnings of information technology. You likely encounter the result of its use in some form every day. LinkedIn, a social media network for professionals, uses the data from its platform to power features such as "People You May Know." If you order or stream movies online, the recommendations that you later receive from that source will be based on the preferences that you've indicated. The offerings you receive from a credit card company will be based on the profile of your financial life that it has developed from past purchases—all thanks to the power of data analytics.

Big data analytics is used routinely by intelligence organizations and private companies to identify an opponent's (or an ally's) strategy, current intent, and motives. Several of these analytics tools exist, and they can provide clues about likely future events for both government and competitive intelligence groups.

Visual analytics is the science of analytical reasoning supported by interactive visual interfaces. A subset of that field—called geospatial analytics, or sometimes just spatial analytics—specifically refers to the application of visual analytics to geospatial problems.

By the beginning of the twenty-first century, powerful and flexible visual interfaces for GIS and big data analytics had been developed, and they were integrated with the extensive set of geospatial data that already existed. This combination allowed analysts to incorporate new intelligence information and draw conclusions while at the same time expanding the existing knowledge base. Many visual analytics tools now exist, and the development experience of one called MapD is typical—though its creator certainly was atypical.

In 2012 Todd Mostak was conducting research in Middle Eastern Studies at Harvard University. His thesis topic was the role of the social media platform Twitter in the Arab Spring, and he was frustrated. His data set included millions of "tweets" (280 characters or less) on Twitter. He had to use the entire data set to run his queries, which would be no great problem if he had access to a powerful supercomputer or a large cluster of server computers. But he had neither. So he spent hours and occasionally overnight to get an answer to a single query. He was familiar with graphics-processing units (GPUs)—the video cards that power computer graphics, so beloved by the online gaming community. It occurred to him that the queries could be analyzed quickly by a massive set of GPU cards working in parallel. In early 2013, he walked just down the road a bit and became a research fellow at the Massachusetts Institute of Technology, specializing in GPU applications to large data sets. Apparently, his efforts bore quick fruit. Later in 2013, he cofounded the start-up company MapD Technologies, Inc., now OmniSci. The company's product, MapD, is a powerful GPU-powered database and visualization platform that can query and display billions of records in milliseconds.[17] One of MapD's applications, called Tweetmap, allows a user to search for the geographical and temporal distribution of words used in Twitter messages around the world—a valuable capability in many types of intelligence, not the least of which is GEOINT.

You can do a lot with GIS and big data. But when those capabilities are combined with the intelligence analyst's tradecraft, creativity, and background knowledge, a new level of analysis is possible. Make no mistake: the addition of visual analytics tools makes that a far faster and more efficient process than the slog likely endured by Charlie Allen's support team of just thirty years ago.

Geospatial Simulation Modeling

Making an estimate about the future—anticipatory intelligence—requires creating a scenario, or model, of the future. And for GEOINT, that usually takes a specific form, called a simulation model.

A simulation model is a mathematical model of a real object, system, or actual situation. Geospatial simulation models, sometimes called geosimulations, describe the variation of one or more phenomena over the Earth's surface in both space and time. They take many forms. For example, they are extensively used in modeling urban environments to identify trends in, among other things, traffic, construction, sanitation, and land use. Important problems that simulation models address include climate change, energy depletion and transition to alternative sources, urbanization, energy, demographic aging, and global migration. The result has been described as a "quantitative revolution in geography."[18] The most detailed of these models simulates the interactions of individuals, households, and parcels of property over time.[19]

The Temporal Dimension

Temporal changes in the geospatial environment occur due to natural causes and human activity. From the earliest times, human survival has depended on understanding both. On a large scale, patterns of floods and droughts carried threats to human survival, and they needed to be understood. On a micro level, knowing the spatial and temporal patterns of game activity was critical for successful hunters. And, obviously, movement patterns of hostile tribes had to be carefully observed.

Chapters 2 and 3 discussed three dimensions of geography—the two horizontal dimensions shown on a map, and the vertical (terrain) dimension. Mapping, and later GIS, can provide snapshots of a geographical region in all three. Such endeavors characterize classical geography, which emphasizes space on the Earth; it is referred to as **spatial geography**. Human activity, however, can't adequately be modeled in snapshots. It needs to be modeled by including the fourth dimension—time—and so received the name **time geography**, or time-space geography. The Swedish geographer Torsten Hägerstrand established the idea of time geography in the mid-1960s, and developed some models and statistical techniques for studying it.[20] His ideas reshaped thinking about quantitative techniques within the social sciences and were to shape later developments in GIS as well.

Geosimulation models built on Hägerstrand's concept are now common in examining human activity in terms of time and space. On the short-term micro level of spatial and temporal coverage, geospatial simulations are both widely accepted and heavily used. For example, in planning for building construction or estimating energy consumption for a region, simulations allow estimating hourly and seasonal energy consumption down to the individual parcel. By combining existing geospatial data with GIS tools, such simulations can provide hourly estimates of electricity and natural gas consumption by time of year or even by day in residential and commercial areas, and the results can be aggregated to give the overall consumption estimate for a large city.[21] Geosimulations can identify patterns of commuting and migration into and out of a community, for example, the changes in public transit routes that depend on the time of day and day of the week. In a temporal GIS, all this information can be gathered, stored, and analyzed.

As an example of how geosimulation works in modeling human activity, the University of Maryland professor Paul Torrens has addressed one life-or-death application: the tendency of panicked crowds to block exits from a burning building. His simulations showed that placing a vertical column in front of the exit, by slowing the fleeing crowd, reduced the congestion and (counterintuitively) allowed more people to escape the building.[22]

Geospatial simulations also are used to produce models of natural phenomena. Earthquakes, floods, ice-sheet movements, and hurricanes all can be modeled. Simulations allow prediction of the aftereffects of natural disasters, such as flooding, fires, structural damage, and disease outbreak and spread. In the medical field, temporal simulations are used to assess the management and prevention of infectious diseases and other epidemiological phenomena. Recall that the study of epidemiology began with the use of cartographic data to identify the source of the London cholera outbreak. One hundred and sixty years later, a different class of geospatial simulation was used to identify optimum locations to build Ebola treatment centers in Liberia during the 2014 outbreak.[23]

Two widely used geospatial simulations predict patterns of crime and describe patterns of military operations over time.

Crime Simulations and Anticipatory Intelligence

In urban law enforcement, geospatial modeling is applied to analyze events through geographic and temporal filters in order to provide anticipatory intelligence for criminal activity. One widely used strategy is to identify "hot spots" (areas within a city that have unusual amounts of crime) so that police can pay increased attention to those communities. The result is typically more arrests than if police resources were uniformly distributed. Action in hot spots also prevents a problem from becoming worse; if left alone, these areas usually transition from minor crimes to more violent types; for example, an area where vandalism is left unchecked is likely to become a high-theft area. Promptly containing and controlling the minor criminal activity in a hot spot prevents a more serious crime situation from developing. Aside from hot spots, certain types of crime are area-associated. Arrests for driving under the influence are most common in areas having a large number of bars or liquor stores, for example.

Temporal patterns also are used in such simulations; certain types of crime occur more often at particular times of the day or week. Assaults are recorded most often between 3:00 a.m. and 7:00 a.m. Conversely, burglars prefer to hit residences during the daytime when residents are generally away at work.

Military Geospatial Simulations

Military geospatial simulations are highly regarded and therefore are widely used. That has much to do with (1) the need for militaries worldwide to anticipate the chances of success for weapons and battle tactics against likely opponents and (2) the relative ease in selecting valid assumptions and inputs for combat simulations (in comparison with those used in the social simulations discussed in the next section).

Military geospatial simulations date back several millennia. They were used to hone strategic and tactical thinking by Sun Tzu, who used color-coded stones to represent opposing armies. Those games led directly to the Chinese game called Go. Sometimes compared with chess, Go is far more complex with a wider variety of possible moves; it is thought to be the oldest board game still widely played today.

Throughout history, Vikings, Celtics, Roman legion commanders, and many military leaders have used tabletops and sand tables to train their subordinates in war-gaming exercises.[24] GIS provided the essential tool to replace the sand table. It allowed militaries to war-game with more elaborate and accurate combat models. The original versions were combat simulation models that required powerful computers and elaborate simulation software. Today, laptops can do the job, and most major militaries now rely on sophisticated geospatial simulations in war-gaming. The Indian Army, for example, maintains and regularly updates a combat simulation model for potential conflicts with Pakistan and China.

Combat simulations also have become familiar parts of popular culture, thanks to the proliferation of online multiplayer war games that are set in past conflicts such as Napoleon's battles, the American Revolution, the Spanish Civil War, and numerous World War II battles.

Crime and military operations geosimulations focus on local geospatial-temporal changes. Concern about regional and global changes led to the development of simulations that look at longer time frames, discussed next.

Long-Term Social, Environmental, and Natural Resources Simulations

Many geosimulations take a long-term view, sometimes of natural changes but mostly of changes due to human activities. For example,

- The last thirty years has seen the rapid growth of megacities—urban areas with populations greater than 10 million. In 2017, there were thirty-one megacities, including fifteen in China and six in India. The UN forecasts that there will be forty-one megacities in 2030.[25] These massive concentrations of people are reshaping global demographics and economic markets in ways that need to be understood, and geosimulations are the primary tool for gaining that understanding.[26]
- Human activity is affecting the world's oceans in ways that are not yet thoroughly grasped. Fishing, climate change, and pollution are believed to be major factors in the changes. But little is known about the cumulative patterns of change, or the primary causes of change, and where the greatest changes are occurring. Geosimulations offer some hope of characterizing the effects of human activity and identifying remedial actions.
- Brazil, beginning in the 1960s, developed several strategies for promoting land use in the Amazon River Basin, including financial incentives for land occupation, agriculture, and pasture. The result has been major infrastructure development and deforestation of about 18 percent of the original land cover, even before the surge in wildfires that occurred in the summer of 2019. Brazil is now trying to assess the effects of development and deforestation on biodiversity, soil structure, hydrology, and local climate.[27]
- Climate changes have had a negative impact on many areas of the world, but not all areas. Over the course of the twentieth century, climate change in the province of Alberta, Canada, resulted in a substantial boost to Alberta's agriculture.[28]

Geospatial simulations, in common with most simulation models, can be difficult to build, and long-term simulations especially so. The main challenge usually is validation: determining that the model accurately represents what it is supposed to, under different input conditions. A second major challenge is ensuring that any results produced by the simulation used valid inputs and were based on valid assumptions. Both are especially difficult to deal with in working with social, environmental, and natural resources simulations. On occasion, neither the inputs nor the assumptions are valid, and that has led to public doubts or suspicions about the results.

Even valid geospatial simulations can run afoul of such suspicions. Perhaps the best-known example was the predictive simulation published in a 1972 book titled *The Limits to Growth*. The book was printed in thirty languages and sold 30 million copies worldwide. It was sponsored by a think tank named the Club of Rome, a group of researchers at the Massachusetts Institute of Technology. The team used several hundred years' worth of data on trends in resource depletion, birth and death rates, population growth, industrialization, pollution, and food supplies. Using that in their simulation model, the researchers predicted a global environmental catastrophe during the twenty-first century.

The book immediately ran into a storm of criticism from other researchers, governments, and industry. The criticisms included unfounded allegations that the report's sponsors had subsequently disavowed it because the conclusions of their report were not correct and that they purposely misled the public in order to awaken public concern.[29]

Despite the criticism, the Club of Rome periodically updates *The Limits to Growth*; and it continues to be widely supported by environmentalists and researchers. Independent studies such as one in 2016 came to the conclusion that "there is unsettling evidence that society is still following the 'standard run' of the original study—in which overshoot leads to an eventual collapse of production and living standards."[30]

While there is no evidence that the Club of Rome team "cooked the books" in their original simulations, it has been done in unrelated simulations. It is easy in a simulation model to get the answer that you want; just select the right inputs and assumptions. And, it is always a bad idea. The Club of Rome book criticism gave predictive simulations an undeservedly poor reputation. It was likely an early factor in fueling what is now a common public distrust of simulation, as has happened with widespread doubts about global warming simulation results today.

Simulation-Based Learning

GIS-based simulations are also used for training and education. They guide students through geospatial, physical, and environmental conditions, processes, and situations, teaching them how to accomplish specific sets of tasks. The training typically immerses learners in a virtual world to do so. The process often involves gameplay—some type of competition in which users attempt to outperform others or challenge themselves. It can include a reward-and-penalty system that is used to assess learning. One widely used example is the flight simulator.

Pilots have long trained using simulators. Flight simulators even made their way into the world of computer gaming when IBM licensed what became known as the Microsoft Flight Simulator in 1982. Flight simulators today incorporate GIS as a way to train pilots about terrain and weather conditions, dealing with runway approaches and navigation.

Chapter Summary

In 2010 the US National Research Council issued a report titled *New Research Directions for the National Geospatial-Intelligence Agency: Workshop Report*. It identified five "cross-cutting themes" that it expected to become increasingly important: fusion, anticipatory intelligence (which it called "forecasting"), human terrain, crowdsourcing (which it called "participatory sensing"), and visual analytics.[31] To date, these expectations have been met. Three of the themes were covered in this chapter and the preceding one as drivers of GEOINT; and they continue to drive its progress. Human terrain (more correctly, human geography) and crowdsourcing have become major factors in sociocultural GEOINT; they are covered in detail in the next chapter.

Notes

1. David E. Kaplan, Kevin Whitelaw, and Monica M. Ekman, "The Inside Story of How a Band of Reformers Tried—and Failed—to Change America's Spy Agencies," *US News & World Report*, August 2, 2004, 40.
2. Charles E. Allen, "Warning and Iraq's Invasion of Kuwait: A Retrospective Look," *Defense Intelligence Journal* 7, no. 2 (1998): 33–44, http://cryptome.org/allen-wiik.htm.
3. Allen.
4. United Nations, "Future Trends in Geospatial Information Management: The Five- to Ten-Year Vision," July 2013, http://ggim.un.org/documents/Future-trends.pdf.
5. "What Is the Difference between Geomatics and GIS?" *GISGeography*, February 18, 1918, https://gisgeography.com/geomatics-gis-difference/.
6. Jon Mills and Rosie Waddicor, "Geo What? Opportunities in Geomatics," *Science*, May 30, 2003, www.sciencemag.org/careers/2003/05/geo-what-opportunities-geomatics.
7. Leanna Garfield, "The CIA's EarthViewer Was Basically the Original Google Earth," *Business Insider*, December 30, 2015, www.businessinsider.com/the-cias-earthviewer-was-the-original-google-earth-2015-11.
8. Matt Alderton, "The Defining Decade of GEOINT," *Trajectory*, March 13, 2014, http://trajectorymagazine.com/the-defining-decade-of-geoint/.
9. CIA, "A Day in the Life of a CIA Cartographer," November 15, 2017, https://twitter.com/cia/status/930926921264828416?lang=en.
10. Robie Mitchell, Patrick Kenney, Jacqueline Barbieri, John Bridgwood, and Stephen Hodgson, "Activity-Based Intelligence in Mixed Reality," *Trajectory Magazine*, January 25, 2019, http://trajectorymagazine.com/activity-based-intelligence-in-mixed-reality/.
11. João Medeiros, "How Geographic Profiling Helps Find Serial Criminals," *Wired*, November 10, 2014, www.wired.co.uk/article/mapping-murder.
12. Medeiros.
13. Medeiros.
14. XS International, "Updated for 2017: The V's of Big Data: Velocity, Volume, Value, Variety, and Veracity," 2017, www.xsnet.com/blog/updated-for-2017-the-vs-of-big-data-velocity-volume-value-variety-and-veracity.

15. The Dark Web is that part of the World Wide Web that is accessible only by means of special software, allowing users and website operators to remain anonymous or untraceable.
16. Cheryl Pellerin, "DARPA Program Helps to Fight Human Trafficking," US Department of Defense, January 4, 2017, https://dod.defense.gov/News/Article/Article/1041509/darpa-program-helps-to-fight-human-trafficking/.
17. MapD Corporation, "About MapD," n.d., www.mapd.com/company/about/.
18. Michael Batty, "Modeling and Simulation in Geographic Information Science: Integrated Models and Grand Challenges," *Procedia Social and Behavioral Sciences* 21 (2011): 10–17.
19. Paul M. Torrens, "Geosimulation," Center for Urban Science + Progress, New York University, http://geosimulation.org/geosim/.
20. A. Pred, "The Choreography of EXISTENCE: Comments on Hägerstrand's Time-Geography and Its Usefulness," *Economic Geography* 53 (1977): 207–21.
21. Shem Heiple and David J. Sailor, "Using Building Energy Simulation and Geospatial Modeling Techniques to Determine High Resolution Building Sector Energy Consumption Profiles," *Energy and Buildings* 40, no. 8 (2008): 1426–36, www.sciencedirect.com/science/article/pii/S0378778808000200.
22. Blake Smith, "Mind of the Mob: The Geosimulation of Crowd Control," Yale Scientific, May 29, 2013, www.yalescientific.org/2013/05/mind-of-the-mob-the-geosimulation-of-crowd-control/.
23. David Brown, "Computer Modelers vs. Ebola," *IEEE Spectrum*, June 2015, 62.
24. Raymond R. Hill and J. O. Miller, "A History of United States Military Simulation," *IEEE Proceedings of the 2017 Winter Simulation Conference*, 2017, www.informs-sim.org/wsc17papers/includes/files/025.pdf.
25. Tanza Loudenback, "Here's How Much It Would Cost You to Live in the 10 Largest Megacities around the World," *Business Insider*, October 20, 2017, www.businessinsider.com/worlds-largest-cities-megacity-cost-of-living-2017-10.
26. Torrens, "Geosimulation."
27. "Forest Governance: Brazil," in *Yale University Global Forest Atlas*, https://globalforestatlas.yale.edu/amazon/forest-governance/brazil.
28. S. S. P. Shen, H. Yin, K. Cannon, A. Howard, S. Chetner, and T. R. Karl, "Temporal and Spatial Changes of the Agroclimate in Alberta, Canada, from 1901 to 2002," *Journal of Applied Meteorology and Climatology*, July 2005, https://journals.ametsoc.org/doi/abs/10.1175/JAM2251.1.
29. Julian L. Simon, *The Ultimate Resource 2* (Princeton, NJ: Princeton University Press, 1996), 49.
30. Tim Jackson and Robin Webster, *Limits Revisited: A Review of the Limits to Growth Debate* (London: All-Party Parliamentary Group on Limits to Growth, 2016), http://limits2growth.org.uk/wp-content/uploads/2016/04/Jackson-and-Webster-2016-Limits-Revisited.pdf.
31. National Research Council, "New Research Directions for the National Geospatial-Intelligence Agency: Workshop Report," 2010, www.nap.edu/read/12964/chapter/2#3.

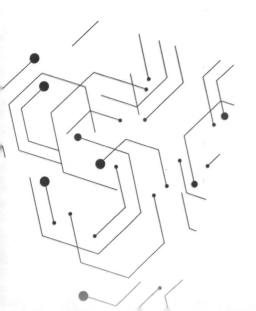

Sociocultural GEOINT

Sociocultural GEOINT deals with **human geography**—the study of people in a spatial and temporal setting. It includes the study of their communities, cultures, economies, and interactions with the environment.[1] It has many other related monikers, including cultural geography, rich ethnography, anthropogeography, sociocultural context, and social domain. But these are simply different terms for knowledge that has been sought since antiquity. Chapter 3 described some of Moses's orders to the spies he sent into Canaan. He also gave them an order to "see what the land is like and whether its people are strong or weak, few or many. Are the cities where they dwell open camps or fortifications?"[2] His order could be summarized as "get me the human geography of that country." The methods for establishing human geography have become more sophisticated over the centuries since then.

Sociocultural Factors in Conflict Resolution

The United States, from its colonial days, has a history of conducting sociocultural GEOINT. From their arrival, colonists had to learn the customs and cultures of the Native Americans. Later, the Lewis and Clark expedition had, as one objective, investigating the language, customs, and laws of the Indian tribes they encountered; that is, of determining the human geography of the lands that they explored. Those ethnographic endeavors in North America proved valuable to the US during the twentieth century, when Americans became more heavily involved in global affairs. Following are three brief examples, spanning almost a century.

Versailles

After World War I, the Allies met in Paris to draw a postwar map of Europe. One of the US objectives in the 1919 Paris Peace talks was to draw national boundaries that would preserve, as much as possible, the national cultures of Europe. But the US government's leaders had no past experience in foreign negotiations of this magnitude. (It was a skill they would learn during the negotiations and become quite good at after World War II.)

President Woodrow Wilson called on the distinguished geographer and director of the American Geographical Society, Isaiah Bowman, to help frame American foreign policy. Bowman pulled together 150 scholars from geography and other related disciplines to collect and analyze information Wilson would need. The study covered, among other things, the language, ethnicity, resources, and historic boundaries of Europe.

During January 1919, Bowman directed the production of over 300 maps per week based on a sociocultural analysis of the human geography of Europe. Bowman's products were presented at the conference, and helped shape the boundaries of postwar Europe. None of the other conference parties had prepared anything that could compare with Bowman's effort.[3]

Dayton

The US was to repeat the Versailles performance seventy-five years later, albeit with much improved technology. In 1995, the Bosnian War had been raging in the former Yugoslavia for three and a half years. From November 1 to 21, 1995, presidents of the warring countries met in Dayton, Ohio, to negotiate an end to the war. The US secretary of state mediated the negotiations, assisted by US intelligence representatives and experts from the Defense Mapping Agency and the US Army Topographic Engineering Center. They provided maps of the former Yugoslavia that included cultural and economic data (or the "human geography").[4] A sample map is shown in figure 15.1. The resulting agreement, known as the Dayton Peace Accords, was signed in Paris on December 14, 1995.

In addition to the maps, three-dimensional imagery of the contested areas permitted cartographers to guide negotiators on a virtual tour of the territory. It was observed that "the power and flexibility of the technology and the technicians gave the political decision-makers the confidence needed to reach agreement. In at least one instance, this three-dimensional experience proved crucial in persuading Yugoslav president Slobodan Milosevic to compromise on a disputed area."[5]

Iraq

The International Coalition mission of liberating Iraq from Saddam Hussein in 2003 and the subsequent occupation of the territory to provide security were met by the Iraqis, initially with a measure of relief and later of anger.

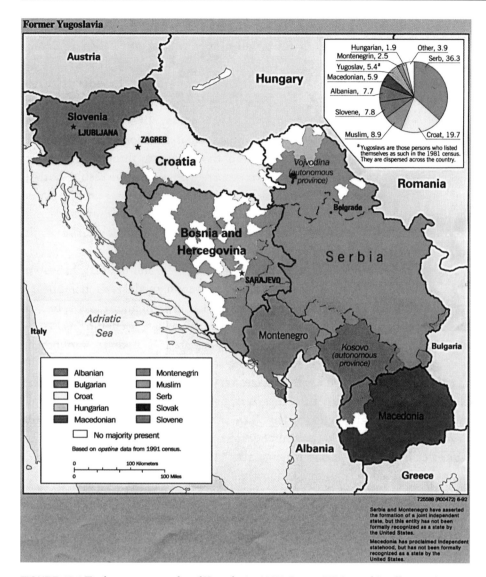

FIGURE 15.1 The human geography of Yugoslavia, 1995. *Source:* US Central Intelligence Agency.

There was an obvious culture clash, and one of the coalition strategies became encapsulated in the tagline "Iraq: Winning Hearts and Minds." In a 2004 article, retired major general Robert Scales observed that the conflict in Iraq required "an exceptional ability to understand people, their culture, and their motivation."[6] In 2005, a *Military Review* article also emphasized the importance of militaries developing an understanding of the local population and culture, popularizing a new term for human geography: the *human terrain*.[7] Subsequently, the US Army developed an anthropology program called the Human Terrain

System. It involved teams of social scientists deployed to Iraq and Afghanistan with the mission of assisting military commanders and staff with an understanding of the local population.

Of course, the project became controversial. Cultural anthropologists were embedded with some of the teams, and the American Anthropological Association didn't like it. In the association's view, the program put at risk both its social scientists and the local people that they surveyed; it also violated, in their view, an ethical rule for "true informed consent" of those whom anthropologists study.[8]

Despite the criticism, the program reportedly had positive results both in military effectiveness and in reducing civilian casualties. Four studies based on interviews with military commanders concluded that the human terrain program resulted in less use of destructive force and more effective use of counterinsurgency operations.[9]

Nevertheless, the US Army officially terminated its Human Terrain System program in 2015. Later, it began a similar program named the Global Cultural Knowledge Network. Military forces now generally consider human terrain modeling to be an essential part of planning and conducting operations in populated areas. The difference is terminology—the name "human terrain" having been replaced by a term that sounds more sophisticated.

Human geography generally changes very slowly, as the preceding examples illustrate. But conflicts, natural disasters, or social upheaval often cause rapid changes within that geography that have intelligence significance. One of the primary tools for quickly recognizing these spatial-temporal changes is a discipline called activity-based intelligence.[10] It requires the use of intelligence fusion, described in chapter 13.

Activity-Based Intelligence

Activity-based intelligence (ABI) is an analytic methodology that involves collecting and analyzing raw intelligence about people, events, and activities in a geographical area. Its purpose is to identify patterns and detect changes in the patterns that have significance for an intelligence customer or, using the term introduced in chapter 1, to provide "situational awareness." ABI has three characteristics:

- It relies on raw intelligence information that is constantly collected on activities in a given region and stored in a database for later metadata searches.
- It employs the concept of "sequence neutrality," meaning that material is collected without advance knowledge of whether it will be useful for any intelligence purpose.

- It relies on "data neutrality," meaning that any source of intelligence may contribute; in fact, open-source may be the most valuable.[11]

The National Geospatial-Intelligence Agency (NGA) has defined ABI as a "multi-INT approach to activity and transactional data analysis to resolve unknowns, develop object and network knowledge, and drive collection."[12]

ABI usually is focused on the activity of a target (person, object, or group) within a specified target area. So it normally includes both spatial and temporal dimensions. At a higher level of complexity, it can include network relationships as well. It differs from conventional intelligence analysis patterns in one important respect: *you can come up with an answer without first specifically defining the intelligence issue.* ABI involves discovery of targets of intelligence interest from observations, rather than identifying a *specific* target and then observing it.[13] In that sense, it's much like the surveillance that big box retailers often conduct: rather than track a specific person within the store, they use cameras to identify persons acting suspiciously. And the observations depend on the fact that targets of interest are identified by their actions, that is, by temporal and spatial activity patterns that indicate hostile or nefarious intent. But where the big box retailers typically react to suspicious activity in real time, ABI is typically focused on the longer term: identifying the target of interest, and then gathering intelligence about it over time.

Though the term "ABI" is of recent origin and is tied to the development of surveillance methods for collecting intelligence, the concept of solving intelligence problems by monitoring activity over time has been applied for decades. It has been the primary tool for dealing with a number of intelligence enigmas.

Because of the ready availability of aerial surveillance, ABI has special value in counterterrorism operations. In the fight against Daesh (ISIS) in Iraq and Afghanistan, vehicle-borne improvised explosive devices became one of the deadlier weapons used by the terrorists against coalition forces. Daesh saw the casualty trade-off as favorable for their side: one Daesh volunteer in exchange for several—perhaps several dozen—coalition fighters. The bomb-carrying vehicles had to be detected and countered in real time; it's often a short trip from the bomb-making building to the target. ABI, relying on video imaging, was a major factor in countering those attacks.

ABI, as these examples indicate, is usually closely related to GEOINT; it can be geospatially referenced, and can rely on imagery, maps, or both. But it does not have to be so. You can get ABI from monitoring COMINT, ELINT, or cyber traffic, where sometimes location information is not available. GEOINT may be needed thereafter to provide more intelligence about the target.

The name ABI comes from US intelligence and the Department of Defense. But ABI has a major role in law enforcement and in commercial and social applications. In those applications, it often carries the name **pattern-of-life (POL) analysis**.

Pattern-of-Life Analysis

We all, individually or in related groups, establish patterns of behavior and repeat them for good or bad in social, professional, financial, and business activities. Pattern-of-life analysis relies on this tendency. It's a fundamental instinct: most animals have established patterns of life, and those are used to predict future behavior of the species.

POL analysis, simply stated, is data collection and analysis used to establish the past behavior of an individual or a group, and from that to determine current behavior or to predict future behavior. It involves the aggregation of data from a variety of sources to develop a profile of past and present human behavior. That profile can be used to predict future behavior based upon recent activity, such as travel, purchases, communications, or criminal acts.

POL analysis is one of the underpinnings of community policing, where officers are assigned to a particular area in order to become familiar with its geography and residents. They observe the normal routines of individuals moving about in the vicinity, and pay special attention to those who break from the typical practices. There are also general tenets regarding locations and habits in public areas. The majority of people do not tend to linger near businesses or public places that are closed. They move purposefully when in or near public buildings. Repeatedly walking the perimeter or interior of a public building or shopping mall, unless it is obviously for physical activity, is likely to draw the attention of building security personnel.

POL analyses are carried out on many different types of data sets in order to spot certain patterns of behavior. Though it is difficult in cyber operations to establish the source of a cyberattack, certain routines can be helpful. The British technology journalist Paul Rubens has noted that

> distinct groups of hackers work at very different times of the day or night. For example, Iranian hackers tend to work during the day (perhaps indicating that many of them are students), while Russian hackers tend to operate in the evening (which suggests that many have daytime jobs and carry out cybercrime as a second job to supplement their incomes). And other patterns provide experts with even stronger indications of where hackers may be from. For example, Russian hacking activity falls away during New Year's Eve (for obvious reasons), while Arab hackers' activity ramps up during the month of Ramadan (when perhaps there is little else to do).[14]

Some may see POL analysis as a euphemism for profiling. They miss the point. Profiling in law enforcement refers to singling out an individual on the basis of racial or ethnic status. POL modeling today uses expansive sets of data points to create profiles based on geographic and behavioral patterns. Interestingly, it can now be done remotely due in large part to the internet of things discussed in chapter 20.[15] It was done successfully, in fact, in the hunt for perpetrators of the 2013 Boston Marathon bombing.

When a bomb explodes, most bystanders flee the scene in panic. In security camera videos of the Boston Marathon bombing scene, two people were seen to observe the chaos and calmly walk away. Those videos led police to the brothers Dzhokar Tsarnaev and Tamerlan Tsarnaev. Tamerlan subsequently died in a shootout with police; Dzhokar was arrested and later convicted of murder, receiving a death sentence.

Intelligence fusion and enhanced surveillance tools and sensors fueled the progress of sociocultural GEOINT. But absolutely nothing could compare to the internet and its spawn, social media, when it came to propelling human GEOINT to its stratospheric heights in the twenty-first century. In 2013 the NGA's midterm strategy called for more use of nontraditional geospatial sources, especially social media, in reporting. Today, social media has become not just an integral part of GEOINT but also essential for managing traditional intelligence collection sources, including HUMINT, SIGINT, and IMINT. The geographic information drawn from social media takes two forms: volunteered and involuntary.

Volunteered Geographic Information

Volunteered geographic information (VGI) is defined as geospatial content provided voluntarily by individuals.[16] It is the "source" in what is referred to as "crowdsourcing," a term introduced in a 2006 *Wired* magazine article.[17] The article describes crowdsourcing as a method of obtaining needed ideas, services, or content by soliciting contributions from a large group of people. In most cases today, it means soliciting online rather than from traditional sources (pre-internet).

As we have seen, like most other components of GEOINT, VGI was collected centuries before the internet was even a thought. Al-Idrisi's *Tabula Rogeriana* (described in chapter 2) was, after all, a product of crowdsourcing. The British Parliament's 1714 Longitude Act was an attempt to solve a problem by crowdsourcing after its usual sources of a solution had failed. Mapmakers long have relied on a form of VGI in preparing their maps.

Despite the long conceptual history, it has been argued that crowdsourcing to support producing GEOINT is the most significant change in the history of cartography.[18] The difference, of course, is information technology, which provides the rich set of source information now available. The internet, as it has done for so many other things, revolutionized the GEOINT discipline for government, the military, commercial interests, and private citizens. It is now used to create, share, and analyze geographic information, bringing together data sourced from many different platforms and also enrolling new volunteers to expand geospatial intelligence. Consider this: hundreds or thousands of people can work on a problem of interest to you, looking at imagery or maps to tag important objects, features, or locations. In commercial applications, the basic idea is to tap into the collective intelligence of the public at large to complete tasks that a company would normally have to perform itself or outsource to a third-party provider.

There is just one problem. As with anything derived from the internet, data quality is a major concern of crowdsourced information that must be dealt with. Volunteered information carries none of the assurances that we rely on in using officially created data. The primary check on the validity of volunteered information comes from the fact that multiple sources can be used to cross-check any given item—validity by consensus, if you will. Wikipedia has demonstrated that crowdsourcing can result in curated knowledge, so long as there are enough sources to work on a specific item of knowledge.

Several web-based applications make use of VGI, including mapping, emergency management, and geosocial networking.

Mapping

Perhaps the best-known crowdsourced mapping effort is OpenStreetMap, founded in 2004 initially for the purpose of mapping the United Kingdom. It originally relied on the massive geospatial data sets produced by government projects like the UK Ordnance Survey. In 2006, the OpenStreetMap Foundation was established as an international not-for-profit organization. Its mission: to leverage crowdsourcing to create and provide geospatial data for public use. Its main product is a freely available and editable world map. By 2015 OpenStreetMap had, with the help of 2.5 million volunteers, digitized more than 130 million buildings and 1.3 million miles of roads.[19]

VGI has also been applied in nautical charting. An extension of OpenStreetMap named OpenSeaMap collects and organizes geospatial data to create a worldwide nautical chart. It is available on the OpenSeaMap website, either for real-time use or for downloading as an electronic chart for offline applications. Commercial and recreational ships collect information about the water depth along their travel routes and upload their depth sounder and GPS locational data to the OpenSeaMap website, where it is integrated with other reporting to maintain and update the chart.

On a smaller scale (relatively speaking), WikiMapia was created in May 2006, and has since then become a widely used mapping website. It features a crowdsourced collection of commercial establishments marked by registered users and guests. By November 2017, it had grown to include almost 28 million place references.

One of the more recent uses of VGI is the result of a partnership between the NGA and the private sector—in this case, DigitalGlobe, now Maxar Technologies. Recognizing that, as an NGA spokesman put it, "the commercialization of GEOINT is leading to exponential growth of publicly available geospatial information," the partnership introduced Hootenanny.[20] It makes use of the open architecture of OpenStreetMap, along with Maxar's Geospatial Big Data Analytics toolkit, to integrate data from satellite and UAV imagery with crowdsourced geospatial data from mobile devices.[21] The term *hootenanny* is not a commonly known expression today; it is an old term for a typically rural social gathering (usually a folk concert) where the audience often joins in the singing. It's an appropriate name for a publicly available crowdsourcing application. Hootenanny features a validation algorithm that relies on achieving a consensus about public inputs.

Emergency Management

Geographic data and tools are essential in all aspects of emergency management: preparedness, response, recovery, and mitigation. Crowdsourcing has proved to be a valuable source in emergency management by virtue of its timeliness. A study of four wildfires in the Santa Barbara, California, area from 2007 to 2009 concluded that "during emergencies time is [of] the essence, and the risks associated with volunteered information are often outweighed by the benefits of its use."[22] The role of volunteers in two of the wildfires illustrate that point:

> In November 2008 the Tea Fire ignited in the hills behind Santa Barbara, and spread extremely rapidly, driven by a strong, hot Santa Ana wind from the northeast. VGI immediately began appearing on the web in the form of text reports, photographs, and video.... Several volunteers realized that by searching and compiling this flow of information and synthesizing it in map form, using services such as Google Maps, they could provide easily accessed and readily understood situation reports that were in many cases more current than maps from official sources.[23]

VGI played an even bigger role in the next big California wildfire:

> In May 2009 the Jesusita Fire ignited, again in the chaparral immediately adjacent to the city, burning for 2 days and consuming 75 houses. Several individuals and groups immediately established volunteer map sites.... The officially reported perimeter of the fire was constantly updated based on reports by citizens. By the end of the emergency

there were 27 of these volunteer maps online, the most popular of which had accumulated over 600,000 hits and had provided essential information about the location of the fire, evacuation orders, the locations of emergency shelters, and much other useful information.[24]

The social media GEOINT response to the wildfires also illustrates the other side of VGI: there were several examples during the fires of false rumors being spread via observer inputs. These likely led to errors on synthesized maps and to unnecessary evacuations; while crowdsourcing mechanisms provided a check on bad data in some cases, they apparently didn't catch them all.[25]

Geosocial Networks

Geosocial networking might be best described as "crowdsourcing for individual benefit." The members of a geosocial network interact based on common interests or employment and a need for location-based information. Labor union members might want to know where job opportunities are located. Members of a professional association might be interested in identifying others in the group with similar interests in their vicinity. Single parents in nearby neighborhoods might want to connect in order to share child care duties.

Geosocial networking enables such contacts, allowing users to interact relative to their current or home locations. It combines web-mapping services with geotagged information about events—local nightspots or restaurants, for example—to enable users to converge for a meeting. Or they can simply visit a local establishment that others in the group have recommended.

There are a plethora of mobile apps customized for specific social group needs. Two examples are Foursquare and Strava.

Foursquare uses geospatial intelligence to provide consumers with information about places to visit near their current locations. The recommendations are personalized based on users' previous browsing history, purchases, or lodging check-in history. It relies on two mobile apps: The Foursquare City Guide app is designed to help the user discover new places, with recommendations from a trusted community of similar users. Foursquare Swarm allows users to share their locations with their friends and create a record of their experiences in their personal lifelogs.[26]

Strava is a geosocial network for athletes. Its fitness-tracking app uses a smartphone's GPS to track when and where a user is exercising. You can use Strava to build an exercise route or acquire another user's route on your phone or GPS device. Its platform features a **heat map**—a graphical representation of data where the individual values are represented as colors, which are then overlaid on the maps—that provides a look at routes that get the most activity around the world.

Geosocial networks also have proved to be an effective tool for planning and organizing protests. On September 22, 2019, members of the #ShutDownDC movement, protesting government inaction on climate change, blockaded streets and disrupted traffic in Washington. The organizers relied on social media to coordinate their actions. They selected specific streets and intersections for their blockades that would cause the most possible disruption, relying on the city's planning department map of "central employment areas" where most commuters go to work.[27]

Intelligence about an individual's locations (or future locations) is available from geosocial networks. In the example of Strava, an astute analyst noted that the map makes it easy for someone to identify the locations of military bases and the routines of their personnel worldwide—a discovery that immediately provoked precautionary warnings to US military personnel.[28] Responding to the threat, on August 6, 2018, the Pentagon banned US military personnel deployed overseas from using the geolocation features of their fitness trackers and smartphones.[29] On another front, users see the social benefits and even consider location information as a safety mechanism, such as letting a loved one know your running path in case of trouble along the way. It is worth flipping the coin to see the potential for harm: those with malicious intent can also be members, gleaning GEOINT to see when particular types of joggers are running at night or in remote areas.

Involuntary Geographic Information

The other type of crowdsourcing has to do with intentionality: when GEOINT draws on internet (primarily social media) material that the source did not intend to provide. This type of information has a name: *involuntary geographic information*, often abbreviated iVGI. The term refers to material where the geographic content either was provided involuntarily or was later geotagged by another.[30]

VGI tends to attract a large volume of people focused on contributing to or completing a project—whether that project's goal is to help map the world or to help to save lives in an emergent crisis. In iVGI, contributors have no overarching goal beyond social interaction. However, their internet posts are used by national, state, provincial, and local government operations; law enforcement; nongovernmental organizations; and in the commercial world for intelligence purposes. Remember that a fundamental principle of GEOINT is to unify as many different sources as possible to create the best understanding of a location and the activities happening there. A social media post, whether the user intends it to or not, provides exactly such a source.

iVGI has the same problem as VGI, in the quality of information provided. Its contributors usually are not experts in the subject being reported, and they are under no constraint to report accurately. The answer is the same for both: it's not a big problem when you have tens of thousands of inputs (see the subsection "Mapping the Syrian Conflict" below). Some individuals may give false or inaccurate reports, as they do in political opinion polls; but that won't affect the final results. It is when there is not a critical mass of inputs about a specific location or topic that accuracy becomes an issue. In either case, geospatial information from trusted sources can be used to cross-check social media inputs. The following examples illustrate, in three quite different arenas, how intelligence can be gleaned from social media users who provide the data.

Disease Outbreak Alerts

Monitoring the spread of a disease, as in the example of the 2014 Ebola outbreak, is a well-known GEOINT capability. But for highly contagious diseases, the critical need is to identify an outbreak quickly and to geotag it. Several web-based applications currently do this; HealthMap and MediSys are two examples.[31]

Yet the first clue of an outbreak is most likely to appear on social media. People talk about epidemics on social media, using key words such as "fever," "virus," "hemorrhage," "cough," and "infection." Some attempts have been made to use this source as an early warning of an outbreak. The problem of information validity and false alarms has to be dealt with, however; words such as "fever" and "virus" especially have nonmedical meanings in social media, as well. Social media–based surveillance may not be able to replace existing methods, but it can supplement them and improve the ability to provide early alerts.[32]

Law Enforcement

Social media monitoring is viewed by some in law enforcement as the next frontier in the use of geospatial intelligence by police.[33] In fact, it's a frontier that's already been crossed.

On August 4, 2011, Mark Duggan was shot and killed by police officers attempting to arrest him in North London. In the period August 6–11, 2011, thousands of people protesting the shooting rioted in London boroughs and in cities and towns across England. Looting and arson were widespread, and five people died in the chaos. During the rioting, many of the rioters did not bother to cover their faces, some even posing for pictures with stolen goods and posting the pictures on social-networking sites. Police forces and investigators used the social websites Flickr and Facebook and the messaging site BlackBerry Messenger to identify and locate looters and vandals for arrest and prosecution. The British public apparently supported the law enforcement use of social media, and it attracted little if any criticism.[34]

The perception of the US public has been otherwise; social media monitoring by government is a controversial issue. In 2016, the American Civil Liberties Union (ACLU) reported that over 500 police forces were accessing user data from Twitter, Facebook, and Instagram using the analytics service Geofeedia and using the data in police operations.[35] The focus of ACLU concern was two incidents subsequent to the deaths of African American men in Ferguson, Missouri, and Baltimore:

- Ferguson, Missouri, August 9, 2014: An unarmed African American teenager named Michael Brown was shot and killed by a white police officer. The shooting started a wave of protests that roiled the area for weeks. Local police used the Geofeedia data to monitor protesters during that time.
- Baltimore, April 2015: Freddie Gray, a twenty-five-year-old African American man, died from a spinal injury after being arrested by police. After Gray's funeral, the city of Baltimore erupted in protests. Baltimore police used Geofeedia data to monitor the unrest. They ran social media photos through a facial recognition system to find protesters with outstanding warrants, in some cases arresting the protesters directly from the crowds. Geofeedia also provided social media posts from a local high school where students were organizing to join the protests. When they were intercepted by police, the students were found to have rocks, bottles, and fence posts in their backpacks.[36]

After the ACLU report was published, Facebook, Instagram, and Twitter all suspended Geofeedia's access to their networks. But Geofeedia is just one of many companies having this level of social media access, so both the US government and local law enforcement will likely continue to use this type of material, perhaps more discreetly in the future.

Facial recognition technology is a powerful tool, even more so when coupled with key word search technology on social media. It is straightforward to do a location-based search for words like "march," "protest," "rally," or the hashtags of protest groups, and from those texts to identify others in the same area who are using the hashtag.

Mapping the Syrian Conflict

Arab nationals are avid social media users. Two-thirds of those users rely on WhatsApp, closely followed by Facebook, with about one-half using YouTube.[37] The Carter Center's Conflict Resolution Program consequently has relied on Syrian social media in a geospatial intelligence project that maintains an interactive map of the Syrian conflict online. The map allows users to watch the evolution of the conflict and the changing fortunes of all sides. Through publicly available Facebook posts, tweets, blogs, photos, and videos, it has documented changes in alliances as government security forces defected and joined rebel groups.[38]

The Carter Center shares some of its Syria maps and reports publicly, making them available to nonprofit organizations, governments, and the news media. Former US president Jimmy Carter, the center's founder, says that the maps provided by the center have been of great use to humanitarian organizations operating in Syria.

As these examples illustrate, geosocial networks can be a source of involuntary information for good or ill. They can be targets for collection to support geospatial intelligence—by law enforcement within a country, by national intelligence organizations against foreign geosocial networks, and even by common criminals who are using them to target victims.

Chapter Summary

The sociocultural side of GEOINT is not new, in one sense; we humans have long studied the geographic patterns of ethnic and cultural groups, a function that much later came to be called the "human terrain" in the military. We've always examined the activities of individuals and groups in what would later be called activity-based intelligence and pattern-of-life analysis. Even crowdsourcing, which took flight with the ready availability of internet social media, was long in use for creating maps and nautical charts, dating at least back to the time of al-Idrisi.

Aided by the internet, the dramatic expansion of social media platforms and users has provided governments, militaries, nongovernmental organizations, commercial entities, and individuals with a vast store of geospatially referenced information during the twenty-first century. It has been applied to a wide range of uses, from good to benign to nefarious. Even in the form of involuntary geographical information, it has become a source of geospatial intelligence that complements the other major sources: HUMINT, SIGINT, maps, and imagery. All these sources, along with the powerful analytic tools provided by information technology, drove the dramatic expansion of GEOINT in governments and commerce. This created the need for government organizations dedicated to providing GEOINT. Chapter 16 tells the story of how the US handled the task, starting with the creation of what was to become the National Geospatial-Intelligence Agency.

Notes

1. Ron Johnston, "Human Geography," in *The Dictionary of Human Geography*, ed. Ron Johnston, Derek Gregory, Geraldine Pratt, et al. (Oxford: Blackwell, 2000), 353–60.

2. Numbers 13:18–19.
3. "Bowman, Isaiah (1878–1950)," "Geography," n.d., http://geography.name/bowman-isaiah-1878-1950/.
4. Timothy R. Walton, ed., *The Role of Intelligence in Ending the War in Bosnia in 1995* (London: Lexington Books, 2014).
5. Gary E. Weir, "The Evolution of Geospatial Intelligence and the National Geospatial-Intelligence Agency," *Intelligencer* 21, no. 3 (Fall–Winter 2015): 53.
6. Robert H. Scales Jr., "Culture-Centric Warfare," *US Naval Institute Proceedings*, October 2004, www.usni.org/magazines/proceedings/2004/october/culture-centric-warfare.
7. Montgomery McFate and Andrea Jackson, "An Organizational Solution for DOD's Cultural Knowledge Needs," *Military Review*, July–August 2005, www.au.af.mil/au/awc/awcgate/milreview/mcfate2.pdf.
8. Scott Jaschik, "Embedded Conflicts," *Inside Higher Ed*, July 7, 2015, www.insidehighered.com/news/2015/07/07/army-shuts-down-controversial-human-terrain-system-criticized-many-anthropologists.
9. Brian R. Price, "Human Terrain at the Crossroads, *National Defense University Joint Force Quarterly*, October 1, 2017, http://ndupress.ndu.edu/Publications/Article/1325979/human-terrain-at-the-crossroads/.
10. Greg Slabodkin, "GEOINT Tradecraft: 'Human Geography,'" *Defense Systems*, October 29, 2013, https://defensesystems.com/articles/2013/10/29/geoint-human-geography.aspx.
11. Gabriel Miller, "Activity-Based Intelligence Uses Metadata to Map Adversary Networks," Defensenews.com, July 8, 2013, http://archive.defensenews.com/print/article/20130708/C4ISR02/307010020/Activity-based-intelligence-uses-metadata-map-adversary-networks.
12. NGA, "Activity-Based Intelligence," n.d., https://usgif.org/system/uploads/3357/original/ABI_Slides_Approved_for_Public_Release_13-231_1_.pdf.
13. Note, however, that you must still identify a general target; the definition of ABI requires identifying an "entity, population, or area of interest."
14. Paul Rubens, "Know Your (Cyber) Enemy," *CIO*, December 14, 2016, www.cio.com/article/3150099/security/know-your-cyber-enemy.html.
15. Lisa Brownlee, "The $11 Trillion Internet of Things, Big Data and Pattern of Life (POL) Analytics," *Forbes*, July 10, 2015, www.forbes.com/sites/lisabrownlee/2015/07/10/the-11-trillion-internet-of-things-big-data-and-pattern-of-life-pol-analytics/#23db17d74eb8.
16. M. F. Goodchild, "Citizens as Sensors: The World of Volunteered Geography," *GeoJournal* 69, no. 4 (October 2007): 211–21.
17. Jeff Howe, "The Rise of Crowdsourcing," *Wired*, June 1, 2006, www.wired.com/2006/06/crowds/.
18. Daniel Sui, Sarah Elwood, and Michael Goodchild, eds., *Crowdsourcing Geographic Knowledge: Volunteered Geographic Information (VGI) in Theory and Practice* (Dordrecht, Netherlands: Springer, 2013).
19. "NGA and DigitalGlobe Open Source Toolkit to Harness the Power of Collaborative Mapping," NGA press release, June 22, 2015, www.nga.mil/MediaRoom/PressReleases/Pages/2015-16.aspx.
20. "NGA and DigitalGlobe."
21. "NGA and DigitalGlobe."
22. Michael F. Goodchild and J. Alan Glennon, "Crowdsourcing Geographic Information for Disaster Response: A Research Frontier," *International Journal of Digital Earth* 3 (2010): 231–41, www.tandfonline.com/doi/full/10.1080/17538941003759255.
23. Goodchild and Glennon.
24. Goodchild and Glennon.
25. Goodchild and Glennon.
26. Elizabeth Stinson, "Foursquare May Have Grown Up, but the Check-In Still Matters," *Wired*, August 9, 2017, www.wired.com/story/foursquare-may-have-grown-up-but-the-check-in-still-matters/.

27. ESRI, "Shut Down DC," n.d., www.arcgis.com/apps/MapSeries/index.html?appid=efcbec197af743749bd5af17cef70f32.
28. Andrew Liptak, "Strava's Fitness Tracker Heat Map Reveals the Location of Military Bases," *The Verge*, January 28, 2018, www.theverge.com/2018/1/28/16942626/strava-fitness-tracker-heat-map-military-base-internet-of-things-geolocation.
29. Ryan Browne, "Pentagon Bans Use of Geolocators on Fitness Trackers, Smartphones," CNN, August 6, 2018, www.cnn.com/2018/08/06/politics/pentagon-fitbit-app-geolocating-ban/index.html.
30. F. Fischer, "VGI as Big Data: A New but Delicate Geographic Data-Source," *GeoInformatics* 15, no. 3 (2012): 46–47.
31. C. Raina Macintyre and Sheng-Lun (Jason) Yan, "Social Media for Tracking Disease Outbreaks: Fad or Way of the Future?" University of South Wales, October 12, 2016, https://newsroom.unsw.edu.au/news/health/social-media-tracking-disease-outbreaks---fad-or-way-future.
32. Macintyre and Yan.
33. Rob Pegoraro, "GEOINT for Policing," *Trajectory*, November 1, 2017, http://trajectorymagazine.com/geoint-for-policing/.
34. Paisley Dodds and Raphael G. Satter, "Police Use Facial Recognition Technology to Nab Rioters," *NBCNews.com*, August 11, 2011, www.nbcnews.com/id/44110353/ns/technology_and_science-tech_and_gadgets/t/police-use-facial-recognition-technology-nab-rioters/#.WtYTX0xFys0.
35. Jonah Engel Bromwich, Daniel Victor, and Mike Isaac, "Police Use Surveillance Tool to Scan Social Media, ACLU Says," *New York Times*, October 11, 2016, www.nytimes.com/2016/10/12/technology/aclu-facebook-twitter-instagram-geofeedia.html.
36. Bromwich, Victor, and Isaac.
37. Northwestern University in Qatar, "Social Media," 2017, www.mideastmedia.org/survey/2017/chapter/social-media/.
38. Kane Farabaugh, "Using Social Media, Carter Center Maps Syria Conflict," VOA, March 14, 2017, www.voanews.com/a/using-social-media-carter-center-maps-syrian-conflict/3764851.html.

16

The Story of the National Geospatial-Intelligence Agency

The prologue to this book introduced the story of two professionals who were separately tackling the pernicious problem of integrating cartography and imagery analysis: Stu Shea and the former director of national intelligence, James Clapper. In this chapter, we go back to the beginning and take a look at how events unfolded in General Clapper's case. (We return to Stu Shea's epilogue in chapter 17.) Recall that after it recovered from the surprise of Saddam Hussein's 1990 invasion of Kuwait, a coalition of thirty-five governments pushed the Iraqis out in the ensuing Operation Desert Storm in 1991. A combination of GEOINT, HUMINT, ELINT, and COMINT helped General Norman Schwarzkopf plan his famous "left hook" invasion through the Iraqi desert west of Kuwait that trapped and shattered the occupying Iraqi forces.

Despite the operational success, US intelligence services were severely criticized afterward for failing to deliver needed intelligence in a timely manner. In fairness, a major part of the problem was inadequate distribution of geospatial intelligence among the coalition forces in the theater, something the intelligence community didn't control. Nevertheless, both Congress and the administration thought that a major reorganization of the intelligence community was the answer. A 1992 study commissioned by the director of central intelligence (DCI), Bob Gates, recommended the creation of a National Imagery Agency that would integrate imagery and mapping. Gates did not like this recommendation at all, nor did the chairman of the Joint Chiefs of Staff (JCS), General Colin Powell.[1] The recommendation died—for the moment.

But pressure was building for change. By the mid-1990s, the long-standing separation between imagery and mapping—the illogical division that Stu Shea had observed two decades previously—was becoming untenable. As National Reconnaissance Office director Jeffrey Harris observed, "We would collect imagery for a six-month period before we even thought about putting it into the mapping process."[2]

Several senior government leaders agreed with Harris. The US intelligence business had changed, in large part due to the lessons learned in the Desert

Storm conflict. But the needs described in chapter 13—especially the need for a complete picture—were still not being well met within the existing structure. And the opponents of the 1992 study conclusions—Gates and Powell—were gone. Secretary of Defense William Perry and DCI John Deutch believed that a unified imagery and mapping agency would function better than two separate agencies had done. The two men, supported by General John Shalikashvili, chairman of the JCS, proposed to Congress that a single agency be created within the Department of Defense (DoD). It would merge two organizations that were key players in the US geospatial intelligence community: the Defense Mapping Agency (DMA) and the National Photographic Interpretation Center (NPIC). The proposed change would incorporate the DMA, with its mapping assets, and NPIC, with its imagery analysis capabilities, to form the core of what would become known as the US National Geospatial-Intelligence Agency (NGA). Because the NGA helped inform many later developments in GEOINT, this chapter tells its interesting story, beginning with the history of the DMA and NPIC.

The Defense Mapping Agency

When the Defense Mapping Agency came into existence on January 1, 1972, it consolidated most of the US military cartography agencies. It incorporated the Mapping, Charting, and Geodesy Division of the Defense Intelligence Agency (DIA), and the mapping organizations of the military services. Those included the Army Map Service, the Air Force Aeronautical Chart and Information Center operations, and the oceanographic and charting services of the US Naval Hydrographic Office. The two principal organizations in this merger had long and interesting histories.

The Army Map Service

At the beginning of World War I, the US Army foresaw a major map shortage as US combat units deployed into the European theater; it therefore created the Central Map Reproduction Plant to meet the need. The shortage, as it turned out, never happened. Static trench warfare required few maps, and European forces provided what was needed. So the Central Map Reproduction Plant produced only 9 million maps during the war. That may seem like a lot, until you compare it with the 500 million maps that its successor, the Army Map Service (AMS), produced during World War II. World War II, however, was a different type of war; it was characterized by rapid force movement around the globe and the need for terrain information by crews of armored vehicles.[3]

The AMS was created in 1942 from existing mapping units, and by 1945 it had prepared 40,000 different types of maps to support campaigns in North

Africa, Europe, the Pacific, and the Far East. The Normandy invasion alone required 3,000 different maps totaling 70 million copies.[4]

Initially, many of the maps that the AMS produced were simply revisions of existing maps. That gradually changed during the war as aerial photography of the war zones came available. AMS cartographers developed skill in using a technique called stereo photogrammetry to produce the large-scale topographic maps that provided ground troops a definite edge in combat. After the war, the AMS continued to develop new types of military topographic maps and map products. However, it soon had to reorient its attention, first in providing maps during the Korean War, and later in the Vietnam War.

The Aeronautical Chart and Information Center

The Map Unit of the Army Air Corps was established in 1928 to produce charts for air navigation. It could at least place in a competition for name and location changes over time. After the outbreak of World War II, the Map Unit moved to Bolling Field in Washington. In April 1944 it was renamed as the Aeronautical Chart Service under Headquarters, Army Air Forces. In August 1952 the organization became the US Air Force Aeronautical Chart and Information Center (ACIC) and moved to Saint Louis. Its products included navigation and planning charts, flight information publications, air target materials, and special products. The ACIC supported US air units in Vietnam, and developed new types of maps that were specialized for the precision targeting that the Vietnam War required.

Merging the ACIC and AMS into the DMA may have solved some issues. There was still just one problem: like its predecessors, it needed imagery to produce maps. And by the time the DMA came into existence, it had a major source, located in downtown Washington. But that source, NPIC, was not part of the Defense Department. It was located within the CIA, and it didn't view mapping as a major part of its mission.

The National Photographic Interpretation Center

In November 1952, the CIA had created its Photographic Information Division to analyze classified photography. The division was formed in anticipation of dealing with photographs from the U-2 program, which was in its early stages. Even then, it was apparent that both the military services and the national leadership needed the strategic imagery intelligence that the U-2 would provide. The new CIA division was designed to meet both needs. The job title given to the assignees foreshadowed later events: they were designated photo *analysts*, not photo *interpreters*, the traditional title. Their role was no longer to just identify facilities and their component parts from photography; it was to do all-source analysis, specifically to "explain what went on in the facilities, how the component parts functioned with respect to each other, pinpoint the bottlenecks or critical control points, and the like. These distinctions in the naming and job titles were intended to convey what Agency planners believed to be the critical difference between photo intelligence . . . and photo interpretation."[5]

In 1953, Arthur C. Lundahl became head of the Photographic Information Division and remained its chief for the next twenty years. Lundahl had been a photo interpreter during World War II and the chief of the navy's Photogrammetry Division after the war. He received extensive support from CIA management in setting up the division. The CIA rented a 50,000 square foot facility in Washington for his team and their film-handling equipment. The facility from the outside could have been from a spy movie. It was "on the top four floors of the Steuart Motor Building, a Ford dealership, at Fifth and K streets, in a run-down area of northwest DC. Between the decrepit working conditions inside the building and the more decrepit neighborhood outside the building, it was the perfect cover location. . . . One of the tenants on the lower floors was the original Toys "R" Us store."[6]

Later on, in the mid-1950s, Lundahl approached senior DoD officials with the idea of a joint photo interpretation center to deal with U-2 photography. The army and navy agreed to participate by contributing personnel, funding, and equipment. In 1958, the center was established, and in 1961 it was renamed the National Photographic Interpretation Center. While the name was, in a sense, a step backward from the 1952 vision of photo analysis, not photo interpretation, the organization was clearly progressive.

Lundahl envisioned imagery analysis to be both an independent form of intelligence and a major contributor to other intelligence products. He began the publication of a memo series to advise all-source analysts about photography related to their issues. Under his direction, NPIC handled the interpretation and analysis first of U-2 and later of satellite photography. Its reporting of Soviet medium-range ballistic missiles in Cuba marked the beginning of the Cuban missile crisis. Its analysts tracked the development of Soviet and Chinese strategic forces during the Cold War and provided detailed reporting of the Soviet incursion into Afghanistan.

The National Imagery and Mapping Agency's Standup

By 1995, DCI Deutch, Secretary Perry, and General Shalikashvili had agreed that an imagery and mapping agency was needed.[7] In November of that year, they delivered to Congress a joint letter of agreement asking for its creation. It would be designated as the National Imagery and Mapping Agency (NIMA).

Congress approved the merger, despite concerns about the level of support that non-DoD customers would receive from the new agency. NIMA came into existence on October 1, 1996, after a year of study and planning.

Two other DoD offices would become part of the new organization:

- Central Imagery Office (CIO). The CIO was the successor to the Committee on Imagery Requirements and Exploitation (COMIREX). COMIREX had been established in 1967 to deal with the increasing customer demand for overhead imagery products. Its job had been to manage national and military imagery collection and analysis. The CIO took over these functions in 1992.[8]
- Defense Airborne Reconnaissance Office (DARO). Formed at about the same time as the Central Imagery Office, DARO's mission was to manage military intelligence requirements for imagery. It developed and managed a system of manned and unmanned aerial reconnaissance aircraft.

As with all mergers, there were the typical problems to be solved. Budget issues were an early complication. The components of the new agency operated under different funding lines, never a good idea. Intelligence and defense operated under different budgetary authorities, and thirteen different congressional committees had oversight of the budgets of NIMA's components. Also, incompatible systems presented a challenge from the beginning. The organizations that now constituted NIMA had different personnel systems, electronic office systems, procurement practices, and systems for creating and distributing products to customers. NIMA had absorbed four organizations entirely: NPIC, the DMA, the Defense Dissemination Program Office, and the Central Imagery Office, along with parts of the Defense Airborne Reconnaissance Office.[9] Not surprisingly, the fundamental challenge was cultural.[10]

A Tale of Two Cities

Logically, the merger made sense. It was a meshing of disciplines that had depended on each other for some time. And the two agencies used similar technologies and drew on similar resources. But at the time of the merger, the major organizations—the DMA and NPIC—were located not just in two different cities, Saint Louis and Washington; they were a 14-hour commute from each other.

Obviously, each organization had its own history and corporate culture. Marrying the cultures would have been difficult under the best of circumstances, which these were not. On both sides there existed a natural reluctance to merge. DMA staff members feared that their mapping support to defense activities would suffer as imagery received all the attention and funding. NPIC analysts and managers saw it as a military takeover that would reduce their intelligence mission; the DMA, after all, was not an intelligence agency.

In short, the organizations supported different customers. The DMA saw the DoD as its primary customer. NPIC products went to the intelligence community. And, much like the two organizations, their customers lived in separate cultural worlds.[11]

These problems could have been handled with some advance preparation and good management. But that isn't what happened. The merger was badly handled at some management levels. Mid-level managers did not expect it to go through, and many did not prepare their employees. Some people felt kept in the dark about what the merger meant. There were stories of seriously poor judgment: a DMA senior official reportedly told an audience in Saint Louis that NIMA meant "Not Interested in Maps Anymore." NPIC staff feared that they were being downgraded from working on their most prestigious product—national intelligence. The DCI's security staff reportedly assigned extra officers to be with him when he visited NPIC because of fear that he would be physically attacked due to his support for the merger. The resistance of managers in all the organizations was so great that for years afterward, the merger was referred to by many senior NIMA managers as "the sewer of standup."[12] The creation of NIMA remains a textbook example of how not to merge independent organizations.

Finally, a different but equally vexing problem was that of the organizational separation being imposed on top of physical separation. NPIC, though located in the DC area, already was physically separated from its CIA and DIA partners, making collaboration between imagery analysts and all-source analysts difficult. The creation of NIMA added to physical separation the challenge of an organizational split.

The Fight to Survive

Washington agency heads and their staffs have a well-used toolbox for gaining and maintaining "market share"—that is, their share of the budget. One of their main tools is the knife. And in the late 1990s, the knives came out. NIMA was predictably struggling. Underfunded and racked with internal conflicts and ongoing turf wars between its mapping and imagery units, and unable to adequately fill all its commitments, NIMA seemed vulnerable. Some national intelligence agencies had lost their organic imagery support, and they wanted it back.

So opponents of the integration supported the creation of the "NIMA Commission" to review the new agency—hopefully, to have it disbanded as dysfunctional. Congress duly asked for such a commission, and it was formed in late 1999. The commission's report, published a year later, identified several problems and made recommendations for fixes, but it reaffirmed the need for NIMA.[13]

The NGA's Standup

While NIMA had survived the immediate threat of dissolution, the next step was to repair the damage inflicted during the standup. A new director did just that. As we know from the prologue, retired Air Force lieutenant general James Clapper became the director of NIMA on September 13, 2001, two days after the 9/11 attacks. Reflecting on the state of NIMA later, Clapper observed that "the first five years of existence, it was simply DMA and NPIC under one tent but they were separate. Even in the production organization, they were separate.... NIMA had not lived up to its intended expectation."[14]

But Clapper had a vision of completing the integration of the two disciplines. He believed that he had a mandate to do it. The NIMA Commission's report had recommended integrating mapping and imagery. Congress, he reasoned, would support him in doing so.[15] And he had the perfect environment in which to make changes. The country had suffered through the images of desperate people jumping from the World Trade Center just before the buildings collapsed. It was a time of immediate need, and, as Clapper observed years later, "never waste a crisis."[16]

There was resistance, of course. The scars from the NIMA standup had not healed. But Clapper had a message for those who didn't get on board: don't let the door hit you on your way out.[17] As he later noted, "Attrition is the most effective management tool. That's why you need time."[18] And he had time. He had agreed to take the job on the condition of being assured five years to do it. The reorganization resulted in the retirements of those who didn't like the integration. By the time Clapper left the job, 60 percent of NGA employees had been hired during his stewardship. For those who stayed, and for the new hires, the director communicated his vision of GEOINT at every opportunity. He encapsulated it in a motto: "Know the Earth; Show the Way."[19] It was embedded in every graphical presentation. Those who remained generally supported the new direction, and the new hires didn't have any attachment to the old ways of doing business.

There was, of course, just one problem. An agency name that continued to define two missions was unsatisfactory. NIMA needed a new name, and the logical one was National Geospatial Intelligence Agency (NGIA). Pronounced as a word, "NGIA" doesn't roll off the tongue well, and some African American NIMA employees also expressed their discomfort about it to Clapper.[20]

Clapper had a solution for that. Observing that only three-letter agencies get respect in Washington, he insisted on renaming NIMA as the National Geospatial-Intelligence Agency (NGA). There was reportedly some distress among NGA staff that their organization's acronym was determined purely as a prestige issue. Clapper saw it differently; in his view, it was essential to his unifying vision. The name change took effect in November 2003.

Reaching Out

The NGA is a member of the US intelligence community; but it is different from its sister agencies that primarily rely on classified means for collecting HUMINT, SIGINT, and MASINT. The NGA has its classified imagery satellite sources, to be sure. But to an extent not shared by other agencies, it also relies on unclassified material produced by commercial entities.[21] And as the government's major provider of maps and charts, much of its product is unclassified as well.

Director Clapper understood that to do its job, the NGA from the beginning had to connect with commercial organizations to obtain its raw product: imagery. Industry also had the needed tools: GIS, data analytics and visual analytics, and simulation modeling software. National Security Agency (NSA) cryptanalysts, CIA case officers, and DIA's MASINT collectors had few counterparts in commercial circles. Such was not the case for the NGA. Much of the expertise it needed was in the private sector. So Clapper strove to form a partnership with that sector. Today, the agency still holds regular government-industry conferences on GEOINT subjects, and it publishes a quarterly magazine, *Pathfinder*, drawing on industry and academic contributors.

Embedded Analysts

Director Clapper also recognized the importance of having imagery analysts deployed to the military's combatant commands, especially the forces operating in Afghanistan and Iraq. He was well aware of the commander's maxim: *If I don't control it, I can't depend on it*. So NGA analysts went to war with the troops.

The NGA needed to cooperate with other agencies for intelligence sharing, as well. So it established partnerships throughout the community, deploying its imagery analysts both to obtain insights for help in their own imagery analysis, and for providing support to others in collection and analysis. Clapper described it succinctly: "exchanging hostages."[22]

In his view, the most important agency with which to collaborate was the NSA. The two agencies could complement each other, and NSA already had a history of embedding its analysts in other organizations. One of the early programs to do this was named GEOCELL.

GEOCELL

In its 2000 report, the NIMA Commission envisioned the integrating of signals intelligence with geospatial intelligence.[23] The success from sharing the NSA's COMINT analysis with NPIC's imagery analysis to uncover Soviet deception during the Cuban missile crisis (described in chapter 13) had illustrated the advantages of such collaboration. As with all matters involving two major agencies, it took a while to take form. But in 2004, the two agencies, the NGA and NSA, began operating GEOCELL.

The NSA had long relied on imagery in targeting its collection efforts. But terrorists posed a new type of threat, and required a much faster response than traditional threats. Military operations in Iraq and Afghanistan required an unprecedented level of collaboration. The GEOCELL program placed a team of NSA and NGA analysts at NSA headquarters with the mission of finding and tracking insurgents and terrorists.[24] The idea was to track individuals by intercepting their telephone traffic. When a suspected terrorist or insurgent began a phone call, an NSA analyst would zero in on its signal, and pass the details to his NGA counterpart—who would immediately use satellite imagery to geolocate the telephone and display its location on a map.[25]

The combination of signals and geospatial intelligence, along with human intelligence, was highly successful in targeting terrorists. In May 2006, the al-Qaeda terrorist leader Abu Musab al-Zarqawi was killed when US F-16s dropped two 500-pound bombs on the safe house where he had taken refuge. The precision strike was the result, according to a US Army spokesman, of "a very long, painstaking, deliberate exploitation of intelligence, information-gathering, human sources, electronic, signal intelligence that was done over a period of time—many, many weeks."[26]

Establishing the Boundaries of GEOINT

Chapter 1 introduced a core definition for geospatial intelligence, and here we add a number of recent ones. Why be concerned about definitions, anyway? *Because they establish boundaries* for a discipline; and in determining budgets, boundaries are important to *all* organizations: governments, militaries, non-governmental organizations, universities, and commercial organizations. In the case of US geospatial intelligence, there are two boundaries in the intelligence community to deal with: the sources of GEOINT, and the nature of the analysis product. GEOINT differs from the other INTs in at least three ways (at least in the US):

- All other INTs are based on exploitation and analysis of raw information derived from specific sources. HUMINT requires the use of a human source. OSINT is based on collection of published material. SIGINT requires collection of some type of signal (including cyber collection, though it doesn't always have a defined signal). But GEOINT requires no specific collection source, though admittedly most NGA products rely on imagery.

- US government publications define HUMINT, SIGINT, OSINT, and MASINT as *products* of collection, processing, and exploitation.[27] Each then has a separate process for creating the product. But, as noted in chapter 1, GEOINT stands alone in being defined as both a process and a product (though IMINT is officially defined as a product, not as a process).[28]
- A closely related issue is that of single-source analysis. In US practice, the five traditional INTs have the responsibility to provide single-source products, not finished intelligence. It is standard procedure for all INTs to draw on outside sources in their analysis process, but in no other INT is it as pervasive as in GEOINT. The only constraint seems to be that GEOINT has a location element.

These differences have resulted in two continuing issues within the intelligence community and the DoD about just what boundaries exist for GEOINT. Let's examine those.

The Source Boundary

Of course, GEOINT has its roots in imagery and mapping; as a consequence, in government practice at least, it hasn't been able to escape the two. But though they usually are part of the product, neither images nor maps are *required* to produce GEOINT. A map is not that useful, for example, when tracking a ship moving on the open ocean. It may be useful in identifying the ship's destination, but that's not the same thing. And if the ship is being tracked using the automatic identification system, for example, neither imagery nor maps are essential—though the result almost always is displayed on an electronic map.

So long as imagery and maps are viewed as essential, the concept of GEOINT remains narrower than it should be. The sources of information should not be a limiting factor, and no single source is required. The emphasis on imagery in the original NGA definition (see chapter 1) is understandable: analyzing imagery is a primary function that intelligence agencies perform. But the implication is that imagery is a requisite component. In fact, geospatial intelligence predates imaging from above, and much still comes from other sources. For example,

- The internet and GPS-equipped devices together probably are the dominant sources of raw geospatial information today.
- Electronic reconnaissance satellites routinely geolocate communications and radar emitters worldwide.
- Geophysical exploration can make use of imagery, but doesn't require it.

- Bathymetry, which is certainly part of GEOINT, doesn't make use of overhead imagery.
- Identifying the sources of underground explosions and earthquakes requires no imagery—only seismic sensing.

All definitions of GEOINT include a geospatial component (everything on the Earth happens at a place and time), but the term has become more far reaching. Let's look for a moment exclusively at the intelligence community. If all intelligence products have some type of geospatial reference or component, where are the boundaries between GEOINT and all other INTs? Even the original NGA definition, which includes imagery, has a gray area. Does it mean only imagery obtained from your own collection systems? Or can it include imagery collection by running a SIGINT operation against an opponent's UAV or imaging satellite downlink, for example? Is the product SIGINT, GEOINT, or both? Also, remote sensing is just one of many ways you can collect imagery. What about a HUMINT operation to steal photographs from a safe or in combination with cyber collection, as happened in the removal of photos in the al Kibar deception?

The answer is that source is not a limiting factor: All INTS can be used to produce GEOINT. Imagery collected from SIGINT, OSINT, HUMINT, or cyber collection still becomes part of GEOINT. And nonimagery material that has a geospatial component can be used to produce GEOINT. The most recent US DoD definition of GEOINT acknowledges this; it says that "any one or combination of these three GEOINT elements (imagery, imagery intelligence or geospatial information) may be considered GEOINT."[29] This leaves the concept almost without a boundary, other than requiring a location and temporal context; "geospatial information" is a term covering an extremely broad area.

But US Defense and intelligence community components are legendary in their fervor for fighting over boundaries. So, of course a dispute arose between NGA leadership and another collection discipline, called Measurement and Signature Intelligence (MASINT). The DIA has functional responsibility for MASINT, while the NGA has functional responsibility for GEOINT. Until the mid-2000s, MASINT included analyzing signatures that were derived from overhead systems, even though some of those signatures were collected from spectral and radar imaging satellites. The products in question weren't imagery; they were the results of measurements and the products took the form of signature diagrams, such as the one from chapter 11, in figure 11.5. But they inevitably had a geospatial reference—they were located *somewhere*.

And so signatures collected from spectral and radar imaging satellites became part of the NGA's responsibilities. In 2005 the director of national intelligence changed the definition of GEOINT to include these types of collection.

What had previously been considered MASINT—the subdisciplines previously called overhead nonimaging IR, and synthetic aperture radar (SAR) MASINT became GEOINT. In the process, they were renamed as "advanced geospatial intelligence."[30]

The Analysis Boundary

US national intelligence collection organizations are tasked with what is called **single-source analysis**—that is, analysis of information from a core collection discipline of intelligence, such as from only human sources, or from only signal sources. For example, the NSA is charged with doing single-source analysis: its job is to process, exploit, and analyze material collected from SIGINT.

In contrast to single-source analysis, a number of national agencies and military service units are charged with producing **all-source analysis**. For example, the CIA, DIA, Department of Homeland Security, and the State Department's Bureau of Intelligence and Research all have the responsibility to provide all-source analysis, meaning that each makes use of all the INTs along with outside sources, at the national level.

From its foundation, the intelligence community instituted an official boundary between these two analysis types. But also since the earliest times, erstwhile analysts have found ways around the bureaucracy in order to do their best work. COMINT analysts, for example, always have had to draw on outside sources of information in order to meet customer demands. So they have often made use of material from other INTs, and referred to such material as **collateral intelligence**. As NIMA evolved to become the NGA, it became standard practice for imagery analysts to draw on not only different imagery sources to get a more complete picture, but on any available source as well—and like COMINT analysts, they also referred to the nonimagery material as "collateral."

The British author Michael Herman in 2001 indicated his approval; he observed that "the single-source agencies now are not pure collectors of 'raw intelligence'; they are also institutionalized analysts, selectors, and interpreters." On the distinction between the two, he noted that it is "intellectually artificial to chop up into parts what is in reality a continuous search for the truth."[31]

Herman was correct. In fact, the most accurate term, introduced in chapter 13—*fusion*—is a word that has a special meaning in the US intelligence community but is also commonly used in NATO and coalition operations. It is a term that indicates what Herman and all intelligence analysts know: though a boundary technically exists to this day, the only way to deal with twenty-first-century challenges is to blur or ignore it. And because intelligence is shared among collection organizations, analysts are able to do so.

Other than the constraint on having some geospatial reference, the NGA has become de facto an all-source intelligence provider, though officially defined as a single-source one. That evolution has not been without consequence. NGA leadership has periodically had to deal with criticism from other US intelligence organizations that it is not staying "within its lane."[32]

The Definition Boundary

The NGA initially defined GEOINT as source-focused, as mentioned above. But as GEOINT has extended its reach into the commercial, scientific, and academic realms, the term has begun to take on an even broader perspective. It's not surprising that other definitions have been proposed, with less focus on the source and more on the process and the product. Darryl Murdock, former US Geospatial Intelligence Foundation director of academic programs, developed a succinct version: "GEOINT is the professional practice of integrating and interpreting all forms of geospatial data to create historical and anticipatory intelligence products used for planning or that answer questions posed by decision-makers."[33]

Pennsylvania State University professor Todd Bacastow has arrived at a more expansive alternative:

> Geospatial Intelligence is actionable knowledge, a process, and a profession. It is the ability to describe, understand, and interpret so as to anticipate the human impact of an event or action within a spatio-temporal environment. It is also the ability to identify, collect, store, and manipulate data to create geospatial knowledge through critical thinking, geospatial reasoning, and analytical techniques. Finally, it is the ability to ethically collect, develop, and present knowledge in a way that is appropriate to the decision-making environment.[34]

Both definitions incorporate the key ideas of an intelligence mission: all-source analysis and modeling *in both space and time* (from "anticipate"). Both model types are frequently used in analysis; insights about social networks, for example, are often obtained by examining them in spatial and temporal ways.

Bacastow's definition views geospatial intelligence as both a process and a product. The NGA's updated definition also defines GEOINT as both, though it places more emphasis on the process than on the product.

It's difficult to think about GEOINT as a single process, because there are many possible processes for producing it. Nevertheless, on its website, the NGA now describes a four-step analytic process for what it terms an "intelligence preparation of the environment." The steps are,

1. *Define the environment:* Gather basic facts needed to outline the exact location of the mission or area of interest. Physical, political, and ethnic boundaries must be determined. The data might include grid coordinates, latitude and longitude, vectors, altitudes, natural boundaries (mountain ranges, rivers, and shorelines), and the like. These data serve as the foundation for the GEOINT product.
2. *Describe influences of the environment:* Provide descriptive information about the area defined in step 1. Identify existing natural conditions, infrastructure, and cultural factors. Consider all details that may affect a potential operation in the area: weather, vegetation, roads, facilities, population, languages, social, ethnic, religious, and political factors. Layer this information onto the foundation developed in step 1.
3. *Assess threats and hazards:* Add intelligence and threat data, drawn from multiple intelligence disciplines, onto the foundation and descriptive information layers (the environment established in the first two steps). This information includes order of battle; size and strength of enemy or threat; enemy doctrine; the nature, strength, capabilities, and intent of area insurgent groups; [and] effects of possible chemical/biological threats. Step 3 requires collaboration with national security community counterparts.
4. *Develop analytic conclusions:* Integrate all information from steps 1 through 3 to develop analytic conclusions. The emphasis is on developing predictive analysis. In step 4, the analyst may create models to examine and assess the likely next actions of the threat, the impact of those actions, and the feasibility and impact of countermeasures to threat actions.[35]

This process delineates an expanded role for GEOINT—a definite change from the NGA's earliest iteration that identified specific sources and omitted anticipatory intelligence. It has several features worth noting:

- It is multidisciplinary; it describes an analytic process much like that followed by any all-source intelligence unit, albeit in a geographic setting.
- It draws on the basic GIS concept of layering different types of information.
- It incorporates the principle of looking at the target from all of the six PMESII factors introduced in chapter 5.
- It emphasizes the importance of anticipatory analysis—the gold standard of intelligence products.

Most important, the process doesn't require that analysis draw on any specific sources. While the context is in support to military conflict, it could be used to describe the GEOINT process for nonconflict issues, with some modification.

The four steps—define the environment, describe influences of the environment, assess threats and hazards, and develop analytic conclusions—can readily be applied to the GEOINT processes discussed in later chapters on local government, nongovernmental, and commercial GEOINT.

Chapter Summary

GEOINT developed within governments, nongovernmental organizations, and the commercial sector over much the same time frame as a combination of need and tools evolved. But one could argue that it was first articulated in the US within the defense and intelligence communities. There, the concept resulted in the merger of mapping and imagery agencies to form NIMA and subsequently the NGA. From these beginnings, the recognition of geospatial intelligence as a valuable process and product quickly expanded both domestically and internationally. Chapter 17 describes some of the other components of this story.

Notes

1. Anne Daugherty Miles, "The Creation of the National Imagery and Mapping Agency: Congress's Role as Overseer," Joint Military Intelligence College, Washington, DC, April 2002, 4.
2. Matt Alderton, "The Defining Decade of GEOINT," *Trajectory*, March 13, 2014, http://trajectorymagazine.com/the-defining-decade-of-geoint/.
3. "WWII History of the Army Map Service," n.d., www.escape-maps.com/escape_maps/history_army_map_service_wwii.htm.
4. "WWII History."
5. US Central Intelligence Agency, "National Photographic Interpretation Center, Volume One: Antecedents and Early Years, 1952–1956," December 1972, www.cia.gov/library/readingroom/docs/CIA-RDP04T00184R000400070001-5.pdf.
6. Jack O'Connor, *Seeing the Secrets and Growing the Leaders: A Cultural History of the National Photographic Interpretation Center* (Alexandria, VA: Acumensa Solutions, 2015), 25.
7. "Standup"—as used in this section's heading—is a term commonly used in the US government to refer to the creation of a new governmental organization.
8. National Reconnaissance Office, "The HEXAGON Story," December 1992, www2.gwu.edu/~nsarchiv/NSAEBB/NSAEBB54/docs/doc_44.pdf.
9. "The Information Edge: Imagery Intelligence and Geospatial Information in an Evolving National Security Environment," final report of the Independent Commission on the National Imagery and Mapping Agency, December 2000, viii, www.dtic.mil/dtic/tr/fulltext/u2/a514975.pdf.
10. "Information Edge."
11. National Geospatial-Intelligence Agency, "The Advent of NGA," January 2015, 18, www.nga.mil/About/History/Documents/NGAAdvent_Dec2014.pdf.
12. Personal communication to the author by former NGA director of research and development Rob Zitz, 2003.
13. Alderton, "Defining Decade."
14. Regina Galvin, "The Genesis of GEOINT," *NGA Pathfinder*, September 10, 2015, https://medium.com/the-pathfinder/the-genesis-of-geoint-aeddd1d3b484.
15. James R. Clapper interview, December 11, 2018.
16. Clapper interview.

17. Clapper interview.
18. Clapper interview.
19. Clapper interview.
20. James R. Clapper, *Facts and Fears* (New York: Viking Press, 2018), 96.
21. The intelligence community's Open Source Center, of course, also relies on unclassified material; but in contrast to other agencies, its product is almost entirely unclassified.
22. Clapper interview.
23. Clapper, *Facts and Fears*, 80.
24. Alderton, "Defining Decade."
25. NSA/CSS, "Remarks by Mr. John C. Inglis, Deputy Director, National Security Agency, at the GEOINT Symposium 2010," November 4, 2010, www.nsa.gov/news-features/speeches-testimonies/speeches/geoint2010.shtml.
26. "'Painstaking' Operation Led to al-Zarqawi," CNN, June 8, 2006, http://edition.cnn.com/2006/WORLD/meast/06/08/iraq.al.zarqawi.1929/index.html.
27. DoD Joint Publication 2-0, "Joint Intelligence," October 22, 2013, www.jcs.mil/Portals/36/Documents/Doctrine/pubs/jp2_0.pdf.
28. DoD Joint Publication 2-03, "Geospatial Intelligence in Joint Operations," July 5, 2017, www.jcs.mil/Portals/36/Documents/Doctrine/pubs/jp2_03_20170507.pdf.
29. DoD Joint Publication 2-03, "Geospatial Intelligence."
30. John L. Morris and Robert M. Clark, "Measurement and Signature Intelligence," in *The Five Disciplines of Intelligence Collection*, ed. Mark M. Lowenthal and Robert M. Clark (Newbury Park, CA: CQ Press/Sage, 2015), 175.
31. Michael Herman, *Intelligence Services in the Information Age* (London: Routledge, 2001), 192–93.
32. Clapper interview.
33. Darryl Murdock and Robert M. Clark, "Geospatial Intelligence," in *The Five Disciplines of Intelligence Collection*, ed. Mark M. Lowenthal and Robert M. Clark (Newbury Park, CA: CQ Press/Sage, 2015), 114.
34. "What Is Intelligence and What Is Geospatial Intelligence," Department of Geography, Pennsylvania State University, n.d., www.e-education.psu.edu/sgam/node/91.
35. National Geospatial-Intelligence Agency, "GEOINT Analysis," n.d., www.nga.mil/ProductsServices/GEOINTAnalysis/Pages/default.aspx.

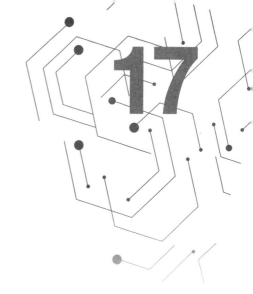

The GEOINT Explosion

The term "geospatial intelligence" and its acronym GEOINT touched a nerve, and not just in the US intelligence community. The consequence has been the GEOINT explosion of the twenty-first century. GEOINT has been widely adopted in national governments (the subject of this chapter); state, provincial, and local governments and nongovernmental organizations (chapter 18); and commercial entities (chapter 19). It was the logical result of a convergence of the tools and need for intelligence about people, their locations, and the events surrounding them. There was just one problem of course, for organizations having GEOINT as a focus: a need for standards.

Chapter 2 introduced the challenge of establishing a prime meridian. It has been located and relocated several times throughout history, starting with Ptolemy's "Blessed Islands." It took an international conference to finally settle on Greenwich in London. The standards problem has plagued cartographers many times since then. When map annotations mean different things to different users, there is great potential for mischief or consequences leading to loss of life.

Imagery has a similar standards challenge. Consider all the different imagery formats you can choose to use just on your computer or smartphone—TIFF, JPEG, GIF, BMP, PNG, as a basic example. If you can transmit only a JPEG file, and your addressee can read only TIFF files, you have a problem. Video files present the same issue—choices include AVI, FLV, WMV, MP4, and MOV. The wide sharing of GEOINT products, just in imagery and video, depends on all parties working to a common standard. And the rapid growth in the field means that new technologies for collecting and processing geospatial materials will regularly arrive on the scene. If the new display technology turns out to be essential to your operations, yet none of your customers have it, things aren't going to turn out well.

By the mid-1990s, even before the term "GEOINT" was coined, the importance of geospatial intelligence was well understood, and governments and nongovernmental organizations around the world were involved in aspects of it. As national and transnational organizations surged in number, most recognized a need to establish standards (for sharing) and to coordinate the introduction of new geospatial technologies; organizations created for that purpose are covered in this chapter. Let's take a look at a bit of the history leading to them—including that favorite of all governmental and military pastimes, renaming and reorganizing as a preferred management tool.

US Geospatial Intelligence Organizations

Chapter 16 told the story of the National Geospatial-Intelligence Agency (NGA) because the agency is the prime producer and functional manager for national GEOINT efforts within the US intelligence community and the Department of Defense. But many US government organizations prepare and disseminate geospatial products. In fact, while the military side of today's GEOINT falls under NGA auspices, creating geospatial intelligence is also a major responsibility of several government organizations. A number of them were producing it prior even to the National Imagery and Mapping Agency (NIMA), and they continue to do so.

The National Reconnaissance Office

We introduced the National Reconnaissance Office (NRO) in earlier chapters. It came into existence in 1961, with the assignment to design, build, launch, and maintain US reconnaissance satellites. From its beginning the NRO operated under a tight cloak of secrecy. Its very existence was classified Top Secret, though articles about it began to appear in the press during the 1970s. It was formed as a triad of satellite development organizations from ongoing Air Force, Central Intelligence Agency (CIA), and Navy programs that became known as Program A, Program B, and Program C, respectively. The three programs remained physically and organizationally separated, and they competed fiercely for the funds available to build intelligence satellites. In September 1992, the NRO's existence was officially acknowledged, and the three programs combined into one at a new headquarters in Chantilly, Virginia.[1] It continues to be a primary contributor of satellite-based imagery and other intelligence products for the US GEOINT effort.

The US Geological Survey

In 1803 the US acquired a vast stretch of central North America from France for 68 million francs ($15 million)—one of the great land bargains in history, even at the 2019 equivalent price of $608 billion. At the end of the Mexican-American

War in 1848, the US added Texas and what is now most of the western US. The government needed a detailed inventory of the vast lands that it now controlled, but that concern had to be set aside while it dealt with the Civil War.

After the war, Congress established the US Geological Survey (USGS) on March 3, 1879, and assigned it the job. Starting from its congressionally directed mission of "classification of the public lands and examination of the geological structure, mineral resources, and products of the national domain," the USGS immediately began topographic and geologic mapping of the vast and largely uninhabited Western territory.[2] It soon found that it had to develop standards for both topographic and geologic mapping, and it continues to do so.

Since its establishment, USGS has produced an extensive set of maps of every part of the US, customized for specific uses. Some maps feature the political boundaries of states, counties, townships, and cities or villages. Others show the means of communication and commerce (past and present): railroads, tunnels, wagon roads, trails, bridges, ferries, fords, and canals. Specialized versions display terrain features: contours, floodplain representations, forests, and sand dunes.[3]

The availability of GIS enabled USGS to undertake the National Map project, launched in 2009, as part of its National Geospatial Program.[4] It repackages GIS data in traditional map form for the convenience of nonspecialist map users. The maps include layers not present on most traditional topographic maps, such as aerial photo and shaded relief images.

USGS today is the largest civilian mapping agency, but it provides a wide range of geospatial information in addition to maps. Many of its publications are intended to help government and industry manage US water, biological, energy, and mineral resources. The USGS product is an essential input to geospatial intelligence conducted by the government and the private sector, though it doesn't fit the definition given in chapter 1; USGS could more accurately be described as a producer of geospatial data and reference materials.

The National Geodetic Survey

The production of nautical charts in the US was originally the responsibility of the United States Survey of the Coast (1807–36) under the tempestuous leadership of Ferdinand Hassler (introduced in chapter 2). During World War II, officers and civilians of its successor organization served in North Africa, Europe, and the Pacific. They were assigned as artillery surveyors, hydrographers, amphibious engineers, and beachmasters during invasions. They conducted reconnaissance surveys for a worldwide aeronautical charting effort.[5]

In 1970, the organization was renamed the National Geodetic Survey (NGS), and it became part of the National Oceanic and Atmospheric Administration (NOAA), within the US Department of Commerce. One of the long-standing

NGS missions is geodesy—the understanding of Earth's shape, orientation, and magnetic field. As part of that mission, for almost two centuries NGS and its predecessors have established geodetic control points throughout the US. These were mostly locations identified by fixing a survey mark in the ground—typically a brass, bronze, or aluminum disk; over 1.5 million now exist across the US. In the past, the NGS also made use of prominent objects such as a water tower or church spire as references; but those markers proved to be less permanent.[6] Today, the NGS relies increasingly on GPS to provide its geodetic reference points. It manages a voluntary network of about 2,000 stations worldwide, called continuously operating reference stations. Each station shares its position data with NGS, which then analyzes the data and distributes the results freely to cooperating nations.[7]

Today, the NGS manages the National Spatial Reference System, which provides the foundation for transportation and communication, mapping, and charting. The system relies on continuously operating reference stations to define the latitude, longitude, and height of points throughout the United States. Surveyors and mapping professionals rely on the system as a positional coordinate standard; that is, when someone marks property boundaries or plans for roads, bridges, and other structures, their information must match that used by others. The NGS also has responsibility for US government oceanographic studies worldwide. But that is a responsibility that it shares with the Naval Oceanographic Office (NAVOCEANO).

The Naval Oceanographic Office

In 1830, the US Navy established the Depot of Charts and Instruments to supply navigational instruments and nautical charts to naval vessels. The depot soon found that the charts needed regular updating, and began to conduct ocean surveys to do that.

The work of chart updating at the depot changed dramatically after Commander Matthew Fontaine Maury took charge in 1842. Maury was the hydrographer of the Navy from 1842 to 1861. During that time, he introduced a novel approach to collecting oceanographic data. He might have studied the success of Muhammad al-Idrisi; in any event, he followed al-Idrisi's example. In what would today be called crowdsourcing, he persuaded shipmasters to provide logs and verbal reports of their voyages to the depot; the reports would then be collated and published for all mariners. Maury's depot was flooded with reports, but it succeeded in providing current charts at a fraction of the cost of the surveys. His approach was subsequently adopted by hydrographic offices around the world.[8]

In 1854, the depot, with its greatly expanded mission, was renamed the US Naval Observatory and Hydrographical Office, but it didn't last very long as such. In 1866, Congress separated it into two units: the Naval Observatory and the Naval Hydrographic Office. In 1962 it became the Naval Oceanographic Office.

During the early years of the Vietnam War, NAVOCEANO discovered that Vietnamese coastal charts were based largely upon World War II data. The river deltas were a particular problem because deltas tend to change in a short period. NAVOCEANO then conducted a three-year geodetic, coastal, and harbor survey of that complex coastline using its survey fleet. NAVOCEANO today operates seven oceanographic survey ships. The ships have no home port; they are continuously at sea, and oceanographers regularly fly to the ships' current ports of call to rotate crews.

The US Central Intelligence Agency

When the Office of Strategic Services (OSS) was established in June 1942, one of its tasks was to deal with a demand for specialized maps for the White House and the military. OSS created a Cartography Section that by the end of 1942 employed twenty-eight geographers. By March 1943 it had become the OSS Map Division, comprising three sections: Cartography, Map Information, and Topographic Models. The division assembled what became the largest collection of maps in the world. Among its specialized mapping products were area maps printed on silk, to be used by downed aircrews for escape and evasion. OSS worked with Britain's MI9 to print versions of these maps on thin paper and smuggle them into German prisoner of war camps in Monopoly board game sets.[9] Other maps like the one shown in figure 17.1 were provided to summarize the battlefront situations for US wartime leadership.

When the CIA was established after World War II, it incorporated elements of the former OSS. One such component was the map division; it continued as the CIA cartography center, retaining the mission of acquiring and creating maps and geographic data that were relevant to national security. The CIA was the first intelligence agency to apply digital technology to cartography. In 1966 a CIA working group borrowed a digitizer and in the course of one weekend compiled a digital map of the world's coastlines and international boundaries.[10] After 9/11, the center produced thousands of maps designed to track terrorist networks and support US military operations; some were used to plan the raid that ended Osama bin Laden's career in 2011.

The CIA entered the imagery analysis field in November 1952, creating the Photographic Information Division. As we know from chapter 16, it later became NPIC and subsequently part of NGA. But the CIA had produced geospatial intelligence since its genesis as OSS, sometimes without using imagery. For example, during World War II, the Germans had very much wanted to conceal the location of their oil refineries; they were priority targets for allied bombing

FIGURE 17.1 Office of Strategic Services 1941–42 map of the Russian front. *Source:* US Central Intelligence Agency.

attacks. But they published the freight tariffs on all goods, including petroleum products. An OSS analyst named Walter Levy used the tariffs, which indicated the price of gasoline delivered at various locations in Germany, to trace back along the German railway network and pinpoint the oil refinery locations.[11] Both OSS and later the CIA conducted many such geospatial analyses, relying on HUMINT and OSINT, well before the CIA's entry into the imagery analysis business.

The US National System for Geospatial Intelligence

With so many US agencies providing geospatial products, sharing of knowledge became essential, and that required some form of standards within the national context. During the lead-up to renaming NIMA, General Clapper put in place the National System for Geospatial Intelligence. It facilitates sharing, collaboration, and standards for geospatial intelligence. Led by the director of NGA, it includes all the US intelligence community organizations and military departments as well as the US Geological Survey. Its partner organizations span industry, academia, and state and local governments.[12]

Five Eyes GEOINT

The globe is a big place, and there are many events of intelligence interest happening on it every day. Also, there are many sources of raw intelligence about those events. The result is a mass of geospatial information that needs to be analyzed and turned into useful intelligence for decision-makers. But no single country has enough intelligence analysts to deal with the challenge posed by the five Vs of big data, introduced in chapter 14. Automated screening tools can help analysts handle the volume, of course. Eventually, however, the raw material needs to be examined by a human. And again, there aren't enough humans with the expertise to cover it all.

The solution to at least easing that problem lies in sharing the analytic load. Countries with a shared intelligence interest and common goals can divide up geospatial analysis problems and share the analytic product among them. That's something NATO members have done for years.[13]

Before NATO, however, an intelligence-sharing arrangement had been set up between the US and the UK. During World War II, the two countries began sharing SIGINT reports. The arrangement was later expanded to add Canada, Australia, and New Zealand to what became known as the "Five Eyes" intelligence services. And the sharing extended beyond SIGINT to include HUMINT and IMINT. These five national services now have a long history of collaborating in IMINT, OSINT, HUMINT, MASINT, and SIGINT products. They do so under a system where cooperation exists *within each discipline*. That is, the SIGINT services deal directly with other SIGINT services, HUMINT services share directly with each other, and so forth. (This setup exists because each service considers that its counterparts best understand how to handle and protect the raw material.) So it's not surprising that the organizational structures and operational patterns have common features within each intelligence discipline of the five.

Within the Five Eyes consortium, GEOINT now has recognition as a coequal INT with HUMINT, SIGINT, OSINT, and MASINT, including a separate organizational structure and a functional manager in those communities. And as with SIGINT and HUMINT, the organizations share directly with each other.

There was certainly no pressure for its Five Eyes partners to adopt the US GEOINT organizational approach, and such pressure would not have been well received anyway. Instead, the move within IMINT services, first to merging imagery and mapping, and later to adopting the GEOINT concept, appears to have been driven by all the Five Eyes partners developing a common view of their profession, and their educational and professional contacts facilitated that. The term "GEOINT" best described what they were doing—though each GEOINT service has its own definition of what GEOINT is, a few of which are illustrated in the following sections. That is, it wasn't a case of "me too"; it was more a case of "that makes sense."

Let's take a brief look at the organizational evolution of each partner before considering some other national geospatial intelligence units. As we will see, the penchant for renaming and reorganizing knows no cultural bounds.

The UK

The UK has a long history in the disciplines that would later merge to become GEOINT. It established the Photographic Interpretation Unit in July 1940. After a series of reorganizations and name changes, in 1953 it became the Joint Air Reconnaissance Intelligence Centre.[14] The UK's map service has an even longer history, dating back to the 1700s. It became the Military Survey Defence Agency in April 1991. The new agency had the role of providing geographic and geospatial support to defense planning and operations.

On April 1, 2000, both units ceased to function as independent agencies and merged into the Defence Geographic and Imagery Intelligence Agency. In making the recommendation for that merger, the UK Select Committee on Defence took into account the advantages that had led the US to create NIMA. It also warned of the consequences of a poorly planned merger—possibly influenced by observing the NIMA experience.[15] In 2016 the organization became part of the National Centre for Geospatial Intelligence, within the UK Ministry of Defence.[16]

The UK definition of geospatial intelligence refers to GEOINT as a product, though it includes the process within: "Geospatial Intelligence (GEOINT) is the spatially and temporally referenced intelligence derived from the exploitation and analysis of imagery intelligence and geospatial information to establish patterns or to aggregate and extract additional intelligence."[17]

Canada

Canada is a pioneer in the use of geospatial intelligence, as represented by its development of the Canadian Geographic Information System (introduced in chapter 6). It continued that leadership pattern with its Canadian Geospatial Data Infrastructure program during the 2000s. The program added a number of geospatial information layers and defined best practices standards for creating and sharing geospatial data.[18]

On the national intelligence side, Canada until recently maintained separate organizations for imagery and for mapping and charting. The Canadian Forces Joint Imagery Centre provided imagery and imagery intelligence. A separate component of the Canadian Forces Intelligence Command was the Mapping and Charting Establishment, with the mission of providing geospatial information to the military and to other government departments. These have subsequently merged into the Directorate of Geospatial Intelligence.

Australia

The Defence Imagery and Geospatial Organisation (DIGO) was created on November 8, 2000, by merging the Australian Imagery Organisation with the Directorate of Strategic Military Geographic Information and the Defence Topographic Agency.[19]

In May 2013, DIGO was renamed the Australian Geospatial-Intelligence Organisation (AGO) within the Department of Defence. It is responsible for the collection, analysis, and distribution of GEOINT to support Australian defense and national interests. The Australian Navy hydrographic services and Air Force aeronautical charting functions merged with the AGO in 2016.[20]

Australia defines geospatial intelligence as a product: "Geospatial intelligence (GEOINT) is intelligence derived from the exploitation and analysis of imagery and geospatial information about features and events, with reference to space and time."[21]

New Zealand

New Zealand's GEOINT effort traces its origins back to the founding of the Royal New Zealand Navy's Hydrographic Office in October 1949. In 2012, it was renamed GEOINT New Zealand, an intelligence component of the New Zealand Defence Force. It provides a range of GEOINT products and services—maps, charts, and imagery products—to New Zealand's military, government, and civil institutions.[22] New Zealand's definition takes a clear process form: "Geospatial Intelligence (GEOINT) is the exploitation and analysis of imagery and geospatial information about features or activities of defence, security, economic, safety or intelligence interest."[23]

Allied System for Geospatial Intelligence

The Five Eyes intelligence services needed a system for establishing standards so that their countries' joint military operations could share geospatial intelligence. In 2009, they created the Allied System for Geospatial Intelligence (ASG). The ASG provides a formal mechanism for sharing the GEOINT product. Of at least equal importance, it provides a system for sharing the workload of providing GEOINT. As noted above, there never are enough GEOINT analysts to address national needs, and being able to count on your partners to cover some of those needs helps all of them immeasurably.

Other National GEOINT Organizations

Many countries have to deal with the emerging GEOINT needs described in chapter 13, and have access to the same tools discussed in chapter 14. So it's not surprising that national GEOINT organizations have proliferated worldwide. GEOINT is now accepted in many countries as a distinct INT, though not necessarily endowed with a distinct organization.

GEOINT organizations have evolved over several decades, but as with all new disciplines, the evolution has encountered its share of challenges. Various countries have faced the need to consolidate their imagery analysis units with cartography units that in most cases have had a long history of existing as separate organizations. Both functions are needed to produce GEOINT, but history and tradition have made their union within a single organization a difficult leap for many countries. Also, history and geography shape each nation's GEOINT requirements. A few of the many national organizations are described here.

France

Outside of the Five Eyes, creating a national GEOINT organization has been less successful in most countries, and France is typical. The cultural problem of merging imagery and mapping or geography is not unique. France has a military geography organization—the Bureau Géographie, Hydrographie, Océanographie, Météorologie—and a separate military GEOINT organization, the Centre de Renseignement Géospatial Interarmées. The two remain organizationally and culturally separate.[24]

China

As noted in chapter 5, China lives in a tough neighborhood. Since the 1950s, it has engaged in conflicts with India, Russia, Taiwan, and Vietnam. It has long been concerned about a possible collapse of the North Korean government. And China has substantial internal threats in Tibet and Sinkiang. An attempt to extend its territorial claims into the South China Sea (building military bases in the disputed Spratly Islands) has increased tensions with Taiwan, Malaysia, the Philippines, and Vietnam, which also claim ownership of the island chain.

Not surprisingly, China has a robust GEOINT effort to support its civil and commercial needs, its increasingly global interests, and at the same time for responding to internal and external threats. Its Belt and Road Initiative, for example, requires geospatial intelligence for land routes in Asia and sea routes across the Indian Ocean. Imagery and mapping responsibilities are divided. Military intelligence is the responsibility of the People's Liberation Army's Second Department. Political and economic intelligence falls within the purview of the Ministry of State Security. And the National Geomatics Center of China—part of the National Administration of Surveying, Mapping, and Geoinformation of China—is responsible for the civil sector. It has the job of compiling and distributing geospatial data and conducting national surveying and mapping efforts.[25]

Russia

Russia has long dealt with external and internal threats, and three organizations provide geospatial intelligence about both. The Federal Security Service (FSB) focuses on internal threats and emphasizes the human terrain. The Foreign Intelligence Service (SVR) is one of two sources for external geospatial intelligence; military GEOINT is handled by the Main Intelligence Directorate of the General Staff (GRU).

But Russia also has economic interests that rely on having good geospatial intelligence, especially in exploiting Siberian and Arctic Ocean resources. Siberia contains about 80 percent of Russia's oil, coal, and natural gas resources.[26] The Arctic Ocean has oil and gas reserves amounting to about 22 percent of the world's total.[27] Both regions have rich mineral deposits, though extraction is difficult. The Federal Service for Hydrometeorology and Environmental Monitoring of Russia (Roshydromet) is responsible for the research and monitoring of these resources. Space observation systems are the responsibility of the Roscosmos State Corporation for Space Activities.

India

India has one of the world's oldest mapping organizations. The Survey of India was organized in 1767 with the assignment of mapping the territories of the British East India Company.[28] It continues the function today of surveying and mapping India's interior.

A separate organization maps India's border areas. India faces external threats from Pakistan and China, having been periodically involved in border clashes with both. And continuing unrest in the northern province of Kashmir has resulted in the Indian Army placing a high priority on current geospatial intelligence. That need is addressed by the Military Survey, which was created as part of India's military intelligence service. It now provides the military with maps of border areas, regularly updated with satellite imagery; creates and updates a digital topographical database; conducts photogrammetric surveys; and develops GIS products.[29]

The Indian military also has a Defence Image Processing and Analysis Centre that handles imagery from military satellites and UAVs. The Indian Space Research Organisation manages the civilian side of GEOINT.

India's interests aren't confined to threats from the land. As noted in chapter 5, India views the Indian Ocean as its preserve, and it is developing an airborne surveillance capability to provide maritime domain awareness in the region.[30] The Indians have good reason for the concern; in 2008, members of an Islamic terrorist group carried out a series of attacks in Mumbai, killing at least 165 people. The 10 attackers used inflatable speedboats to land on the coast near Mumbai.[31]

Japan

Japan shares with the UK all the geographical advantages of being an island nation. But geography also conveys disadvantages. The country's location on multiple geological fault lines has led to a history of dealing with earthquakes and consequent tsunamis. To deal with the threat, Japan now has an extensive warning system that may be the most sophisticated in the world. The country has a network of seismic and pressure sensors on the ocean floor that feeds data to a center where simulation modeling determines the danger posed by a tsunami.[32]

Japan does not have a centralized intelligence organization. Its intelligence functions are divided among five independent organizations. The Cabinet Intelligence and Research Office is responsible for geospatial intelligence and reports directly to the prime minister. A subdivision, the Cabinet Satellite Intelligence Center, operates a network of imaging satellites that includes the IGS-Optical and IGS-Radar series.[33] In 2018, Japan had seven of these reconnaissance satellites in orbit, with plans for a total of ten. The satellites' primary missions include monitoring North Korean missile facilities and assessing damage from natural disasters.[34]

The Open Geospatial Consortium

What was done for standards and sharing within the US by the National System for Geospatial Intelligence and among the Five Eyes by the ASG is also done globally by the Open Geospatial Consortium (OGC), though with more emphasis on standards than sharing. The OGC is an international organization with the goal of developing open standards for the international geospatial community. Since its beginning in 1994, the OGC has facilitated collaboration by more than 500 commercial, governmental, academic, and research organizations worldwide.[35] The OGC has allowed the NGA to take the lead in defining standards for some geospatial technologies.[36]

Transnational GEOINT Organizations

Transnational organizations have also developed as part of the GEOINT explosion. They are largely focused on advancing the state of knowledge and education about geospatial intelligence generally. The three dominant ones hold well-attended annual conferences that attract geospatial intelligence practitioners from all over the world.

The US Geospatial Intelligence Foundation

The story of the US Geospatial Intelligence Foundation (USGIF) is best told by revisiting our friend Stu Shea from the prologue. By the turn of the century, Shea had reason to be pleased. His vision from twenty years past of integrating cartography and imagery had come to fruition with the standup of NIMA. And in the interim, he had organized and led computer mapping and geospatial programs for government clients. His team had put in place the first GIS at the National Security Agency and the CIA. Stu could have relaxed.

But, he was still bothered: "The knowledge sharing isn't going far enough. It has to be more than the intelligence community working together on GEOINT. The academicians, professional trade associations, technology companies and commercial imagery satellite operators all have unique roles to play and contributions to make. And, it should go beyond the government level, anyway."[37] Stu reasoned that NIMA alone couldn't make that happen—though it might be willing to help if someone took the lead.

So in 2002 Stu approached an old friend with the idea for an inclusive GEOINT event. John Stopher was the budget director for the US House of Representatives' Permanent Select Committee on Intelligence, and had the job of monitoring NRO and NIMA programs. An electrical engineer with a background in advanced imaging systems, Stopher well understood the issues around GEOINT. The two men were not in an elegant bar and had neither a drink nor a cocktail napkin at hand. Instead, they crafted the basic details one afternoon, literally sitting on the trunk of John's used BMW (that he had just bought) in the Defense Intelligence Agency parking lot. They then began to contact others in government, industry, and academia with the idea of a geospatial intelligence conference.[38]

In 2003 the idea bore fruit. The director of NIMA, General Clapper, was initially skeptical about the idea, but he agreed to support a conference. The undersecretary of defense for intelligence, Steven Cambone, agreed to be the keynote speaker. The GEO-INTEL 2003 conference was held in New Orleans in October. The organizers planned for 200 attendees. Over 1,200 showed up.

Buoyed by the conference's success, the organizers began planning for an annual conference and a nonprofit organization to manage it. And in January

2004, the USGIF came into existence, with Stu Shea as president. Its vision: fostering geospatial intelligence tradecraft in government, industry, and academia, and encouraging collaboration among them all.[39]

Later that year, the second annual conference, renamed GEOINT 2004, was again held in New Orleans, and this time there were over 1,800 attendees. (The conference name was changed because, as it turned out, Intel Corporation had the copyright on the name "INTEL.") GEOINT 2005, in San Antonio, attracted 2,300. Membership and conference participation continue to grow. The 2018 conference in Tampa had 4,200 attendees from around the globe—making it the world's largest event centered on GEOINT.

The USGIF has continued to broaden its mission, and today it can best be described as a facilitator of GEOINT as a discipline. It sponsors events and programs, such as tech days focused on specific geospatial technologies, scholarships, and college and university accreditation programs designed to promote GEOINT education.[40] Despite its name, the USGIF is an international organization.

The Geospatial World Forum

The Geospatial World Forum is an annual conference that addresses topics from every type of geospatial intelligence. The participants include more than 1,500 members from more than 500 organizations—government, academic, private sector, and transnational—from more than seventy countries.[41] Recent venues have been India, Switzerland, and Portugal. In April 2019, the forum held its eleventh conference in Amsterdam.

The conferences emphasize futuristic themes, and some of its major programs—smart cities, the environment, location analytics, and business intelligence—are topics discussed in later chapters. The forum also is committed to creating a collective and shared vision of the global geospatial community; it has the ambitious goal of making "geographic information a common language for 7 billion people around the globe."[42]

Defence Geospatial Intelligence

Defence Geospatial Intelligence is Europe's largest annual GEOINT conference. It is held annually in London and attended by military, government, and industry geospatial intelligence professionals from more than forty-five nations at last count. The conference presentations cover new developments in collecting, integrating, managing, analyzing, and sharing geospatial data on topics that range from public safety to national and global security.[43]

Chapter Summary

The GEOINT explosion was an almost inevitable outcome of the pressures to more effectively combine mapping, imagery, and other sources of geospatial information for government use. While the NGA has functional management of GEOINT within the Defense Department, a number of agencies in both the military and civilian sectors of the US government produce or contribute to producing geospatial intelligence. They include the US Geological Survey, the National Geodetic Survey, NAVOCEANO, and the CIA. Sharing of intelligence among these agencies is handled under the US National System for Geospatial Intelligence.

A similar structure exists for sharing GEOINT among the Five Eyes national intelligence services: the UK, Canada, Australia, New Zealand, and the US. All five have organizations dedicated to providing geospatial intelligence, and they cooperate both in dividing up collection responsibility and in sharing the resulting product.

GEOINT also now serves a wide range of national and transnational government interests. Most world powers also produce geospatial intelligence for both military and civil purposes and cooperate in establishing standards (though generally not in sharing intelligence.) Transnational organizations such as the USGIF, the Geospatial World Form, and Defence Geospatial Intelligence hold annual conferences to facilitate the sharing of geospatial knowledge and technologies.

But the development of GEOINT is not the exclusive province of national governments. Geospatial intelligence has also exploded among state, provincial, and local governments and nongovernmental organizations—the subject of chapter 18.

Notes

1. Bruce Berkowitz, "The National Reconnaissance Office at 50 Years: A Brief History," 2nd ed., July 2018, www.nro.gov/Portals/65/documents/about/50thanniv/The%20NRO%20at%2050%20Years%20-%20A%20Brief%20History%20-%20Second%20Edition.pdf?ver=2019-03-06-141009-113&timestamp=1551900924364.
2. Mary C. Rabbitt, "A Brief History of the US Geological Survey," US Geological Survey, 1975, https://pubs.er.usgs.gov/publication/70039204.
3. Rabbitt.
4. US Geological Survey, "National Geospatial Program," n.d., www.usgs.gov/program/national-geospatial-program.
5. National Oceanic and Atmospheric Administration, "National Geodetic Survey," n.d., www.ngs.noaa.gov/web/about_ngs/history/index.shtml.
6. National Oceanic and Atmospheric Administration, "National Geodetic Survey."
7. National Oceanic and Atmospheric Administration, "CORS," June 26, 2018, www.ngs.noaa.gov/CORS/.

8. National Park Service, "Matthew Fontaine Maury," n.d., www.nps.gov/people/matthew-fontaine-maury.htm.
9. US Central Intelligence Agency, "The CIA Museum: OSS Escape and Evasion Map," n.d., www.cia.gov/news-information/featured-story-archive/2010-featured-story-archive/oss-escape-and-evasion-map.html.
10. US Central Intelligence Agency, "The Mapmaker's Craft: A History of Cartography at CIA," November 10, 2016, www.cia.gov/news-information/featured-story-archive/2016-featured-story-archive/mapmakers-craft.html.
11. Walter Laqueur, *The Uses and Limits of Intelligence* (Somerset, NJ: Transaction, 1993), 43.
12. Matt Alderton, "Governing GEOINT Growth," *Trajectory*, June 7, 2017, http://trajectorymagazine.com/governing-geoint-growth/.
13. Arndt Freytag von Loringhoven, "Adapting NATO Intelligence in Support of 'One NATO,'" *NATO Review*, August 9, 2017, www.nato.int/docu/review/2017/Also-in-2017/adapting-nato-intelligence-in-support-of-one-nato-security-military-terrorism/EN/index.htm.
14. "Unit History: Joint Air Reconnaissance Intelligence Centre," n.d., www.forces-war-records.co.uk/units/3536/joint-air-reconnaisance-intelligence-centre.
15. UK Select Committee on Defence, "Fifth Report: The Defence Geographic and Imagery Intelligence Agency," n.d., https://publications.parliament.uk/pa/cm199900/cmselect/cmdfence/100/10003.htm#n10.
16. Matt Alderton, "Eyes Wide Open," *Trajectory*, November 5, 2018, https://trajectorymagazine.com/eyes-wide-open/.
17. UK Joint Doctrine Publication 2-00, "Understanding and Intelligence Support to Joint Operations," August 2011, www.gov.uk/government/uploads/system/uploads/attachment_data/file/311572/20110830_jdp2_00_ed3_with_change1.pdf.
18. "Canadian Geospatial Data Infrastructure 2005–2009," report by GeoConnections, Natural Resources Canada, for Ninth United Nations Regional Cartographic Conference for the Americas, New York, August 10–14, 2009, https://unstats.un.org/unsd/geoinfo/rcc/docs/rcca9/crp/9th_UNRCCA_econf99_CRP5.pdf.
19. "Australian Geospatial-Intelligence Organisation," n.d., www.defence.gov.au/AGO/.
20. "Australian Geospatial-Intelligence Organisation."
21. "Australian Geospatial-Intelligence Organisation."
22. New Zealand Defence Force, "GEOINT New Zealand," September 22, 2017, www.nzdf.mil.nz/about-us/geoint/default.htm.
23. New Zealand Defence Force.
24. Ret. Col. Frédéric Hernoust, Thierry G. Rousselin, David Perlbarg, Nicolas Saporiti, Jean-Philippe Morisseau, and Ret. Gen. Jean-Daniel Testé, "GEOINT on the March: A French Perspective," in *2018 State and Future of GEOINT Report*, 2018, http://usgif.org/system/uploads/5489/original/2018_SaFoG_PDF_Final.pdf.
25. United Nations, "National Geomatics Center of China," n.d., www.un-spider.org/links-and-resources/institutions/national-geomatics-center-china.
26. US Energy Information Administration, "Country Analysis Brief: Russia," n.d., www.eia.gov/beta/international/analysis_includes/countries_long/Russia/russia.pdf.
27. Hobart M. King, "Oil and Natural Gas Resources of the Arctic," n.d., https://geology.com/articles/arctic-oil-and-gas/.
28. "Survey of India," n.d., http://www.surveyofindia.gov.in/.
29. Lt. Gen. P. C. Katoch, "Modernisation of Armed Forces: Military Survey and GIS," *Geospatial World*, July 25, 2013.
30. Rahul Bedi, "Indian Navy Extends Maritime Surveillance Capability with Additional Naval Air Squadrons," *Jane's 360*, January 22, 2019, www.janes.com/article/85875/indian-navy-extends-maritime-surveillance-capability-with-additional-naval-air-squadrons.
31. CNN Library, "Mumbai Terror Attacks Fast Facts," September 18, 2013, www.cnn.com/2013/09/18/world/asia/mumbai-terror-attacks/index.html.

32. Josie Garthwaite-Stanford, "New Tsunami Warnings Could Get More People to Safety," *Futurity*, January 22, 2019, www.futurity.org/tsunami-warning-sensors-1961732/.
33. Deyana Goh, "Japan Launches Reconnaissance Satellite IGS-Radar 6," Spacetech Asia, June 13, 2018, www.spacetechasia.com/japan-launches-reconnaissance-satellite-igs-radar-6/.
34. Adrienne Harebottle, "Japan Places Eighth Reconnaissance Satellite in Orbit," *Via Satellite*, June 13, 2018, www.satellitetoday.com/government-military/2018/06/13/japan-places-eighth-reconnaissance-satellite-in-orbit/undefined.
35. OGC website, http://www.opengeospatial.org/.
36. *Priorities for GEOINT Research at the National Geospatial-Intelligence Agency* (Washington, DC: National Academies Press, 2006), 20.
37. Stu Shea, interview, August 7, 2018.
38. Stu Shea, interview.
39. Stu Shea, interview.
40. US Geospatial Intelligence Foundation website, http://usgif.org/.
41. Geospatial World Forum website, https://geospatialworldforum.org/.
42. Geospatial World Forum website.
43. "DGI 2020," https://dgi.wbresearch.com/.

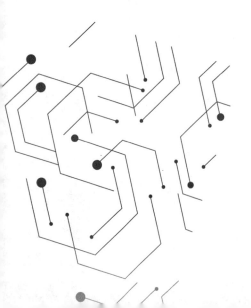

Non-National Geospatial Intelligence

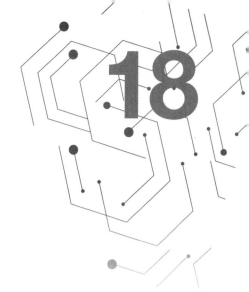

In 2008, a coalition of nongovernmental organizations (NGOs) joined together on an ambitious plan: to ensure that clean water was available to every resident of Liberia by the end of 2020. To do that, they would need to map the country.

Clean, drinkable water is in short supply across much of Africa, but few if any states are worse off than Liberia. Two civil wars—the last one ending in 2003—have devastated much of the country's infrastructure. And in 2014 an outbreak of Ebola swept across the country—with poor hygiene and dirty water helping it spread.

The consortium of NGOs, aided by the Liberian government and commercial entities, collaborated to gather data on the demographics of every village in Liberia. One hundred fifty Liberians were organized into teams and trained in the basics of collecting data for GIS input. The teams traveled around the country on motorcycles, cataloguing population, existing water resources, locations, and incidence of waterborne bacterial infections.

Next, the survey results were layered on a map that tracked the progress of the effort for follow-up action, for example, to manage the installation of water filters for villages where hand pump wells weren't an option.[1] The project appears to be on track to hit its 2020 goal: By 2018, more than half the targeted households had water filters in place. Diarrheal disease (primarily caused by contaminated water) is a leading cause of childhood death in Liberia. In the 2007–17 period (the most recent available data), that mortality rate dropped by 10 percent.[2] In households with the filters installed, the drop was even more dramatic: diarrheal incidence dropped from 36 percent prefilter to 2.9 percent postfilter.[3]

Does this seem like an application of geospatial intelligence? Perhaps not at first glance. The idea that any type of intelligence is conducted by state, provincial, or local governments (other than in law enforcement) might seem surprising. But many functions of non-national governments and organizations deal with geospatial intelligence. It is in a different context than traditional national and military intelligence operates, where the definition is more about uncovering official secrets. Chapter 16 points out the broader view that has emerged since GEOINT has extended its reach into other realms. Murdoch and Bacastow's definitions, for example, make no mention of concealment (though they may assume it). The basic elements of geospatial intelligence explored in chapter 1 apply in almost all arenas. There, we said that GEOINT means analyzing situations and creating products that

- are anticipatory—deal with the future,
- involve some type of human activity,
- draw on knowledge of the Earth (its surface, whether on land or sea, and the objects on or beneath that surface), and
- provide an information advantage to a decision that someone must make.

With that introduction, let's look at some examples—with the understanding that this is only a sampling of the geospatial intelligence activities conducted by non-national governments and entities.

State/Provincial and Local Government

State or provincial and local governments seek geospatial information (and have done so throughout history) for many purposes. Emergency responders have to know the locations of fires, criminal activity, or natural disasters in order to help. Records keepers and tax authorities have to identify the owners of property. State governments have many of the same needs for geospatial information that national governments have, such as security, emergency response, environmental assessment, natural resource management, conservation, education, and regional planning.

And in most cases, they have the same tools for gaining and analyzing geospatial information that national governments possess. For example, they had access to aerial photography beginning in the 1930s, and made extensive use of it for land surveys. Landsat and SPOT accelerated those applications, and the tools of GIS have enabled local governments to expand the types of GEOINT that they use. Local government geospatial programs now support a wide variety of functions, including those noted in the next subsections.

Community Planning

All governments at the local level have to acquire and understand geospatial information to carry out their responsibilities. They must plan educational facilities locations and define school boundaries. They provide utilities and oversee their provision; asset management, call-before-you-dig programs, forecasting demand, and responding to outages all require geospatial knowledge. They have to plan for community development, regulate zoning, and analyze land use. In recent decades, new issues have entered the planning process: environmental protection, protecting endangered species, and stormwater management, for example.

Urban planners have the same issues, along with a few that are of particular interest to cities. Planning for and managing capital improvements such as mass transit, port facilities, and parks; dealing with traffic congestion; protecting critical infrastructure; and planning for special events or emergency evacuation require tapping into geospatial information—down even to mundane problems, such as dealing with solid waste.

The experience of Thurston County in Washington State is typical of what community planners do today. In 2016, the county needed a new solid waste management plan to help it deal with an overload of its existing waste transfer facility. The study it commissioned assessed the shortcomings in the existing system. But the study team also did what is now common practice: prescriptive geospatial intelligence. It astutely analyzed the expected future demographics of the region—where employment and population growth were likely to occur over the next twenty years. Based on those results, the county's land use planning, and the estimates of expected waste generation, the team identified the optimal locations for building future transfer stations.[4]

Citizen Interaction

The tools of GEOINT have changed the way that individuals and businesses in an area interact with their local governments. Are crime patterns, code violations, or real estate prices in your locale a major concern? Do you want to know the schedule for debris pickup, snow plowing, inspections, or other non-emergency public services in your area? Would you like to see where the snow plows are at the moment, what they're seeing, and to get an area map of road conditions and lane closures? Or check the public transportation schedule in your area? All that, and more, is available on map and imagery displays provided by governments in some localities. Increasingly, you can interact directly with these services by simply entering your physical address and then being informed of when, for example, to put unwanted items on the street curb for pickup.[5]

For people who are considering opening a business or planning to move an existing one, local governments are quite interested in and ready to help with GEOINT on demand. They care about business retention and attraction. One online mapping application lets business owners browse through available buildings to find one that meets their needs, using parameters such as zoning, square footage, and building type. And, of course, the same prospects are able to tap into Google Street View or a drone-created video to look at the property from the street and from above.[6]

Citizen interaction is a two-way street. In the US and Canada, citizens have for some time been able to dial 311 for nonemergency reporting about incidents and public service outages. Increasingly, it's possible to file such a report, providing location information automatically, in reporting from a mobile device.

Assessing Real Estate

Both local governments and realtors use geospatial information for establishing property values—one for revenue projection and assessing taxes, the other for buying and selling properties. A wide range of geospatial concerns affect values: crime statistics in the vicinity, school locations, air pollution, and proximity to noise from roads are a few examples.

Thanks to current technology, tax assessors today have an easier job than any time in the past. Using traditional field inspections, they had a difficult and sometimes hazardous job. It was time-consuming because they had to obtain permission to access the backyards of properties. Gated communities presented an additional layer of restricted access. And they had to deal with assessors' worst nightmares: they were subject to dog attacks and occasional threats from gun-carrying homeowners. Aerial imagery has eliminated these problems. Today, assessors use high-resolution imagery to identify new construction, improvements, demolitions, and other changes that could easily be missed in a field inspection. They rely on the same change-detection techniques that imagery analysts use in national intelligence. Setting aside peace of mind, the results in efficiency are astounding, with assessors finding ten to twenty new structures each day, instead of the one or two found using the old method.[7]

Emergency Response

Emergency dispatch and management depend on quickly knowing the location of the fire, crime, suspicious activity, natural disaster, hazardous material incident, water or gas line break, or stricken person. Less time-sensitive responses might track a disease outbreak or pollution spill. Though the universal 911 help number in the US and Canada is a nationally sponsored program, the response is up to state and local governments from where the call is placed. Until the 1990s, only about 50 percent of the population had access, and for those who had it, the systems relied on the caller for location information. It's a different story now. In North America, the trend has been to make accurate locations of callers available to 911 call centers. It's been a challenge, because the 911 system is big and complex; the changes involve new hardware and software, and coordination among many public safety and governmental elements. The first enhanced 911 rolled out, requiring that call centers receive the positions of callers to within 300 meters. The next generation 911 (NG911) expanded the types of calls (e.g., from medical alert devices) that could be sent to call centers with geolocation information.[8]

GEOINT is especially central in dealing with crisis response to wide-area disasters such as those caused by hurricanes, flooding, and wildfires. During 2017 it played a key role in response to hurricanes Harvey, Irma, and Maria, and to the California wildfires. During 2018, when Hurricane Florence hit the Carolinas and Virginia, extensive geographic information was available online for disaster response teams and the public from numerous sources:

- An interactive map of affected regions that included estimated rainfall, winds, and power failures (*New York Times*).
- A journal to provide GEOINT for FEMA leaders to use in decision-making, with some information available to the public (FEMA/ESRI).
- GEOINT for commercial organizations and first responders from DigitalGlobe's Open Data program (DigitalGlobe).
- Interactive maps of the hurricane, GIS assistance, and apps (ESRI).
- A heat map showing the density of service requests in hurricane-hit areas, based on social media posts (Humanity Road, US Coast Guard, and others).
- A weather analytics app providing real-time updates on road conditions (Harris Corporation).
- A map showing storm surge warnings (National Hurricane Center).
- A crowdsourced map showing the locations of FEMA shelters (OpenStreetMap).[9]

Law Enforcement

Applying GEOINT in state and local law enforcement obviously relies on knowing where and when incidents happen. And the internet of things provides police with a new and enhanced set of automated sensors that pinpoint activities of interest. Police forces also operate their own sensor systems, in the form of GPS anklets worn by offenders, license-plate-reading sensors, in-car video and body cameras, and the ShotSpotter gunfire-detecting system described in chapter 7.[10]

State and local law enforcement routinely deal with denial and deception, and they consequently conduct COMINT, IMINT, clandestine HUMINT, and cyber operations to obtain intelligence. And GEOINT long has had a role in doing more than detecting and geolocating crimes as they happen. Since the early days of GIS, geospatial analysts in major police departments have mapped and analyzed crime patterns in their areas. Their product feeds law enforcement initiatives that have many names: intelligence-led policing, information-led policing, problem-oriented policing, and community policing.

Using the power of GIS, crime analysts overlay disparate data sets to draw their conclusions: census demographics, lighting conditions, and locations of pawn shops, transit stops, ATMs, liquor stores, and schools. They often can identify the underlying causes of crime and help law enforcement administrators devise strategies to deal with the problem. They use it to catch individual perpetrators by geographic profiling, as Kim Rossmo did in helping catch the Leeds rapist discussed in chapter 14. They assess crime trends and identify where crime is likely to occur in the future. The predictions (often courtesy of the tools cited in the previous paragraph) are used to deploy police units both to deter crime and to catch perpetrators—frequently, in the act.

Nongovernmental Organizations

Nongovernmental organizations and academic researchers often operate in some rough areas of the world, where maps and other geospatial details are in short supply and not uncommonly out of date. This section touches on a few of the many NGOs both inside and outside the US that make heavy use of GEOINT.

Drone Adventures

Drones are the latest tools for quickly and safely obtaining imagery in difficult-to-traverse regions. And they are superb for getting the work done. It's a specialty for the nonprofit Drone Adventures, whose tagline is "Mapping the World with Drones." Founded in 2013, based in Switzerland, and staffed heavily by volunteers, the organization partners with companies, individuals, institutions, and other NGOs. Its mission generally is to apply drones for humanitarian purposes, including nature conservation, preserving culture, and cleanup after natural and human-made disasters. These include projects as diverse as nature conservation in Africa, imaging sea life in the Seychelles at the request of the SaveOurSeas Foundation, imaging in the aftermath of the Fukushima nuclear reactor meltdown, and archaeology research in Turkey.

Chapter 5 discussed the use of thematic cartography, pointing out its application by radical Islamic groups to argue for a new caliphate that would reclaim territory lost to non-Muslims many centuries ago. Something like that has been used on a much smaller scale by local governments to promote governmental or special interests. It typically takes the form of producing maps that misrepresent or omit geographical features, often with the objective of dispossessing owners and taking control of their property.

An example is the Lima enclave called Barrios Altos, a deprived area in the historic center of the city. Lima government agencies have produced a series of thematic maps that depict Barrios Altos as a poor zone, overcrowded, with high criminality, and at risk of physical collapse. The maps were used by a special municipal body in charge of strategic vision for the renovation of the area; it argued that nearly half of Barrios Altos should be demolished and renovated by private investment firms.[11]

Enter Drone Adventures. One of its projects during 2014 involved what is called "counter-mapping"—imaging Barrios Altos and developing a series of maps to present the other side of the story. Local NGOs opposing the demolition used the drone imagery as a tool in following up with on-the-ground investigations and in discussions with residents. The mapping products were then subsequently used to mobilize residents and to fight the proposed demolition.[12]

Satellite Sentinel Project

Chapter 13 pointed out that one of the issues driving the evolution of GEOINT is human rights and war crimes. It was a special concern of the Satellite Sentinel Project, launched in 2010 with the purpose of trying to prevent a full-scale civil war in the Sudan; its tagline is "The world is watching because you are watching." Using DigitalGlobe's openly available satellite imagery, analysts looked for evidence of war crimes and mass atrocities. Aided by the imagery, researchers on the ground were able to investigate further and ask questions. Their objective was to provide an early warning system. By focusing world attention and international responses on human rights and human security violations, they hoped to deter the sort of mass atrocities that have characterized conflicts in the former Yugoslavia, Syria, and Iraq. In South Sudan, they uncovered evidence of alleged atrocities and quickly published stories that war criminals would have preferred to keep hidden. In 2015, the project's job was taken over by an organization named the Sentry, an investigative initiative seeking to dismantle the networks of perpetrators and facilitators that benefit from Africa's deadliest conflicts.[13]

Doctors Without Borders

One of the best-known NGOs worldwide is Médecins Sans Frontières, better known in the US as Doctors Without Borders (DWOB). Headquartered in Switzerland, this humanitarian organization has the mission of helping anywhere in the world by delivering emergency medical aid to people affected by conflict, epidemics, disasters, or exclusion from health care. In 2017 DWOB had missions in more than sixty countries.[14]

DWOB requires GEOINT for planning deployments, monitoring conflicts, and keeping track of disease spread, among other activities. During the

2014 Ebola outbreak in West Africa, it found that some of the hardest-hit areas had never been mapped before—making it difficult to plan how to combat the virus. DWOB dispatched a GIS expert to Sierra Leone to provide GEOINT about the disease's incidence and its spread over time. It then was able to do more than simply treat new cases; its teams were able to concentrate on containing the disease. The GEOINT products also allowed teams to find sick people who were quarantined and get them access to water, food, and health care.[15]

The International Anti-Poaching Foundation

The International Anti-Poaching Foundation (IAPF) is a nonprofit organization that targets the illegal trafficking of wildlife, primarily in Africa. Trafficking in African elephants, rhinoceroses, and gorillas has developed into a major criminal enterprise, and African park rangers are hard pressed to stop the poaching. Rhinos are especially vulnerable because of the value of their horns—which in 2017 fetched prices of up to $100,000 per kilogram in Vietnam, making them more valuable than gold by weight.

The IAPF mission includes training park rangers in military-style tactics for defending against the poaching. Drones that can operate at long ranges and provide day/night surveillance of wilderness areas have become the principal IAPF technology for catching poachers.[16]

Monitoring North Korea's Nuclear Program

During 2018 and 2019, the US and North Korea were in negotiations to ease the tensions surrounding North Korea's nuclear program. The US goal was to obtain verified denuclearization. North Korean leader Kim Jong Un wanted a formal end to the Korean War and the removal of sanctions. In 2018 North Korea unilaterally suspended nuclear and ballistic missile tests, ostensibly halting further weapons development during the negotiations.

In the past, any denuclearization agreement verification would have depended primarily on classified reporting from the US intelligence community. There now are independent sources of information about the North Korean program, thanks to the dramatic improvements in commercial satellite imagery, along with the work of some NGOs in making use of it. And the picture they present is disturbing.

Jeffrey Lewis is a director at the James Martin Center for Nonproliferation Studies in Monterey, California. His team has been using satellite imagery—the best of which may approach the quality of that used by national intelligence agencies—to track the progress of North Korea's nuclear program. After a July 2018 tipoff that the North Koreans were continuing their missile manufacturing, Lewis's team conducted a geospatial intelligence analysis effort that a veteran of the National Geospatial-Intelligence Agency would probably admire. According to a *New Yorker* article, the team

> identified the factory in question and used historical commercial-satellite imagery to rewind the clock. They discovered images of supply trucks, including bright red trailers that had traditionally been used to transport intercontinental ballistic missiles, traveling in and out of the facility. By matching details from North Korean media photos with those of satellite images, they were also able to confirm that it was one of the missile factories where Kim had, in recent years, personally examined missiles and launch vehicles. As Lewis told Warrick, the facility "is not dead, by any stretch of the imagination."[17]

Another NGO, the Center for Strategic and International Studies (CSIS) in Washington, has a stellar fifty-plus-year history of providing strategic studies for policymakers. CSIS used satellite imagery to detect changes at North Korea's ballistic missile bases over time, supplemented by interviews with defense and intelligence officials and North Korean defectors. CSIS published a satellite imagery history, along with an all-source analysis demonstrating that work was ongoing to improve the operational status of sixteen such bases.[18]

Disease Prediction

Chapter 6 described Dr. John Snow's pioneering effort to trace the source of an 1854 London cholera outbreak by creating a map of the area, pinpointing cases of cholera, and overlaying the location of water pumps—thereby narrowing the source to a single water pump. This layering of information over a map allowed Snow to identify relationships, and in turn develop insights into a disease source. Since then, the tools of GEOINT have been applied to tracking disease outbreaks and further to predicting and responding to such outbreaks. As mentioned above, those tools were used in West Africa to track the spread of Ebola, to plan transportation routes for health care personnel, and to understand weather patterns that would affect both. But as with many GEOINT topics, the real payoff comes when predictive models are used to warn of outbreaks well in advance. Two examples:

- In May 2017, the Jutla Research Group, a West Virginia University–sponsored research team, used GEOINT modeling to predict an outbreak of cholera in Yemen. Cholera is a waterborne bacterial disease, and outbreaks occur under specific conditions: hot, dry seasons in coastal regions where good sanitation and clean water are in short supply. Using satellite imagery, the team monitored temperature patterns, water storage, precipitation, and population movements in Yemen, feeding the data to their model. The model results indicated that a cholera outbreak was likely to occur soon in West Yemen along the Red Sea. Within a month, the outbreak struck densely populated cities along the country's west coast; over 350,000 Yemenis suffered moderate to severe cholera symptoms.[19]

- In the fall of 2017, a team of researchers from Duke University and Johns Hopkins University began using a different model to predict in detail malaria outbreaks in the Peruvian Amazon. The research team makes use of NASA's Land Data Assimilation System (LDAS). It incorporates data from several NASA satellites (Landsat, Global Precipitation Measurement, and Terra and Aqua) to provide information on precipitation, temperature, soil moisture, and vegetation around the world. The team's malaria forecasting tool uses that information to pinpoint areas where prime breeding grounds for the anopheles darlingi mosquito overlap simultaneously with human populations. Predicting where these mosquitoes will flourish then relies on identifying areas with warm air temperatures and calm water, such as ponds and puddles, which the mosquitoes need for laying eggs—data that the LDAS provides. The project is being continued in other parts of Latin America.[20]

Predictive models in both examples accurately provided warnings of disease. But the LDAS project was more remarkable because of its exquisite detail, indicating the direction that geospatial modeling is taking. It used the most current predictive model, which is called *agent-based*; it is so named because it models the behavior of every agent, or every human, mosquito, and malaria parasite within a given area—a significant application of the data analytics and visual analytics described in chapter 14. The model uses the results to assess the probability of when, where, and how many people are expected to get bitten and infected with the disease. It also can provide prescriptive intelligence, simulating possible results from any one of several actions, such as providing bed nets and protective sprays or administering preventive antimalaria treatment.[21]

Chapter Summary

GEOINT, defined in the broadest sense, is not produced only by national governments and their military arms. Many functions of state, provincial, and local governments and nongovernmental organizations practice and make use of geospatial intelligence. These entities are typically not concerned with uncovering official secrets, as national governments are. But they have many geospatial interests that are identical to, or overlap with, those of national governments.

Non-national governments rely on geospatial information to support emergency responders, to aid with taxation, for transportation and utility planning, and, more recently, for resource management. They make use of the same tools as national governments for collection and analysis of geospatial data, such as GIS, satellite, and drone imagery.

NGOs use GEOINT both for planning and in conducting day-to-day operations, again relying on many of the same tools. Some NGOs also produce much the same sort of GEOINT as the national intelligence agencies, for example, by monitoring conflict areas for evidence of war crimes, reporting on nuclear proliferation, and for disease prediction.

Not surprisingly, commercial interests have many of the same concerns, often with a slightly different twist. The next chapter deals with those.

Notes

1. GIS Cloud, "NGO Maps Out Entire Liberia and Brings Clean Water to Every Single Resident," *GEO Awesomeness*, n.d., http://geoawesomeness.com/ngo-maps-entire-liberia-brings-clean-water-every-single-resident/.
2. Institute for Health Metrics and Evaluation, "Liberia," n.d., www.healthdata.org/liberia.
3. Sawyer Products, "The Last Well to Provide Liberia with Clean Drinking Water by 2020," June 15, 2018, www.prnewswire.com/news-releases/the-last-well-to-provide-liberia-with-clean-drinking-water-by-2020-300667007.html.
4. FLO Analytics, "Solid Waste System Assessment," n.d., www.flo-analytics.com/case-studies/solid-waste-system-assessment/.
5. Ben Miller, "7 Ways Local Governments Are Getting Creative with Data Mapping," *Government Technology*, January 25, 2016, www.govtech.com/7-Ways-Local-Governments-Are-Getting-Creative-with-Data-Mapping.html.
6. Miller.
7. "5 Advantages Every Assessor Should Have in the New Year," EagleView, January 11, 2018, www.eagleview.com/2018/01/assessor-advantages-2018/.
8. 911.gov, "Next Generation 911," n.d., www.911.gov/issue_nextgeneration911.html.
9. US Geospatial Intelligence Foundation, "The GEOINT Community Responds to Hurricane Florence," *Trajectory Magazine*, September 13, 2018, http://trajectorymagazine.com/the-geoint-community-responds-to-florence/.
10. Rob Pegoraro, "GEOINT for Policing," *Trajectory Magazine*, November 1, 2017, http://trajectorymagazine.com/geoint-for-policing/.
11. Rita Lambert and Adriana Allen, "Participatory Mapping to Disrupt Unjust Urban Trajectories in Lima," September 8, 2016, www.intechopen.com/books/geospatial-technology-environmental-and-social-applications/participatory-mapping-to-disrupt-unjust-urban-trajectories-in-lima.
12. Lambert and Allen.
13. "Satellite Sentinel Project," n.d., http://www.satsentinel.org/.
14. Doctors Without Borders, "Opening & Closing Projects," n.d., www.doctorswithoutborders.org/who-we-are/how-we-work/opening-closing-projects.
15. Doctors Without Borders, "Maps Save Lives," March 17, 2017, www.doctorswithoutborders.org/article/maps-save-lives.
16. Damien Mander, "Rise of the Drones," *Africa Geographic*, February 2013, https://iapf.org/wp-content/uploads/2017/03/riseofthedronesjan2013.pdf.

17. Doug Bock Clark, "How Civilian Firms Fact-Check North Korea's Denuclearization Efforts," *New Yorker*, February 26, 2019, www.newyorker.com/news/news-desk/how-civilian-firms-fact-check-north-koreas-denuclearization-efforts.
18. Joseph Bermudez and Victor Cha, "Undeclared North Korea: The Kumchon-ni Missile Operating Base," September 6, 2019, https://beyondparallel.csis.org/undeclared-north-korea-the-kumchon-ni-missile-operating-base/.
19. Andrew Foerch, "Forecasting Disease from Space," *Trajectory Magazine*, January 5, 2018, http://trajectorymagazine.com/forecasting-disease-space/.
20. NASA, "Using NASA Satellite Data to Predict Malaria Outbreaks," September 13, 2017, www.nasa.gov/feature/goddard/2017/using-nasa-satellite-data-to-predict-malaria-outbreaks.
21. NASA.

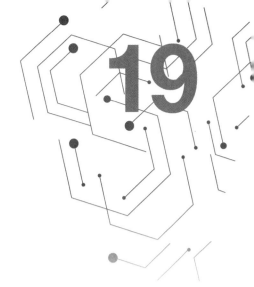

Commercial GEOINT

In 2018, Keith J. Masback, then executive director of the US Geospatial Intelligence Foundation, wrote,

> It's time to talk about geospatial intelligence in the commercial and consumer marketplace. It's time to share lessons learned and to crosswalk best practices. It's time to develop GEOINT practitioners native to the commercial sector. It's time for GEOINT to be taught in business programs at the undergraduate and graduate level. From precision agriculture, to oil and gas exploration, high-velocity logistics, marketing and retail, smart cities, the Internet of Things, and autonomous vehicles, GEOINT is a key competitive differentiator. Those who apply it will have a decided advantage. Those who fail to apply it will ignore the power of GEOINT at their own peril.[1]

Masback was right. Geospatial knowledge was a commercial necessity from the time that ancient traders needed to know how to get their goods to their destination, while avoiding natural hazards and bandits or pirates. The fundamental perils of delivering products haven't disappeared, but commercial firms now rely on geospatial intelligence for many more purposes. Sophisticated tools aid in locating faraway customers, identifying marketing and sales approaches, tracking the movements of products in their supply chains, locating sales and distribution points, and assessing the geospatial operations of their competitors, among others.

Commercial entities are much like nongovernmental organizations in their acquisition of intelligence. There are legal restrictions on the methods they can use in most countries, but they can collect information using IMINT, some types of HUMINT, and OSINT—relying heavily, of course, on the internet. Much of the geospatial data that companies need is readily accessible via the most basic of devices, such as smartphones. And the information is available in levels never seen before. Companies have learned how to use it primarily for geospatial business intelligence, both strategically and in day-to-day operations.

Before diving into our discussion, a distinction needs to be made. Business intelligence (BI) and competitive intelligence (CI) are two different topics, related only insomuch as both concern the commercial world. CI is a term that refers strictly to the collection and analysis of information that a commercial competitor keeps concealed in an effort to gain competitive advantage. The distinction will become clearer throughout the following sections, beginning with BI.

Geospatial Business Intelligence

Business intelligence may be best described as market research. It involves transforming data through analysis into information that will help corporate executives and managers make informed strategic and operational decisions. Because the field consistently makes use of geospatial knowledge, the term "geospatial business intelligence" has come into vogue in the last decade. It is generally defined as "geospatial information that allows companies to make better decisions."[2] Updated definitions note that it is the process of combining traditional business intelligence technologies and methods with GIS. Or, if you prefer, geospatial BI adds to existing tools the spatial analysis and map visualization features that lead to better business decisions.[3]

In 1992, the author and database expert Carl Franklin estimated that about 80 percent of corporate data had a spatial component.[4] Imagine what the figure must be almost thirty years later. Steven Fleming, at the University of Southern California's Spatial Sciences Institute, said in 2017 that "most of the world's data is geo-tagged; I think it's 85 to 90 percent. We know where a banking transaction starts and where it ends. We can track digits. We can certainly track where people are."[5] How to make use of this morass? With the tools of location analytics.

Location Analytics

Location analytics depends on drawing insights from geospatial data relationships to solve problems. In the BI context, "location" can mean an address, a service boundary, a sales territory, or a delivery route. It involves choosing different data sets spatially or chronologically, layering them on a map display, and drawing conclusions from the display. The underlying premise is that location directly affects the kinds of business insights that you might discern from different sets of information.[6]

Take the case of Earth Fare. It was the second-largest natural foods chain in the US, having more than thirty stores located in ten eastern states. The company's goal was to expand and open new stores in the same region. To do that, it relied on location analytics to understand who its competitors were, the market covered by their current stores, how far customers would travel to visit an Earth Fare, and, most importantly, to identify specific cities and land parcels within those cities where a new store could be most profitably located.[7]

In the process of examining data geographically, companies discover relationships between data types that would not have otherwise been obvious. Consequently, location analytics provides insights that businesses can use to increase their profitability.[8] Statistics, tables, and charts are limited in their ability to support this sort of in-depth analysis. But when map displays are added to the mix, things change dramatically; executives now can visualize the information in ways that fit their analytical style, and that visualization has power. It has become as easy to do as clicking on a website and entering a zip code. The executive then can get a quick view of demographics, behavioral patterns, and spending habits in a region.[9]

Getting to that view, however, requires a lot of work by the geospatial BI analyst. These analysts have something in common with their counterparts in government agencies: the challenge of sifting through the mass of data to find the right part to layer on a display. But when they have the tools to do that, location-referenced intelligence—heavily dependent on mobile devices—is a game changer in business intelligence. Location-based apps on smartphones are used every day to check public transportation schedules, locate friends, find restaurants and entertainment venues, and request ride services. And as any savvy consumer knows, businesses are tapping into that online traffic.

There are many factors for an analyst to consider in looking at this data stream, and it's easy to overlook critical ones. The Earth Fare study, for example, examined socioeconomic and demographic data. It investigated land use, transportation, and drive-time information for possible store locations, and the supporting infrastructure for a natural foods vendor.[10] There are many other possible perspectives for businesses to pay attention to in analyzing geospatial data. It may help to integrate the data using all of the PMESII perspectives discussed in chapter 5, for example:

- **Political.** Some areas are more favorable to business; local governments offer incentives to companies to encourage relocation and expansion. The regulatory climate is a key factor. For example, in some European countries, regulations make it practically impossible to fire an employee and therefore unattractive for enterprises.
- **Military.** Again, "military" in this context refers to law enforcement. It might include examining potential threats from criminal and gang activity; or the attitude of local law enforcement in protecting business interests.
- **Economic.** Income, purchasing power, and spending habits of local residents or tourists are economic factors to consider. Expensive natural food stores

may not find success in lower-income areas, for instance. Government taxation of business is also a factor.
- **Social.** This includes demographics, recreational patterns, educational levels, religious values, and more. Businesses that cater to young adults or families with small children select areas that fit that demographic profile.
- **Infrastructure.** Issues here include parking, transportation layout and hubs, risk and natural hazard information, utilities, supplier locations, and maintenance facilities.
- **Information.** Mobile phone coverage, availability and usage, the internet, and local news media are relevant factors.

In 2017, the online retailing giant Amazon began the search to establish a location for its second headquarters. The search was a massive effort, and in a brilliant stroke, Amazon made effective use of crowdsourcing to gather the geospatial information it needed by conducting a contest or competition among locales. Social and traditional media heavily covered the excitement among local governments and individuals as they began sales campaigns to win over the company. Amazon considered over 200 major US cities (including its current home, Seattle) before the field was narrowed to just twenty. In the process, it relied heavily on the cities being considered to provide the details it needed to make a decision.

And finally, there were two cities. On November 13, 2018, Amazon announced that it would establish secondary headquarters in two locations: the Crystal City area of Arlington, Virginia, and the Long Island City, Queens, neighborhood of New York City.[11]

The announcement created an unexpected backlash and a storm of criticism from political and labor leaders in New York when details emerged about the incentive package that Amazon would receive. In response, on February 14, 2019, Amazon dropped its second bombshell. It would abandon the New York location and establish its second headquarters in Arlington.[12]

Geography certainly was a major factor in Amazon's decision to locate its second headquarters on the opposite side of the country. But its search criteria included all the PMESII factors. Looking at the major factors in Amazon's decision process:

- *Political* certainly included garnering strong support from local political leaders, measured by the attractiveness of the incentive package they would offer. The Arlington package was worth up to $573 million; the New York package, up to $1.525 billion.[13] The Washington area also had the political advantage of being close physically to the national power structure.

- *Military* was a factor, again in the context of the quality of law enforcement: Detroit, Saint Louis, and Memphis have the highest violent crime rates, so they probably fell off the list rather quickly.[14] In crime rates, Queens and Arlington both are near average for all cities and towns in the US.
- *Economic* had several aspects, the size of the incentive package being probably most important. But New York was clearly attractive because of its status as the nation's financial center. Considerations about the regions' economic attractiveness for employees were less appealing, in both cases. Both cities have a high cost of living, and New York's tax structure is not designed to appeal to well-paid technical staff.[15]
- *Social* factors centered on the availability of a high-tech labor force. New York has the highest number of tech professionals (despite its tax structure), with the Washington area coming in second.[16] An important factor in Amazon's final decision was antibusiness sentiment and especially labor opposition in New York; the company has a history of opposing labor unions.
- *Infrastructure* included issues such as transportation and education. Mass transit was a requirement Amazon established early on. And the existence of top-tier universities was clearly important to provide a tech talent pool.
- *Information* certainly was important for Amazon; it depends on the internet for most aspects of its business. But the way the local media views Amazon was also an important factor, and in the end it, along with the political and social opposition it generated, turned out to be decisive.

Three Business Types

Joe Francica, the Pitney Bowes managing director for geospatial industry solutions, has described three business types that particularly rely on geospatial knowledge:

- Those that have a long history of using location intelligence, because it is essential to their operations. These businesses are rooted in geography. Telecommunications and transportation are two well-known examples.
- Those that didn't start out as being dependent on geospatial information, but are today. Three of these are retail, insurance, and agriculture.
- Those that critically depend on geospatial intelligence. For them, location intelligence *is* the business. Ride-hailing companies like Uber or Lyft fit that profile.[17]

In each of these industry types, geospatial business intelligence mirrors the use of GEOINT in national intelligence in one way; it must take both a strategic, long-term perspective as well as an immediate or operational view. Let's look at a few of the many applications using each view, starting with strategic GEOINT.

Strategic GEOINT

Chapter 5 described the evolution of strategic thinking geospatially. Like governments, businesses have to develop their fundamental strategies with an eye on geospatial realities and shaping those to their advantage. Strategic GEOINT contributes to a planning process that includes selecting the optimal locations for a business.

Telecommunications

Spatial analytics enables the telecommunications industry to assess the strength of the current infrastructure and also analyze competitors' network coverage. For mobile phone companies, the most critical question is, Where should we place our cell towers? The company wants to find locations that will provide the best network coverage at the lowest possible cost. It might also want to know where to add new towers to its network to boost weak signals or where expected population increase means adding towers to meet demand. Cable companies delivering telephone, internet, and media services have to look at demographics, hazards to facilities, and supporting infrastructure. Terrain is a major factor in both cell phone tower siting and planning to lay trenches for fiber-optic cables. For both, the regulatory climate has to be considered.

Transportation

In strategic planning, airlines obviously have to plan routes and establish hubs that will most efficiently handle their customer traffic. They then must identify the mix of aircraft needed to support those routes and all the supporting infrastructure. Marine passenger and cargo companies have to do much the same planning, considering the ports they are expected to serve. Long-haul trucking companies have to plan to deal with terrain (the road network) and fleet service infrastructure, among other things. Companies like United Parcel Service and FedEx optimize their delivery routes in the planning stage.

Retail

Though online shopping has seemingly begun to dominate the retail sector, there are still plenty of stores with physical locations. Also, it's not yet easy to dine at a fine restaurant, get a haircut, or fill up your gas tank online. Strategic planning for retail outlets is the same as with a natural foods store: it begins with thinking about location. Whether it's a restaurant, barbershop, or filling station, you need to understand the impact of location on profitability. What are the automobile and foot traffic patterns? What competitors are already in the area or are expected? How does the location affect stock delivery, store management, and inventory management? You also want to know not just the customer traffic patterns and demographics of the likely customer set; what about your potential employees? That was a major criterion in the Amazon location decision discussed above. Resort areas are a favored place to locate a factory outlet mall, for example. An area having lots of retirees is the logical place for assisted living facilities—which, in turn, is a logical place for stores that provide geriatric products. Conversely, activewear and trendy shops prefer to locate in an area that is heavily populated by millennials (who don't shop exclusively online).

Once you've established a physical location and the associated demographic segments, the next issue is which products should be offered there. **Geographic segmentation** refers to your target audience having preferences that are specific to a location or region. It's a standard marketing approach for businesses that have a wide demographic customer set. People buy grits in Arkansas but not in Maine. Thong swimsuits do well in California but not in Iowa and definitely not in Muslim communities. McDonald's hamburgers are made with beef in most places, but lamb is used in India because of religious preferences; and your burger will have chili sauce in Cancún but will have teriyaki in Tokyo.

Insurance

Insurance companies are about reducing the risks of claims and establishing pricing to cover those. They need to have a clear picture of all the possibilities, which usually vary with the location of the property (or person) being insured. The more information about geospatial variation in risk that an insurance company can obtain, the more flexible it can be in pricing insurance. Real property, automobile, and health and life insurance all rely on geospatial information in assessing risks and pricing individual policies. The insurers also use it in aggregate to measure overall risk by understanding the location and temporal distribution of their policyholders and policyholder assets.

Real property insurance, for example, requires establishing specific details associated with a property address: the number and type of buildings on the property, their age, construction, residential or commercial usage, tax or sale value.[18] The sales products used by realtors, especially the aerial views taken from drones, are valuable inputs to an insurer's estimates.

Both natural and social risks have to be taken into account. Is the property located in a hazardous zone—more than normally subject to earthquakes, floods, volcanic eruptions, hurricanes, tornadoes, or wildfires? Obviously, areas that experience frequent hurricanes or floods will have higher insurance premiums. Areas subject to hillside erosion and landslides require special analysis of the terrain: an insurance company will examine the soil, subsurface structure, and slopes to identify locations that would be affected by a major landslide.[19] Social risks to property are affected by demographics: crime rates, poverty, and population density.

Automobile insurance long has been priced based on the automobile's location and on distance driven daily. Demographic segmentation is a factor as well (both young and old may pay more for insurance). Urban area drivers also tend to pay more. But insurance companies additionally look at details of the commuter routes traveled, including proximity to hazards, emergency services, and crime. Normal traffic patterns are of special interest; areas of heavy traffic and congestion are examined using the heat maps introduced in chapter 15.[20] In chapter 14, the display of the likely residence of the Leeds rapist, shown in figure 14.2, is an example of a heat map. The radar displays of storms shown on TV are heat maps; intensity is shown by color, with red usually being the most intense.

Life and health insurance obviously are priced based on the age and health of policyholders. But location has become a factor to be considered as well. Demographics, care facilities available, where people live or have lived, all affect their likely future need for health care and their expected mortality rates.

Agriculture

GEOINT has several different roles in the business of agriculture, ranging from global views such as expected crop yields to managing individual farms and ranches. The strategic view begins with a need for farmers and futures traders in agricultural products to understand likely prices of those products next month or next year. The result is crop forecasting on a large scale, supported by GEOINT tools. Governments also use the tools in agriculture; in fact, it was a national-level debacle that led to their widespread use by the US.

In 1971, the USSR was facing a major wheat crop shortage that had to be covered by massive imports of wheat. Had that fact been generally known, it would have caused a major surge in the price of wheat worldwide and would have resulted in the Soviets having to cover the shortfall at substantially higher prices than the existing market offered. They avoided that outcome in what became known as "the Great Grain Robbery."[21]

Soviet representatives contacted major US grain exporters and arranged secret meetings to purchase the wheat without revealing the buyer. At the same time they entered into negotiations with US officials for an agricultural trade deal, while concealing their impending shortfall and their negotiations with exporters.

In July and August 1972, the USSR purchased 440 million bushels of wheat for approximately $700 million. At about the same time, the Soviets closed a trade deal under which the US agreed to provide them with a credit of $750 million for the purchase of grain over a three-year period.[22]

In the 1970s, monitoring crop developments worldwide was the job of agricultural attachés from the US Department of Agriculture's Foreign Agricultural Service. They routinely reported on expected Soviet crop yields based on visual observations—requiring frequent visits to crop-producing areas. In this particular case, Soviets steered the attachés to a few selected fields that were producing high yields—ensuring that their visitors had no chance to view other fields en route.[23]

If the US government had been aware of the Soviet shortfall, it certainly would not have extended the $750 million credit for grain purchases. It would have forced the Soviets to accept substantially higher prices for the grain that it did purchase. And it would have avoided the later political and economic consequences: The Soviets had acquired their wheat at bargain prices; the US government had funded the purchase; and US consumers had to deal with sharply higher food prices. Grain prices that year reached 125-year highs in Chicago. The result was a political embarrassment for the US administration.

The Great Grain Robbery drove the US government to begin monitoring global agricultural output via satellite imagery—first visible imagery, then infrared and spectral imagery. Monitoring allowed the US Central Intelligence Agency in 1981 to forecast a wheat shortage in the USSR that would require the Soviets to purchase wheat on the international market.[24] Analysts quickly figured out that such methods could be used for crop forecasting generally, and satellite imagery now is used by many countries for that purpose.

Ride-Hailing

Start-up companies often introduce disruptive technologies or business practices, and many have done so in geospatial commerce. The ride-hailing companies in the last decade have done exactly that. In 2009, Uber Technologies moved to disrupt the taxicab industry by launching its service, and in the process it became an archetype for the on-demand economy. Its business model has been followed by other ride-sharing companies around the world and for startups in other industries as well.

The key was to skillfully use geospatial knowledge in strategic planning to take on a vulnerable target: the taxicab industry. Riders in major cities were generally dissatisfied with the taxi service and response times, but they mostly had no feasible alternative way of getting to their destinations. Uber offered a more convenient, efficient, and enjoyable travel experience. And it chose to begin its operations with a keen eye on geospatial factors. Uber launched its service in San Francisco, a city with expensive, poorly rated taxi service and a large tech-savvy population that is comfortable with using smartphones for almost any purpose. As Uber (and other ride-sharing services) expanded, they made their initial moves into urban areas having the right demographic and geospatial characteristics and segments: lots of major sporting and theater events, active nightlife scenes, and poor taxi service.[25] The ride-sharing services also considered the political climate in targeted areas—specifically, the regulations on taxicab services and how they could be avoided in different localities.

Operational GEOINT

Commercial GEOINT isn't just about gathering and analyzing geospatial data for strategic planning. It's also about

- Running business operations on a day-to-day basis. Tracking where assets are; communicating with employees; and dealing with suppliers and customer demand issues.
- Interacting directly with customers. Customers have become quite competent at using mobile devices to engage with the world around them, to locate the things they need in that world, and to use the geospatial information those devices provide, using electronic maps and layering information on them to make decisions.[26] Businesses have tapped into that capability to enable two-way communication with customers.

Let's touch on operational GEOINT in commerce, using the same six categories that we just covered for strategic GEOINT.

Telecommunications

Once a cable or mobile phone system is in place, day-to-day operations have a strong geospatial element. Cable providers have to handle service calls and area outages and know the locations of customers having problems. Cell phone companies have to constantly check the quality of service for each tower. Both have to deal quickly with attacks on their networks from natural or human-made causes.

Transportation

Airlines use spatial analytics to continuously track in-flight operations in detail. Airport, meteorological, and fleet data are monitored in real time and used to reroute flight paths as necessary. Marine shipping companies and trucking companies also track the movements of their assets, and respond to problems that ships or trucks encounter en route. Companies such as UPS, FedEx, and Amazon track package delivery schedules, sharing the information with customers, and interacting with customers on changes, cancellations, and delays—most often, in real time.

Retail

Retail chains, having a network of geographically dispersed stores, obviously make more use of geospatial information than do sole proprietorships. The chains have to monitor multistore operations and respond to unexpected demand or lack of demand, and they have to deal with suppliers to get goods to the right stores. But both chains and sole proprietors make increasing use of geospatial information in communicating with customers. One of the rapidly expanding uses is in **geofencing**.

A geofence is a virtual perimeter created around a real-world geographic area. The perimeter could be a circle of a specific radius around points such as a store, an outlet mall, or a restaurant. It could be the boundary of a specific area such as an entire neighborhood.

Once the geofence is established, it can be used in many ways. For example, a location-aware device such as a smartphone entering or exiting the geofenced area triggers a message if the device has the proper app loaded. The message can go in one of two directions:

- It can trigger an alert to the smartphone user. A geofence around a retail area sends an instantly targeted advertisement—encouraging the user to stop in; it displays special short-duration deals or pushes a coupon to the smartphone, for example. A geofence around a nearby competitor could push an invitation to come visit the geofence operator's establishment instead.
- It can trigger a text message or alert to the geofence operator, often including the location of the mobile device, for example, to monitor activity in secure or sensitive areas.[27]

In retail, a variant of the geofence is to use an in-store or in-mall beacon to communicate, using Bluetooth technology, with shoppers. It allows users to navigate through a large mall or store, and to receive offers from retailers based on their location. Some hotels use it to automatically check in arriving guests.[28]

Geofencing has many applications other than smartphone apps. It is used to track vehicle fleets, vessels in the shipping industry, and livestock; monitor employees in the field; keep track of the organization's property; and automate timecards. The Federal Aviation Administration is moving to geofence drones to warn the drone operator, obtain the operator's identity, or even to bring the drone down.[29]

Insurance

GEOINT is valuable for property insurers at the beginning of the insurance process (assessing risk and pricing accordingly), as previously discussed. But it is also used operationally in handling claims and increasingly in reducing possible claims in advance.

One of the newer applications of GEOINT for property insurers is in planning for claims before a major disaster so that they are handled more quickly and efficiently afterward. Insurance companies can use predictions of a path and intensity of hurricanes, major floods, and wildfires to determine the number of claims adjusters to send in. Some such events, for example, tornadoes and earthquakes, are not currently so predictable. But after an event, a GEOINT analyst can accurately pinpoint the path taken by a destructive tornado, or the area hit by a flood or wildfire, and compare that with predisaster imagery for claims estimates.

Increasingly, GEOINT applications help the insurer do more than just react to disasters. They can provide the detailed knowledge necessary for companies to interact directly with insured individuals, and thereby to reduce claims in advance of predictable major disasters. The major GIS provider ESRI offers insurers that capability. Suppose you are an agent responsible for mitigating potential claims for the damage expected in an approaching hurricane. You might use ESRI's tool to do the following:

1. Call up a map display of your policyholders in the affected region.
2. Locate the majority concentration of policyholder residences.
3. Fine-tune the display to show where the highest-value homes are located.
4. Overlay the areas that are expected to be severely hit with different colors representing different levels of flood or wind damage.
5. Identify in those areas the homes that will probably be most affected by the hurricane and what the potential claim value will be.
6. Notify the policyholders that are most severely at risk to take immediate action to protect their families and property.[30]

A logical future development for this process would be for the insurance company, in advance of a storm, to have agreements with the insured and with local companies to have protective measures provided in the event that the homeowner is unable to do so.

Operational geospatial intelligence also has a role in the auto insurance industry. Insurers appear to be moving to obtain GEOINT for the purpose of setting individual insurance rates. Major insurers now provide devices that track location, speed, driving time, and braking and acceleration events for the purpose of establishing or changing rates.[31]

Agriculture

One of the major contributors to agricultural production is termed "precision farming" or "precision agriculture." It relies on the tools for obtaining and analyzing geospatial information. The concept developed because different parts of a farm or ranch will have different physical properties—terrain features, soil type, nutrients, and moisture content, for example. So a combination of GPS, GIS, and remote sensing can tell farmers exactly where, when, and how much of expensive products like fertilizers, pesticides, and herbicides to apply. It can help farmers to more efficiently use water resources that are increasingly difficult to obtain. For ranchers, it helps in managing livestock. Precision farming has proved to be highly successful in large-scale farming that is the norm in many countries.

The spectral and visible sensors that support precision farming are mounted on satellites, UAVs, and GPS-equipped harvesters. They are used to identify gaps, water shortages, weed growth, and crop health in fields. Multispectral and hyperspectral sensors are especially prized as sensing tools because of the detail that they provide. For ranchers, sensors monitor fence lines and the status of livestock food and water; some ranchers also outfit their herds with internal sensors to track their livestock's current location and state of health.

Ride-Hailing Companies

Day-to-day operations of a ride-hailing company are a continuous process of receiving ride requests, identifying the driver and putting driver and passenger together at the same location, and handling the payment and evaluations of each party afterward. The company has to accept scheduling rides from 30 minutes to 30 days in advance, and must shuffle expected demand with drivers available in an area. It also adjusts fares by time and location to level demand (a technique known as surge pricing), for example, during major sporting matches or holidays such as New Year's Eve. All these actions require operational GEOINT.

Geospatial Competitive Intelligence

In contrast to business intelligence, competitive intelligence operates more like national intelligence. CI analysts have to deal with deception and denial, for example. However, they are far more restricted (compared with national intelligence organizations) in the methods they are allowed to use for collection. In most countries, CI practitioners operate under stringent legal and ethical restrictions. Still, CI is about intelligence in the narrowest sense: determining for commercial purposes knowledge that a competitor wishes to conceal. It concerns the actions of gathering and analyzing intelligence about competitors and their products to support commercial decision-making for competitive advantage.

GEOINT has several possible roles in CI. It can include locating a competitor's facilities and determining their plant capacity, use rate, efficiency, product mix, capital investment, hours of operation, and identifying plant suppliers and logistics support. It can be used to identify the most threatening competitors by looking at their business locations in relation to your own, and the geographical relationship of both to customers.[32]

The geospatial intelligence tools available for CI vary widely. As in national intelligence, open source and especially internet searches can be the major contributor. Social GEOINT, the subject of chapter 15, provides a rich source of information. Monitoring sites such as Facebook, LinkedIn, Instagram, and Twitter can yield information on geospatial patterns that have competitive significance. Geospatial information that is openly available by international regulation is another source: the automated identification system described in chapter 13 can be used to track a competitor's ships and thereby identify trade patterns.

The reconnaissance and surveillance tools used in government GEOINT may also be useful. For example, drones can obtain detailed images of a competitor's factory. These types of intelligence tools, including drones, are legal in the US, but their use in competitive intelligence is prohibited by some countries. Satellite imagery is widely available and is routinely exploited to assess a competitor's construction, mining, or factory production operations. And CI professionals are borrowing a trick that national intelligence analysts have long used: counting autos in an installation's parking lot as an indicator of activities being conducted there. Counting autos observed in satellite imagery of a clothing retailer's parking lots has allowed predicting the retailer's likely income, well before actual numbers were available.[33]

The terms "deep web" and "dark web" are often used interchangeably. They are technically differentiated, but both are based on the concept of privacy. The deep web refers to the expansive part of the internet that is not indexed and therefore not normally visible or accessible from typical search engines. The dark web emphasizes anonymity. Access-restricted commercial databases, websites, and services make up much of the deep web, but they can be searched (again, not with conventional search engines). And they contain geospatial references that can be used in competitive intelligence. Companies offer services that include deep web searches to support competitive intelligence. For example, one CI company explains how it can be done:

> Flightaware will give you access to all the flight plans or a specific aircraft. The site was built to allow you to track commercial flights, but you can also enter any tail number and track a private aircraft or a helicopter (of course, you'd have to know that tail number—check the Federal Administration Registry by searching by company name to get those). This is certainly useful when tracking a flight you are taking, but some of our clients have used it to track competitors—a mining company tracked helicopters to check prospections, a corporate office tracked the corporate jet of a competitor to anticipate mergers and acquisitions.[34]

Watchers of Amazon's second headquarters decision-making were able to do the same thing by observing Amazon CEO Jeff Bezos' corporate jet flights in 2018. The *Puget Sound Business Journal* identified Boston and Washington as the most likely winners. A career site for tech professionals added New York to the list, based on the geographical distribution of its members.[35] Two of those cities, of course, wound up as winners—at least, temporarily.

Chapter Summary

One of the three threads running through the history of GEOINT cited at the beginning of this book is commerce. Geospatial intelligence in business predates even the development of the original Silk Road. And the same tools that facilitated the GEOINT explosion within governments are now being applied to create a commercial revolution as well.

Geospatial business intelligence often begins with a decision on where to locate your company. The PMESII perspectives discussed in chapter 5 all can play an important role in that decision. Next in line are the other geospatial factors to consider in strategic planning and in daily operations. While businesses have a long history of doing those things, in the past it was painstaking work often done by reading hard copy materials and by making repeated trips to scout locations, demographics, and overall viability. Now commercial concerns can draw on a volume of online material, even employing crowdsourcing, to make sophisticated projections based on all sorts of variables in near real time, and to react quickly to opportunities and threats.

This is true for firms not just in planning their own operations, either; observing another firm's geospatial activities allows companies to gain a substantial competitive edge. Open source, in the form of social media, is now a valuable source for competitive intelligence, at least as much so as in government intelligence.

Notes

1. Keith J. Masback, "Geospatial Intelligence: The Totally Made-Up Term That's Changing the World," *Space News*, July 24, 2018, https://spacenews.com/op-ed-geospatial-intelligence-the-totally-made-up-term-thats-changing-the-world/.
2. "What Is Geospatial Business Intelligence?" 3dlaser mapping, April 28, 2016, www.3dlasermapping.com/blog-post/what-is-geospatial-business-intelligence/.
3. Thierry Badard and Etienne Dubé, "Enabling Geospatial Business Intelligence," *Technology Innovation Management Review*, September 2009, www.timreview.ca/article/289.
4. Carl Franklin and Paula Hane, "An Introduction to GIS: Linking Maps to Databases," *Database* 15, no. 2 (April 1992): 17–22.
5. Rob Pegoraro, "The Vanguard of Commercial GEOINT," *Trajectory*, January 31, 2018, http://trajectorymagazine.com/the-vanguard-commercial-geoint/.
6. ESRI, "Location Analytics: The Next Big Step in Business Analysis," *ArcNews*, Fall 2012, www.ESRI.com/news/arcnews/fall12articles/location-analytics-the-next-big-step-in-business-analysis.html.
7. FLO Analytics, "Retail Expansion Strategy: Natural Foods Chain," n.d., www.flo-analytics.com/case-studies/retail-expansion-strategy-natural-foods-chain/.
8. Hugo Moreno, "Location Intelligence: Mapping the Opportunities in the Data Landscape, *Forbes*, January 19, 2017, www.forbes.com/sites/forbesinsights/2017/01/19/location-intelligence-mapping-the-opportunities-in-the-data-landscape/#794034441bc6.
9. ESRI, "Location Strategy for Business," n.d., www.ESRI.com/location-strategy.
10. FLO Analytics, "Retail Expansion Strategy."
11. Monica Nickelsburg, "Amazon Announces HQ2 Cities, Splitting Second Headquarters as Extraordinary Contest Concludes," *GeekWire*, November 13, 2018, www.geekwire.com/2018/amazon-announces-hq2-locations-dividing-second-headquarters-extraordinary-contest-comes-close/.
12. Irina Ivanova, "Amazon Cancels Plans for New York City HQ2," CBS News, February 14, 2019, www.cbsnews.com/news/amazon-long-island-city-amazon-cancels-plans-for-new-york-city-hq2-today-2019-02-14-live-updates-breaking-news/.
13. Nickelsburg, "Amazon Announces HQ2 Cities."
14. Faraz Haider, "The Most Dangerous Cities in the US," *WorldAtlas*, July 26, 2019, www.worldatlas.com/articles/most-dangerous-cities-in-the-united-states.html.
15. Henry Grabar, "Your City Will Lose the Contest for Amazon's New HQ," *Slate*, September 8, 2017, https://slate.com/business/2017/09/your-city-will-lose-the-contest-for-amazons-new-hq.html.
16. Sissi Cao, "Jeff Bezos' Private Jet Records Suggest Winners of Amazon HQ2," *Observer*, April 2018, https://observer.com/2018/04/jeff-bezos-private-jet-records-suggest-amazon-hq2-winners/.
17. Pitney Bowes, "The Future of Location Intelligence: An Expert's Outlook," n.d., www.pitneybowes.com/us/location-intelligence/case-studies/the-future-of-location-intelligence—an-expert-s-outlook.html.
18. Moreno, "Location Intelligence."
19. Julio Ochoa, "Geographic Information Systems (GIS) and the Insurance Industry," April 4, 2017, http://blog.julio8a.com/gis-and-the-insurance-industry/.
20. Ochoa.
21. Dan Morgan, "The Shadowy World of Grain Trade," *Washington Post*, June 10, 1979.
22. US Department of Agriculture, "Exporter's Profits on Sales of Wheat to Russia," February 12, 1974, http://archive.gao.gov/f0302/096760.pdf.

23. Bob Porter, *Have Clearance, Will Travel* (Bloomington, IN: iUniverse, 2008), 34.
24. CIA Intelligence Memorandum, "USSR: A Third Consecutive Crop Failure," August 1981, www.foia.cia.gov/sites/default/files/document_conversions/89801/DOC_0000498196.pdf.
25. Nevada Small Business.Com, "What Startups Can Learn from Uber about Marketing," n.d., https://nevadasmallbusiness.com/how-start-ups-can-learn-from-ubers-marketing-strategies/.
26. Hugo Moreno , "The Where Factor: Location Intelligence and the Competitive Edge," *Forbes*, April 22, 2015, www.forbes.com/sites/forbesinsights/2015/04/22/the-where-factor-location-intelligence-and-the-competitive-edge/#7178a8814cfb.
27. Sarah K. White, "What Is Geofencing? Putting Location to Work," *CIO*, November 1, 2017, www.cio.com/article/2383123/mobile/geofencing-explained.html.
28. Matt Alderton, "Inside Game," *Trajectory*, November 2, 2015, http/trajectorymagazine.com/inside-game/.
29. White, "What Is Geofencing?"
30. ESRI, "Insurance Risk Management Demo," n.d., http://webapps-cdn.ESRI.com/Apps/StepByStep/.
31. Interview with Wilmington, NC, GEICO local agency owner Patrick Punzalan, October 10, 2019.
32. Christophe Othenin-Girard, Claude Caron, and Manon G. Guillemette, "When Competitive Intelligence Meets Geospatial Intelligence," in *Proceedings of the 44th Hawaii International Conference on System Sciences, 2011*, https://pdfs.semanticscholar.org/f437/c39d065c1198388a1e3c0f2bdc2f4947f457.pdf?_ga=2.159277818.1524240487.1570796194-1550076764.1570284265.
33. Grant Burningham, "How Satellite Imaging Will Revolutionize Everything from Stock Picking to Farming," *Newsweek*, September 8, 2016, www.newsweek.com/2016/09/16/why-satellite-imaging-next-big-thing-496443.html.
34. Estelle Metayer, "Tapping into the Invisible Web for Competitive Intelligence," *Competia*, November 9, 2010, www.competia.com/tapping-into-the-invisible-web-for-competitive-intelligence/.
35. Cao, "Jeff Bezos' Private Jet Records."

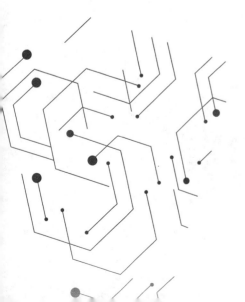

The Road Ahead

In 1949, George Orwell published *1984*—his dystopian picture of a future characterized by mass surveillance and control of human behavior by repressive governments. One of the book's key features was a world where television cameras watched everyone—in their homes, on the streets, at work—where news was censored, and where thoughts were controlled. The book, always popular, remains a staple taught in US high school classes today.

Dystopian futures are, of course, all the rage in books and media productions, and for good reason. They excite and engage the audience. A movie that details nothing other than a sunny future is a guaranteed bust. In fact, most measures of the human condition (income, education, freedom, literacy, and health) have markedly improved during the last 200 years, despite a sevenfold increase in the world's population. But surveys indicate that only a small fraction of people around the world (6 percent in the United States) seem to think that the world is getting better.[1]

Sixty years after Orwell's book, there's both bad and good news. Many of the surveillance technologies the author envisioned have become part of daily life. And some countries are making use of them to control dissidents. But many other tools are making life difficult for dictators. There are already numerous ways to defeat censorship and to organize against repression and governmental corruption, with new ones continuing to come online. Social media was used to great effect during the Arab Spring that began with mass protests in Tunisia in 2010, spreading across the Middle East and toppling governments. In 2019, despite the Venezuelan government's best efforts to censor both mass media and social media, protest organizers found ways to share news about what was happening where within the country.

The tools of geospatial intelligence can be used for good or ill. The same technologies used by nongovernmental organizations to save lives—satellite imagery, cell phone geolocation, social media, and others—also are used by drug and human-trafficking organizations, pirates, and illicit arms dealers to cause harm for profit. But projecting how existing tools are likely to be used, individually or in new combinations, for either purpose is difficult and over the long term tends to err on the side of conservatism. With this caution, let's give it a try anyway.

Predicting the Future

The Danish physicist Niels Bohr once reportedly observed that "it's difficult to make predictions, especially about the future." And predictions by experts in many fields over many decades have completely missed the mark. If businesses could with some confidence identify important developments in technology and anticipate their impact, Xerox would likely dominate the hardware side of the personal computer business (its research group developed the graphical user interface that Apple and then Microsoft adopted). A company called Digital Research Intergalactic would probably dominate the software side (it had the operating system that IBM wanted for a personal computer, before switching to Microsoft). Steve Jobs and Bill Gates might be minor footnotes in the history of computing. And, most cameras, including the one in your smartphone, would probably be built by Kodak. There is a reason that the intelligence business has moved away from the term "prediction" in favor of "anticipatory." Nevertheless, this chapter attempts to address some of the emerging developments and potential future trends in geospatial intelligence.

The preface noted the three drivers on the road to GEOINT: war, commerce, and taxes. For much of the last century, war (with its need for military intelligence) was the primary driver. Commerce arguably is the current major driving force.

For individuals, key factors shaping the future of GEOINT are the forces (basically, the needs), and the enabling technologies. The psychologist Abraham Maslow's theory (put forth in 1943), explains that human needs form a hierarchy, as illustrated in figure 20.1. From the earliest times, geospatial knowledge was important for the basic survival needs identified as physiological (i.e., food and shelter) and safety (or personal security). Those continue to be the priority in parts of the world ravaged by war and famine, and people continue to migrate to regions that offer better prospects for food, shelter, and safety. On a fundamental level, those factors continue also to stimulate geospatial information development even in stable, developed regions—such as applications for avoiding dangerous areas, for locating food, and for finding a place to sleep.

The higher level needs on the hierarchy—social belonging, esteem, and self-actualization—have resulted in an abundance of geospatially focused social-networking applications. Though there are giants such as Facebook and Twitter that dominate, the smartphone has thousands of applications aimed at the self-actualization need. This trend is unlikely to disappear.

The Road Ahead

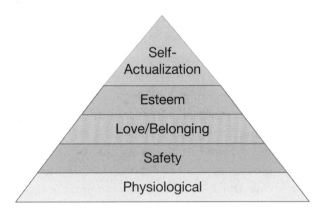

FIGURE 20.1 Maslow's hierarchy of needs. *Source:* FireflySixtySeven, Wikimedia Commons.

Maslow never intended his framework to be applicable to governments. And at first glance, it might seem an odd fit. But governments long have been formed based on the promise that they would provide basic physiological and safety needs for the governed. Once those requirements are met, democratic governments usually attempt to address the higher needs of their constituents—educational and recreational, for example. Furthermore, survival is a basic priority for the government itself. At a higher level, all government leaders crave esteem—to be respected. And at the top of the hierarchy, self-actualization could be thought of as control of its territory for many if not all governments. So governments naturally want the knowledge that supports control—driving the major trend to assemble and integrate geospatial data from more sources, with emphasis on time-stamped geospatial data—because temporal patterns are essential for prediction.

This book began with a brief overview of cartography's storied history. But cartography also appears to have a bright future—though certainly not one that Eratosthenes or Ptolemy would recognize.

The Future of Cartography

There is an entertaining video making the rounds on the internet where an old rotary telephone is put in front of ten-year olds; viewers then watch the various ways that they try and fail to operate it. It is easy to envision a time when a paper map stymies children, too. During the latter part of the twentieth century, of course, paper maps were largely displaced by digital map displays. Both maps and digital displays provide information. But those still display mostly static information. It's possible now to immerse yourself in the map and observe detailed changes over time within it—with confidence that what you are observing is "ground truth." Two major trends already in progress are dynamic cartography and virtual reality, both providing an exquisite level of geospatial detail.

Dynamic Cartography

Dynamic cartography—applying graphic techniques to inject life and motion into maps—combines geospatial information from all possible sources to create and update digital maps in real time. Topography, human geography, imagery products, and details about building interiors all are layered throughout the data so that users can interact with the display using data analytics. One company describes its dynamic cartography product as having change detection embedded in it.[2] The trend is to provide even more details, aided by surveillance technologies and crowdsourcing. Dynamic cartography can provide detail beyond just where individuals live and work, all the way to where they are at the moment.

Virtual Reality

Another approach is to dispense entirely with map displays and represent the world as we visualize it every day—that is, to apply the techniques of virtual reality that were introduced in chapter 6. According to traditional definitions, this isn't cartography; there is no true map involved. In virtual reality, we create a computer image of the real world. Closely akin to it is augmented reality, whereby we place an image on a picture of the real world. (Pokémon GO, also covered in chapter 6, is an example of augmented reality.) The two ideas, along with three-dimensional visualization, have been described as "the next big thing in the GIS industry."[3]

In 2003, San Francisco–based Linden Lab introduced an online virtual world named Second Life. By 2013, Second Life had approximately 1 million regular users. It resembles massive multiplayer online role-playing games, except that it has no script to be followed and no objective. It just presents a virtual world that individuals (represented by avatars) can enter and act within.

It's not difficult to envision a Second Life in the real world. Your avatar might go into a supermarket or building supply store to pick out the items you want from an actual video, place them in a virtual cart, and have them delivered in reality. And if your friends happen to be using the same app to shop at the same store, you can interact with them socially, without having to think twice about not having showered or brushed your teeth. Your avatar, after all, is always perfectly dressed if you wish him or her to be.

The combination of dynamic cartography and virtual reality has a range of potential geospatial intelligence applications. They could be used together, for example, to monitor compliance with nonproliferation treaties by an "on-site" inspector who moves virtually around a nuclear reactor facility. On a tactical level, in urban military operations, it would be a clear advantage to conduct a virtual tour of an area before actually moving troops into it.

Crowdsourcing

Chapter 15 discussed the crowdsourcing of geospatial information. The trend to combine crowdsourced data with official data, as is being done in the Hootenanny project and in dynamic cartography, will continue. On a broader scale, we're likely to find new ways to use collective GEOINT. Crowdsourcing still has much room for growth, and it will be exploited in new applications that are now difficult to foresee. The key issue still is the credibility of the data obtained. We need to find ways to better curate the material that goes into a collective cartographic product. And the solution to this problem may lie in the blockchain—one of the tools we'll turn to next.

The Tools

Sensors are one big part of the future. The technology just keeps getting better, providing smaller and more capable sensors and those that will make use of new sensing methods. But they're not the only part. The communications essential to support sensors keep improving as well. And the analytic power provided by software continues to make new applications possible. Improved display technology allows analysts to visualize even more complex results of their analysis.

Blockchains

A blockchain is a technology that allows the creation of a digital ledger of transactions or events that is then shared among participants in a distributed network. It relies on cryptography to allow each participant in the network to enter transactions on the ledger securely without the control of a central authority. Once software is deployed on a blockchain, programs run automatically and are accessible to any of the participants. This design makes them basically autonomous and uncontrollable by governments.

Blockchains are best known as the technology underpinning Bitcoin cryptocurrency, as that was the initial use of the term. Commercial enterprises are now using blockchains to make and verify transactions on networks instantaneously. Once an item is entered into a blockchain ledger, it is very difficult to remove or change. If a participant wants to make a change, others in the network have to independently verify it. If a majority of participants agree that a transaction is valid, then it will be approved and a new block will be added to the chain. It creates a record of "truth" that can't be changed unless the change is agreed to by the rest of the network. Blockchain technology allows transactions between people, none of whom trust the others. The key feature is that a majority of the entire network must confirm the validity of each transaction.[4]

A central approving authority—the vulnerable point in previous online transactions—doesn't exist.

So how do blockchains fit with geospatial intelligence? The history associated with a location is fairly easy to blockchain. And it's critically important to get the history of land transactions right. So blockchain-based land registries are beginning to appear and will likely proliferate.[5]

The current information about a location poses a more difficult problem, because it has to be asserted and then agreed to by independent sources. But you can establish levels of trust for different sources—assigning higher trust to government or academic sources, for example. And you can combine inputs from unchecked sources with current data from IoTs and UAVs, for example.

Blockchains also enable secure data sharing to present a common geospatial picture for both governmental and business intelligence uses. Within the US government, projects already are under way to apply blockchains geospatially:

- The Food and Drug Administration is testing its use in tracking pharmaceutical supply chains.
- The State Department started a pilot project in 2018 for a blockchain registry of international workers, the goal being to prevent forced labor.
- The Centers for Disease Control and Prevention in 2018 tested a blockchain system for tracking public health issues.
- The Department of Homeland Security is examining blockchain technology for tracing the origins of raw material imports.[6]

Finally, blockchains with an associated geospatial tag (known as location-enabled blockchains) are likely to appear in the future to deal with criminal use of the blockchain. Cryptocurrencies such as Bitcoin, for example, are used by criminals to anonymously extract ransom for kidnappings, for ransomware attacks, and as payment for illicit transactions conducted using the dark web. But if the currency's digital wallets are geotagged, then the likelihood of identifying a perpetrator goes up dramatically.[7]

The Internet of Things

The internet of things (IoT) refers to a network of physical objects—devices, appliances, vehicles, buildings, and other items. These objects have embedded electronics, software, and sensors; and because they are connected to the internet, they are able to collect and exchange data. And they can be sensed and controlled remotely via the internet.

Experts estimate that 8.4 billion such "things" were connected to the internet in 2017, with the number expected to reach 20.4 billion objects by 2020.[8] They include a wide class of devices and systems, ranging from garage door openers to devices that power smart homes. On a larger scale, the IoT enables intelligent transportation systems and smart cities. Each "thing" carries a unique identifier that is recognized by the internet. These network-connected devices automatically collect and exchange data, allowing people, enterprises, and governments to be more efficient and productive.

We now have smart meters in our homes. We have them on streetlights. We have traffic sensors on the road. They all generate data that include locational information. The most common, however, are the sensors that observe scenes and provide both location information and tracking of objects and people. Video surveillance cameras are common in stores, banks, homes, and government installations, monitoring activity both within and without. All these data are being collected and made available for analysis, thanks to the IoT.[9]

Why is this important? Because spatial analytics depends on having accurate location information, and lots of it. The more completely and accurately that objects and people can be located and tracked, the better the analysis that can be done.[10] And better analysis includes the ability to predict the future locations and actions of the objects and people being tracked.

The potential of IoT to support spatial analytics has not been lost on military forces. It is included in the design of the US Army's Nett Warrior system, which provides detailed situational awareness to combat units from a network of sensors—including, of course, the IoT. Military leaders are eyeing ways to include tapping into civilian IoT devices to assist their forces in the most difficult of modern combat environments: urban warfare.[11]

IoT devices are installed by companies to optimize logistics and to track the locations of equipment—and, of course, for building security. Self-driving automobiles belong to the general class of IoT as well. The cars will rely heavily on mobile creation of and access to real-time information among people, objects, and the environment surrounding them. This requires access to internet bandwidth on a scale not currently available.

The IoT has many benefits but carries the risk of misuse when someone gains unauthorized access. One recent addition to the IoT is of special concern to privacy advocates. These are the voice-activated personal assistants that connect to the internet—Google Home and Amazon Echo, among others. The devices must continually listen for their users to ask a question or issue a command ("Alexa, play some Elvis songs"). And if hackers can get into the personal assistant—(which isn't that difficult, and has already happened with IoT baby monitors), depending on the user's ability to enforce data security—they can listen in on private conversations.[12]

Hacking a large-scale IoT isn't just a possibility; it happens routinely. In 2017, a Moscow-based site was livestreaming video from over 500 security cameras in Canada. The cameras were focused on homes, industrial workshops, parking lots, and daycare facilities, among others. The hackers relied on the cameras having been installed with default passwords unchanged.[13]

Personal Sensors

Personal sensors already are used to keep track of fitness-related activity, sleep patterns, and current health statistics, including measuring vital signs. They will only get better. For example, they could potentially also sense the environment around you and correlate that to your health readings. Think of the way a self-driving vehicle can sense what's in the immediate area. We mostly don't hang around in dangerous areas such as hazardous chemical or ionizing radiation environments, but we occasionally are subject to pollens or other unhealthy airborne conditions (smog and cigarette smoke). And the history of your interaction with the environment at various locales can tell much to a physician.

Personal, wearable air sensors are already available. Tied to a location sensor, they give a continuous record of the air quality you are breathing in a given area. Their real power comes when they are shared via crowdsourcing. They can then provide real-time data on temperature, carbon dioxide, nitrogen dioxide, and particulate levels in the air, and can even detect and geolocate toxic chemical leaks.[14] We currently rely on government-provided environmental monitoring stations to warn of airborne hazards, but that doesn't give an accurate picture to someone who's living close to an interstate highway or near an industrial facility.

Remote-Sensing Technologies

Chapters 8 and 9 described remote-sensing platforms, followed by a discussion of sensors in chapters 10 through 12. Drones appear to be the future in aerial platforms: they're comparatively inexpensive to buy and operate, and they can carry a wide range of sensors. And though single drones currently dominate, drone networks offer a wider area of coverage and the potential for using a variety of sensor types.

Unmanned aerial vehicles (UAVs) now are cheap, widely proliferating, and equipped with high-quality sensors. Depending on whom you believe, the commercial and military drone markets are expected to reach anywhere between $23 billion to over $100 billion annually by 2022, up from just $6 billion in 2014.[15]

The technology used in drones is advancing, and that will drive the availability of drones with new uses and more sophisticated capabilities. For example, most commercially available drones today operate within line-of-sight of their controllers, because they need to maintain a command link to someone on the ground. But as they gain payload capability and are able to operate at longer ranges, they will be able to operate well beyond line-of-sight. That implies either flying a preprogrammed course and returning with stored video, possibly using artificial intelligence to search for something of interest during the flight, or possibly sending a second drone aloft as a communications relay to the distant drone.

Drone-equipped video surveillance cameras already are used for monitoring activity around many civilian and government facilities worldwide. They produce a vast amount of video that has to be transmitted, stored, visualized, and analyzed in real time. And when equipped with multiple cameras, like the Gorgon Stare drones described in chapter 10, they can offer both wide area video coverage and high spatial resolution.

One obvious development is smarter drones that can sort through the massive video intake, and select only video showing something of interest to the drone operator—an intruder in a secured area, for example. Such UAVs increasingly are able to react to observations of interest, moving to obtain more detail, calling in additional drone support, and switching from reconnaissance to surveillance.

Another advance enabled by smart drones is the use of drone swarms. This is more than a simple matter of launching many drones—something that is not uncommon now. What makes a group into a swarm is its ability to coordinate behavior. The capability was developed from studies of the way insects such as ants exhibit swarm behavior in nature. Drones are already flying that use artificial swarm intelligence relying on communications between the individual drones and on their interactions with the environment. Individual drones in the swarm position themselves autonomously, flying in formation without being told explicitly where to go. The drone operator manages the swarm instead of controlling any one drone.[16]

While drones are proliferating, so are satellites—very capable smallsats, in particular. We are moving toward a world of satellite swarms, put in place by companies such as Planet Labs and EarthNow (discussed in chapter 9) that can provide good-quality imagery of every point on Earth every day—something that is beyond what drone swarms are likely to do. It will take machine learning to handle the volume of imagery produced, but the impact in areas such as business intelligence, competitive intelligence, and agriculture will be huge. For example, companies will certainly find ways to use knowledge of the numbers and types of merchant ships in every world port, or the locations and usage of every oil storage tank worldwide. Farmers and commodity traders alike will find uses for detailed information about the condition of crops and need for irrigation or fertilizer on every farm.[17]

Spectral sensing is another remote-sensing technology with a bright future. Hyperspectral sensing is currently used for applications such as identifying chemical emissions, pollution, agricultural problems, and invasive plant species.

However, ultraspectral sensing appears to be the new frontier. It promises to provide insights about a target that hyperspectral imagery cannot, though its potential is still being investigated. But it requires a more complex imaging spectrometer and, more importantly, trained analysts with sufficient resources to extract knowledge from the vast information content in the imagery. Nevertheless, both UAVs and satellites are likely to carry ultraspectral imagers in the future.

Autonomous Vehicles—and More

The self-driving car receives frequent and global media attention. Less well known is the progress on autonomous farm vehicles, heavy trucks, ships, and even airplanes. But in 2019, autonomous ships were being designed in several countries, and autonomous passenger-carrying electric-powered aircraft were in testing.

All of these vehicles require extensive geospatial information for navigation, and knowledge of immediate surroundings for collision avoidance. This means that they carry an array of sensors—radar, sonar, LIDAR, GPS, and units that keep track of movements and acceleration. And, of course, artificial intelligence to make sense of it all, and communication with some controlling or emergency backup center. This means that we can see a new set of mobile sensors gradually becoming available to add to the data banks that GEOINT analysts draw on.

Wireless Communication

Several advantages have driven the development of wireless communication. It is inexpensive and quickly deployed, compared with either cable or satellite communication. The receivers can be located almost anywhere, without concern about wiring issues. IoT-type sensors can be battery powered, allowing them to be quickly emplaced. The early problems with wireless technology—slower speeds, interference, and vulnerability to hacking—are being solved. The most recent development, the fifth-generation wireless systems, abbreviated 5G, is a major advancement on those fronts.

The 5G systems represent a set of improved wireless network technologies that began to deploy in 2018. The major changes are that, first, 5G relies on a denser network, with each cell covering a smaller area, and, second, the cells operate at higher frequencies. The result is that each cell has a range more like that of a home router rather than a traditional cell phone tower.[18]

The redesign means that 5G has several advantages in comparison with existing 4G networks: more than ten times greater speed, with the ability to connect many more smart devices and sensors simultaneously. The result is to enable small, inexpensive, low-power devices to connect to the network.[19]

Smart cars can become even smarter because they will be able to instantly connect with everything else on the road. As they proliferate, these networks promise to open new dimensions in geospatial intelligence—for example, supporting communication between vehicles and between vehicles and infrastructures. Public safety organizations will have more access to geospatial information in video and data forms, for example.

Agent-Based Modeling

Chapter 18 introduced the concept of an agent-based model for disease prediction. In its most general form, **agent-based modeling** is a type of simulation that is applied in the study of complex adaptive systems. This usually means that humans are part of the system, since we are both complex and adaptive. In one GEOINT context, a specific simulation would model the actions of individual animals (humans included) and discrete objects in time and space. The advantage of agent-based modeling for GEOINT is that the models not only identify patterns; they also are used to explore the *causes* of those patterns.[20] And it is especially effective when paired with a GIS display.

According to ESRI's Kevin Johnston, agent-based modeling in a GEOINT context allows us to address a wide array of problems, such as:

- developing corridor connectivity networks for wildlife movement;
- anticipating potential terrorist attacks;
- analyzing traffic congestion or producing evacuation strategies;
- planning for the potential spread of disease, such as bird or swine flu;
- understanding land use change;
- optimizing timber tract cutting;
- exploring energy flow on electrical networks; and
- performing crime analysis to deter future impact.[21]

Simulations on Demand

Chapter 14 discussed geospatial simulations. They're used extensively by businesses, law enforcement, military forces, and government organizations generally. On any type of smart media, you now can get snow, flood, and hurricane damage forecasts, and traffic and election result predictions, all based on simulations. We get the product of a simulation routinely in mapping routes on Google Maps and similar apps.

Online games use simulation heavily, so why not employ it for useful applications? Those could easily be made accessible to individuals who would customize them to their specific interests and locations. They're in fact increasingly available for businesses and individuals, and a richer variety of uses are likely.

A fairly obvious application would provide predictive warning about any type of threat affecting individuals. Simulations such as those for cholera or malaria outbreaks, discussed in chapter 18, could readily be available to individuals, and give results that are specific to a user's location.

Simulations, combined with the technologies for handling big data, have the power to dramatically change the way that geospatial intelligence is done within intelligence communities. The National Reconnaissance Office has a program that appears to be leading the way; it is an analytic tool called Sentient. According to an official, Sentient "ingests high volumes of data and processes it; ... catalogs normal patterns, detects anomalies, and helps forecast and model adversaries' potential courses of action."[22] Based on these types of forecasts and models, Sentient would then train satellite-based collection sensors on the targets of interest to monitor events as they are happening.

Pulling It All Together

At the US Geospatial Intelligence Foundation's GEOINT 2017 symposium, National Geospatial-Intelligence Agency director Robert Cardillo estimated that, if his organization continued to manually analyze imagery, it would require 8 million analysts to look at all the industry's imagery generated in the next decade.[23]

That, of course, won't happen—just as it did not happen with the telephone. In 1900, the exponential growth of telephones led to estimates that by 1920 the entire US population would be working as telephone operators. But automated switching (with the dial phone and later the push-button phone) managed to handle the increasing volume (in the process, ironically, turning all telephone users into "operators").

A similar process is at work in geospatial intelligence today, combining the tools of

- both remote sensing and close-in or in situ sensing (e.g., personal sensors, IoT, and autonomous vehicle sensors);
- artificial intelligence or machine learning;
- wireless communication and networking technologies;
- the visualization and sourcing capabilities described in the preceding section, "The Future of Cartography"; and
- the modeling and simulation systems now being fielded to perform a different type of geospatial intelligence, whereby an issue of intelligence significance can be identified and located and difficult questions about it can be answered in real time. The result of connecting directly to sensors allows a user to observe and analyze a situation as it is evolving.

The key to success happens in the "back end" of the process. Geospatial information can be collected and communicated to a data center, but then something must be done with it. The tools that will do that "something" have many names—machine learning, artificial intelligence, and deep learning being three popular ones. A suite of these that can manage and analyze the volume of information is essential. Chapter 14 discussed some of those that have shaped and will continue to shape GEOINT.

One tool that likely will have the biggest impact in the future actually has been around for over six decades. It developed as an academic discipline called artificial intelligence (AI) in 1957, and since then it has seen several alternating waves of optimism and disappointment. In consequence, other terms—such as, more recently, machine learning or deep learning—have been applied to AI. There are differences between the three terms, but the distinctions matter mostly to information technology professionals. Whatever the name, we're talking about software that writes new software based on its experiences. It's something many of us encounter daily in the form of personal assistants on our smartphones, computers, or stand-alone devices. These basic AI applications (e.g., virtual personal assistants) learn our voice and habits over time. At a more sophisticated level, machine learning (the term we use here) finds patterns in large volumes of noisy data and, from repeated patterns that it observes, builds models to apply to new situations.[24]

Several companies already are applying machine learning to deal with the massive volume of satellite imagery. The algorithms train themselves to identify objects such as vehicles, aircraft, ships, structures, and animals based on repeated observations. Early results are promising: using commercial satellite imagery, deep-learning tools have been able to identify the locations of surface-to-air missile sites in a 90,000-square-kilometer area of China, taking 42 minutes to make a search that would have taken imagery analysts about 60 hours.[25] They show promise of being able to make tough predictions about, for example, likely growth in sales by a business at a specific location, agricultural productivity during the next growing season, and the rate of growth of major cities.[26]

Applications of GEOINT

Many GEOINT applications that are still seen as futuristic are already being applied but are poised for dramatic growth. They will need to deal with a major challenge that won't go away quickly: how to handle the massive amounts of geospatial information collected from all sources—for example, in building the "smart cities" discussed next. Centralized (cloud) real-time processing may not be up to the task, given the combination of storage, bandwidth, and computational power needed. Distributing the processing throughout the network may be necessary.[27]

Smart Cities

The idea of a smart city first appeared in the late 1990s, and it has caught on globally since then. Several large urban areas around the world have adopted a smart city strategy. As successes mount, the concept is likely to be applied in smaller cities, and eventually to smart villages in rural areas as well.

The basic concept is for city government to collect data from individuals, sensors, and IoT devices throughout the city, aggregate it, and analyze it in real time, with the intent of better managing virtually any aspect of public life. This could include managing transportation systems, utilities meters, public lighting, power plants, water supply networks, waste management, law enforcement, information systems, schools, libraries, hospitals, and other community services; tracking city assets; managing fault and disaster (fire, flood, and environmental) insurance; and traffic and parking management. Much of the focus has been on the health of the infrastructure, but it's a straightforward step to monitoring the city's economy and the health of its citizens. And while the information is collected, organized, and analyzed by the government, it is possible to share it with residents and visitors.

The tools described in chapter 14 and in this chapter have powered much of the smart cities trend: IoT, personal sensors, smartphones, and drones for collecting data, and GIS, data analytics, and visual analytics for organizing and analyzing the resulting big data. Major drivers of the next generation of smart cities are likely to include the 5G network, augmented reality, and machine learning.[28]

Each year, market researcher Juniper Research ranks the top smart cities in the world, based on their level of integrating technologies and services in four areas: productivity, mobility, health care, and public safety. In 2017, Singapore topped the list, followed by London and New York.[29]

There are many issues to be considered, but one stands out: Who controls access to the information that is collected? The collector. And control of information confers power. This is a major factor to watch in the case of Singapore, which is both a city and a country. Nominally a democracy, since 1959 the city has been ruled by the People's Action Party. The move to a smart city is likely to strengthen the Singapore government's ability to monitor the activities of its residents. Citizens are not protected against government surveillance, even in their homes. Doctors are required to submit patient information to a national health care database. The government has broad powers to limit citizens' rights and to stifle critics and activists. Citizens are guaranteed security, education, affordable housing, and health care. In exchange, they are expected to surrender many individual freedoms and civil liberties. Severe penalties are assessed for conduct such as chewing gum in public or littering. Internet usage is monitored, and media outlets must be registered with the government. The People's Action Party already has a history of using intimidation and legal means to ensure its control, and it has diverted infrastructure upgrades away from parts of the city that voted against it. So Singapore could become a poster child for how repressive governments can use smart city technologies to maintain political control. The point is reinforced by China, which already uses smart cameras to, for example, identify jaywalkers for police attention and to surveil ethnic minorities for detention and interrogation—evoking images of Orwell's *1984*.[30]

The smart city has other potential drawbacks. It relies on technologies that are vulnerable to hostile action in the form of a cyberattack against the sensors or wireless network or in collecting intelligence—for example, by hacking into surveillance cameras. Still, if it's worth the risks to have smart cities, why not have smart countries? (Of course, Singapore is really there already, being both a city and a country.) Technologically advanced countries already pull in geospatial data about crime, population, health, housing, and funds transfers. Countries having substantial international interests track movements of their citizens, commercial activity, trade, social unrest, and transnational criminal activity. Most, however, place the data in separate compartments. That separation is changing. The need to forecast developments or provide warnings drives data consolidation. France has been pushing such consolidation to deal with terrorist threats—sharing geospatial and other information among commercial entities, government agencies, and counterterrorism units.[31]

Business

Businesses have several needs that are now, and will be even more so in the future, addressed by geospatial intelligence. The obvious ones include what many larger commercial enterprises routinely do already: identify how existing business locations are performing, and where changes are occurring. They rely on inputs from IoT, location sensors, mobile devices, and social media to make such assessments. The tools allow businesses to collect georeferenced temporal data about almost any event or thing. They can track personal data, and the data can be accessed by businesses to profile customers and sell products.

The analytical tools for geospatial analysis have to be able to make sense of all the incoming data—to identify new business opportunities, cost savings, and potential partnerships, for example, or to provide warning of events such as changing political situations and weather that could affect business locations. Companies want to be able to, for example, know how hurricanes might affect their trade routes and warehouses.[32] Tools already exist to do these things, but they have considerable room for growth. One author notes that "in the near future, industries are expected to benefit from emerging technologies like geospatial intelligence fusion, crowdsourcing, human geography, visual analytics and forecasting. And of course there are new evolving technologies like machine learning and artificial intelligence algorithms, which can give feedback on the location and behavioural pattern of people to build models for better business intelligence."[33]

Health

There are areas in the world where people live longer, happier, and healthier lives than average. These places are characterized by several features, including

- lots of green spaces and sunshine;
- good air quality;
- access to whole foods and a corresponding absence of fast food establishments;
- ready availability of fitness centers, pools, and gyms; and
- a suburban existence, rather than either urban or rural life.

In consequence, life expectancy varies by location; in the United States, it varies by zip code.[34] Over time, fine-grained geospatial statistics may allow more precise measurements of ideal places to live. On a day-by-day basis, personal sensors will allow better measurements of the current health and exposure of individuals. Long term, the ability to track exactly where people have lived and places they have visited can be an input to diagnoses by medical staff.

On a larger scale, public health threats are getting better monitoring, especially on the issue of food and water contamination. In 2018, a total of 197 people in the US became ill as a result of eating romaine lettuce contaminated with E. coli bacteria. The specific grower, processor, and distributor could not be identified beyond locating the source to the farming region around Yuma, Arizona. So romaine lettuce quickly disappeared from supermarket shelves and restaurants, though most of it was uncontaminated.[35] Future tagging and tracking systems, probably using blockchain technology, are likely to allow more accuracy in identifying a source of contaminated food and removing from markets only the food from that source.

Climate Change

You read about it all the time. Climate change is happening now, and it will continue to happen. And the predictions about its effects are mostly dire. Geospatial intelligence has a role to play in both making and assessing these predictions. All the platforms, sensors, technologies, tools, and organizations described in this book will be involved in dealing with the effects of climate change in coming decades.

The primary effect, of course, is global warming. If the predictions hold up, global warming will be apocalyptic. But it won't be the apocalypse of a meteor strike, global thermonuclear war, or zombie invasion. It will take the form of a long-term, slow-moving apocalypse. Like most geological and geographical changes, it will take its time. A global temperature increase of between 2° and 3° C, which is expected to occur during this century, will have these effects—among others:

- **Migration.** The last two decades have been characterized by mass migrations within countries and across national boundaries, with consequent societal impact. Much of that movement was sparked by conflicts in regions such as Syria. Global warming has been estimated to have been one factor in such conflicts. But it is likely to become the driving force in future migrations, even in areas without conflicts. The World Bank in 2018 estimated that as many as 143 million people in areas near the equator would be forced to relocate within their countries by 2050.[36] It didn't estimate how many more would attempt transborder migration.
- **Flooding.** Over 90 percent of urban areas worldwide are coastal and thus are vulnerable to ocean rise. Venice may be the first city to go underwater. New York will get its share of higher water. Across the river, New Jersey's Meadowlands will become New Jersey's wetlands.[37] But even inland around much of the globe, flash flooding is more likely due to climate change.[38]
- **Water.** The combination of several factors—including pollution, climate change, and mismanagement—has created a steadily worsening water shortage worldwide. Globally, over 1 billion people now lack access to drinkable water. In India alone, more than 500 million people are estimated to be without access to drinking water by 2030.[39]
- **Agriculture.** The southern and southwestern United States will be hit with massive crop failures,[40] but the US Midwest will likely see a continued increase in yields for its primary crops, corn and soybeans—at least until 2050. After that, the future looks a lot grimmer, with models showing that yields will "fall off a cliff."[41] The future of agriculture looks even worse for Latin America and Africa.

Most scientific and environmental studies on the effects of climate change, as noted above, take a dire tone. But there will be winners and losers. Some areas will be especially hard hit (the Amazon River Basin, for example). The Arctic Ocean will become more ice-free, with a severe impact on its existing animal and plant life. But it, along with Siberia and northern Canada, will be more accessible for resource exploitation. While 76 percent of the world's population will have higher mortality rates, warming will save lives in northern regions.[42]

And, we humans are not dinosaurs. We have demonstrated an incredible ability to adapt to catastrophe, and technologies will undoubtedly be developed to help avoid, disrupt, or mitigate some of the effects listed above. Geoengineering, agricultural genetic engineering, and precision agriculture technologies are rapidly rising disciplines where engineers specialize in studying climate change. Let's take a brief moment to consider the three.

Geoengineering to deal with climate change effects will take many forms. Artificially generating clouds to cool parts of the planet is an option being investigated. Engineering glaciers to slow their melting is under consideration.[43] Capturing rainwater from monsoons in India could help abate that country's impending drinkable water shortage. All these options depend on geospatial intelligence, especially based on remote sensing, for projecting and measuring the effects of mitigation and adaptation. Current platforms and sensors are keeping pace and improving all the time. The Arctic is already receiving much attention; remote sensing is providing policymakers and environmentalists with extensive details on deforestation, coastal erosion, glacial changes, and new construction there.[44]

Agricultural genetic engineering is demonstrating the potential to develop new crops and animals that can thrive, or at least survive, in a hotter world.[45] Genetically engineered corals may wind up replacing the existing coral reefs that are being lost in warmer oceans.[46]

Precision agriculture technologies can improve the world food outlook, and, like the others, they depend on fine-grained geospatial intelligence—in particular, from remote sensing. In Africa, herders already rely on airborne and spaceborne sensors to identify sources of water for their herds; farmers use them to identify agricultural problems that can be addressed by fertilization or pruning.[47]

The Dark Side Applications

Many governments, organizations, and individuals will continue to apply the new and improved tools of geospatial intelligence, as they develop, for unjust ends and illicit profit. Both government and private misuse was noted in the introduction to this chapter. Lone-wolf hackers aren't likely to go away. And organized criminal groups and terrorists have developed considerable skill in the use of existing tools and—because their survival depends on it—have shown the ability to quickly adopt cutting-edge tools and methods as they become available.

Mexican drug-trafficking organizations have perhaps the most sophisticated geospatial intelligence operations of all nongovernmental organizations, for two good reasons: they need them, and they can afford them. In the case of the United States, a trafficker wants to maximize the likelihood that a shipment will get across the Mexican-US border and to stateside distributors. At the same time, the trafficker wants to keep the transportation costs as low as possible. Much planning goes into selecting routes that avoid towns on the Mexican side controlled either by other drug-trafficking organizations or by unfriendly town mayors. Then the geospatial planner has to identify the best border crossing points to use, and rendezvous points stateside. The planning must be done both spatially and temporally because US interdiction patterns change frequently.[48]

International arms traffickers have similar needs for detailed geospatial intelligence, many of which are the same as those of legitimate businesses. They have to plan transportation routes and handling facilities, such as remote landing strips and warehouses. They use advanced technologies to maintain current knowledge of where law enforcement is operating. Traffickers plan their distribution networks for suitable governmental environments—that is, for locations where regulatory and oversight systems are weak or where local officials can be either bribed or blackmailed. And they prefer regions that have a lax financial system so the large amounts of money moved by the trafficker will not be seen as suspicious.

National-Level GEOINT

On October 26, 2019, Daesh (ISIS) leader Abu Bakr al-Baghdadi found himself in a difficult position. A series of explosions and the rattle of gunfire earlier warned him that American forces had at last located him and were closing in. His compound in the Syrian village of Barisha, Idlib Province, was surrounded. After US Delta Force troops blew a hole in the compound walls, al-Baghdadi had fled into a tunnel under the compound, along with two small children. He could hear troops coming down the tunnel, ordering him to surrender. For him, that wasn't an option. With his children surrounding him, he triggered the detonator on his explosive vest.

The raid that killed al-Baghdadi had been long in the planning stage. It depended of course on GEOINT, including intensive coverage of eastern Syria with satellite and UAV imagery. But al-Baghdadi had seen what US imaging and SIGINT systems could do against his organization, and he had successfully eluded imagery and SIGINT searches for years.

His capture was a sterling example of intelligence fusion (of INTS and of international coalitions), which was introduced in chapter 13. HUMINT from several sources was key in this instance. US intelligence already knew from HUMINT that many Daesh troops had fled to Idlib province as their last holding in Syria collapsed. The wife of an al-Baghdadi aide and one of al-Baghdadi's couriers had been captured in Iraq earlier in 2019 and interrogated. They gave their interrogators names and locations—enough leads so that Iraqi and Kurdish intelligence officers could establish al-Baghdadi's pattern of travel. A critical source that the Kurds had recruited was a trusted confidant of al-Baghdadi.[49] With the help of these sources, along with satellite and UAV imagery, US intelligence began surveillance of the routes al-Baghdadi used and identified his movement pattern—a classic example of pattern-of-life analysis.[50]

The al-Baghdadi raid was an exemplar of GEOINT in a combat situation. But it also points the direction that all GEOINT is taking at the national level. During the second half of the twentieth century, much of the emphasis in national intelligence was on building the platforms and sensors to collect SIGINT and IMINT, to locate and assess targets of intelligence interest—the subjects of chapters 7 through 12. Countries will continue to build these platforms and sensors for collection to address the continuing demand described in chapter 13: a broad range of issues that national governments want to know more about. National GEOINT organizations will have to answer more questions in more detail. They will have to measure more things, monitor more trends, and identify more threats.

But, as chapter 19 pointed out, commercial sources now provide very good imagery from satellites and UAVs in a quality that almost matches that of the most advanced government collectors. Increasingly, those products provide continuous video of a scene and imaging in spectral bands. More IoT sensors are being installed every day, and some governments will tap into the ones installed in their territory. They can reach outside their territory as well; the 2017 incident where a Moscow-based site livestreamed video from security cameras in Canada illustrates how easy it is to obtain geospatial intelligence from IoT sensors. Finally, the internet (social media in particular) is an abundant source of geospatial intelligence.

With so many readily available sources, the emphasis in national GEOINT has to shift away from building more collection means. Newer and better collection no longer provides much of a competitive edge, because most countries have access to good-enough geospatial intelligence. The edge will go to the countries that can most efficiently identify and assess material in the massive body of data that is already available or will become available. An intelligence service has to be faster than others at tapping into *all* intelligence sources, traditional and nontraditional; identifying signatures and patterns of activity that are of intelligence interest; locating and tracking targets; and monitoring changes. So the tools for handling massive volumes of data, such as AI and machine learning, provide a decisive advantage.

The bottom line? The intelligence advantage goes to the country that can get there first. US intelligence, in concert with its allies in Syria and Iraq, demonstrated this advantage in the al-Baghdadi raid.

The Challenge of Ubiquitous GEOINT

Privacy is increasingly difficult to achieve. Governments, businesses, and individuals know where we live and quite a bit about us. They are able to learn where we are and what we're doing at the moment. There exists a trade-off between having all the conveniences offered by technology and tools such as geotagging—while protecting our privacy. We love the convenience and the enabling features of GEOINT. We are at best ambivalent about giving up our privacy for that convenience.

Yet for any measure, there usually is a countermeasure. To every offense there develops a defense. Several technologies already allow us to counter intrusive features of GEOINT. They'll become more accessible, and more will develop in the future. And they can be especially effective when employed by groups, instead of individuals, in a form of guerrilla warfare in cyberspace. Let's look at a few of them.

Video Surveillance

Chapters 9 and 10 described the ambitious plan of EarthNow to create a network of satellites that would provide continuous video surveillance of almost the entire Earth with 1 meter ground resolution. The concept appears to be achievable—if not by EarthNow, then eventually by others. The challenges that this degree of surveillance pose both to nefarious actors and to ordinary citizens also have to be considered eventually. Individuals may be more inclined, for example, to undertake certain activities only at nighttime or when clouds provide a protective cover.

Closer to home, video surveillance is becoming more sophisticated, thanks to the power of artificial intelligence. It is now possible to conduct searches of video for specific persons, objects, license plate numbers, or suspicious activities—from body cameras or fixed surveillance cameras. And the technologies to do this are rapidly improving while their expense is decreasing.[51]

Drone Proliferation

As drones have proliferated, they have become readily available to those with malicious intent: narcotics traffickers, terrorists, and industrial spies, among others. In the United States, the use of drones to observe and counter police operations is on the increase. During 2017, criminals used several such drones to observe and disrupt a hostage rescue operation conducted by the Federal Bureau of Investigation on the outskirts of an unnamed large US city.[52]

Close encounters between drones and commercial aircraft are increasing. After a video surfaced in February 2018 of a drone flying above a Frontier Airlines flight on approach to Las Vegas, pilot and air traffic control groups called on the US Congress to impose tight restrictions on drone operators.[53]

Not surprisingly, countermeasures have been developed to deal with the drone threat and are commercially available (though not in the United States). A 2018 report identified 230 antidrone products being sold by 155 manufacturers in 33 countries.[54] Typically, countermeasures rely on jamming the uplink signal that controls the drone, though some versions can take control of the drone and force it to land; others jam the GPS signal on which the drone depends. In response, drone operators interested in surveillance will likely be able to buy drones that can be preprogrammed to fly a specific course and return, with no uplink that can be jammed. Then the only defense will be a laser or kinetic takedown, for example, by drone-killing drones or by trained raptors such as eagles (which have been very effective at taking down errant drones). But counterdrone measures will have some difficulty dealing with drone swarms—except by hacking into the collective intelligence of the swarm.

But all these countermeasures are also available to the "bad guys," and so drone-based security systems will need to eventually deal with the threat of being blinded or otherwise taken out of action.

The proliferation of drones to deliver commercial packages and medical supplies also opens up an opportunity for intelligence gathering: they could be potentially used for clandestine reconnaissance over denied areas. Surveillance cameras and SIGINT receivers now can be very small and lightweight, and delivery drones have to carry heavy packages anyway. So it's possible to add the camera or receiver in the drone design, but the device's concealment or disguise would have to be very good to avoid detection.

Location History

Location history can have two different meanings, one useful and another that is both useful and problematic. In the future, the history of a place can readily be available at a tap or click of a smart device. That is already true of real estate. Go to any county website, and with one tap, you can determine who previously owned a dwelling, who owns it now, the price history of the property, its current valuation, and its tax records. Depending on the website, if it's a business, you can determine what it has for sale. You can find out the crop history of a tract of farmland. For a mine or an oil well, you can ascertain what has been extracted. That would be valuable in identifying the past use of a facility that would indicate the possibility of surface or groundwater pollution, for example.

Finally, location history can be applied on a larger geographical scale, aggregating it to produce a bigger picture of trends in commerce and the use of resources.

The other meaning of location history refers to the movement history of a person or mobile object. That's routine now for many vehicles. Trucking companies make wide use of radio frequency identification tags to keep track of their fleets. Movement history of individuals is recorded on smartphones: they can provide a detailed history of places frequently visited. You can take control of your smartphone's movement tracker and delete its history. But turning off location history in your account settings won't necessarily stop a history record from being created. Opening Google Maps, pulling up a weather update, or doing a browser search creates a history that is stored.[55] And cell phone towers retain location information about all mobile phones connecting through any tower. In the United States, law enforcement requires a warrant to obtain such location information, but that's not a requirement in most countries. Tracking a person's movements by other means—ubiquitous surveillance cameras, for example—remains legal in most countries.

And China has taken movement history of vehicles one step farther into the realm of real-time surveillance. Electric vehicle manufacturers selling in China are generally required to install in their cars GPS-equipped tracking devices that connect via internet to the manufacturer—something Tesla already does with all its vehicles, for example. But all such tracking data in China is also provided to Chinese government–backed monitoring centers that can then easily identify the vehicle and track its movements.[56] In a country such as China that has few protections on personal privacy, the potential for abuse is obvious.

Geofencing

Geofencing was introduced in chapter 19. It's used by businesses and individuals for a wide range of purposes. You can set up a geofence or obtain alerts based on ones that others have set up.

Geofencing already is used to track or clock employees, provide area security, and warn of intrusions. It can work with IoT devices and social networking to trigger actions or reminders based on a person's location. It can readily be set up around a residence to send reminders to occupants as they're leaving home or to warn of hazards in an area.

It is one of those tools that will continue to benefit from synergy with other tools. According to a 2017 press release, the geofencing industry is expected to grow over 27 percent by 2022. The article cites "technological advancements in use of spatial data and increasing applications in numerous industry verticals" as driving factors.[57]

Countering Surveillance

Much surveillance depends on IoT devices, and it is relatively easy to defeat individual devices that depend on wireless technology. IoT networks that make use of an extensive set of connected devices can introduce multiple points of vulnerability in the overall surveillance network.

Encryption and Authentication

Encryption has been around for millennia. But location-based encryption is relatively new. As its name suggests, it is a secure form of encryption that relies on geospatial information. A message recipient can decrypt the cipher text only when his or her smartphone or computer is in a specific location that the message sender selects. That is, the sender encodes a set of geographic coordinates into the encryption keys for added security.[58]

Location-based authentication is a similar concept. It relies on obtaining the GPS location of a smartphone user, for example, to confirm the identity of the user. Both, however, could be compromised by location-faking apps, discussed next.

Concealing and Faking Location

A recent trend that could gain popularity is location faking software for smartphones. Want to avoid geofencing? Make an employee monitoring system think you're in the factory instead of at the local Starbucks? Defeat a distance-based travel toll system? Or conceal a traffic violation from your insurer? Simply download an app to your smartphone and set it up. Then all you have to do is select where you wish to seem to be, and hit the option "simulate location."[59]

If that trend should spread, large numbers of people will be able to conceal their location and to control who gets to know where they are. Location details could gradually disappear from some data, and the locational data banks would have to deal with credibility problems.

It also can be desirable to hide your computer's internet protocol (IP) address, as it can be used to establish your location and to learn more about you. There are several reasons why you would want to hide your IP address. The obvious one is to protect your privacy. When you visit a website, its server makes a record of your IP address along with such information as the items you click on and the time you spend looking at those items. It uses that to build a profile of you. Another reason is to evade censorship or geographic restrictions. Many governments (e.g., Singapore) now block certain types of content, and some companies also make their websites unavailable in selected countries. Evading these geographic roadblocks isn't difficult if you can appear to be located in another country.[60]

But how to do that? There are several current choices. The deep web browser Tor enables you to conceal both your location and IP address. You also can set up a virtual private network (VPN) or rely on a proxy server (an intermediary server with a separate IP address). Then websites that you visit can see only the IP address of the VPN or proxy server, not your actual IP address.[61]

Defeating Blockchains

Much of the future work in assuring the reliability of geospatial data appears to depend on the use of blockchains. But blockchains, though difficult to hack, are not immune to it. Relatively unimaginative distributed-denial-of-service attacks and targeting cryptographic vulnerabilities can do the trick. But there are more subtle ways to attack a blockchain, if your purpose is to introduce deception.

Blockchains depend on the participation of large numbers of users to prevent hacks. But crowdsourced information about a particular location is unlikely to be a large set. If you can ensure that you control 51 percent of the crowdsourced inputs about a particular site, you can control the geospatial picture there. It's not just a theoretical possibility. In August 2017, a group of hackers called "51 Crew" attacked blockchain clones Shift and Krypton. The group took control of more than 51 percent of the network in an attempt to collect ransom, and it disabled both blockchains when the ransom wasn't paid.[62]

Chapter Summary

This chapter has attempted to touch on the future, and to identify some of the directions that GEOINT will take, along with a number of benefits and downsides of those directions. Undoubtedly, it will miss the mark in some areas. Once a field that evolved in an understandable and manageable fashion, geospatial intelligence now moves explosively. Governments, businesses, and individuals alike face the challenge of keeping up with it.

Notes

1. Steve Denning, "Why the World Is Getting Better and Why Hardly Anyone Knows It," *Forbes*, November 30, 2017, www.forbes.com/sites/stevedenning/2017/11/30/why-the-world-is-getting-better-why-hardly-anyone-knows-it/#15649b317826.
2. BAE Systems, "Dynamic Cartography Moves Maps into the Future," June 7, 2017, www.baesystems.com/en-us/article/dynamic-cartography-moves-maps-into-the-future.
3. Dominik Tarolli, "The Next Big Thing in the GIS Industry: 3D, AR and VR," *Geospatial World*, April 25, 2017, www.geospatialworld.net/article/technology/gis-3d-ar-and-vr/.
4. Steve Norton, "The CIO Explainer: What Is Blockchain," *Wall Street Journal*, February 2, 2016, http://blogs.wsj.com/cio/2016/02/02/cio-explainer-what-is-blockchain/.
5. Arup Dasgupta, "The Game Changer of Geospatial Systems—Blockchain," *Geospatial World*, September 2017, www.geospatialworld.net/article/blockchain-geospatial-systems/.
6. Matt Alderton, "Locating the Blockchain," *Trajectory*, no. 2, 2019, 27.
7. William Tewelow, "Bitcoin, Blockchain and GIS Could Change the World," *Geospatial Solutions*, March 14, 2018, http://geospatial-solutions.com/bitcoin-blockchain-and-gis-could-change-the-world/.
8. "Gartner Says 8.4 Billion Connect 'Things' Will Be in Use in 2017, Up 31 Percent from 2016," Gartner Group, February 7, 2017, www.gartner.com/newsroom/id/3598917.

9. Pitney Bowes, "The Future of Location Intelligence: An Expert's Outlook," www.pitneybowes.com/us/location-intelligence/case-studies/the-future-of-location-intelligence--an-expert-s-outlook.html.
10. Julian Ereth, "Geospatial Analytics in the Internet of Things," Eckerson Group, August 14, 2017, www.eckerson.com/articles/geospatial-analytics-in-the-internet-of-things.
11. Vivienne Machi, "Internet of Things to Provide Intelligence for Urban Warfare," *National Defense*, January 22, 2018, www.nationaldefensemagazine.org/articles/2018/1/22/internet-of-things-to-provide-intelligence-for-urban-warfare.
12. Bill Brenner, "Know the Risks of Amazon Alexa and Google Home," *nakedsecurity*, January 27, 2017, https://nakedsecurity.sophos.com/2017/01/27/data-privacy-day-know-the-risks-of-amazon-alexa-and-google-home/.
13. Patrick Cain, "Russian Site Streams Live Video of Canadian Living Rooms, Daycares," *Global News*, May 8, 2017, https://globalnews.ca/news/3434362/russian-site-streams-live-video-of-canadian-living-rooms-daycares/.
14. Richard Schiffman, "A Cheap Pollution Sensor Will Keep You Off the Dirtiest Roads," *NewScientist*, September 26, 2017, www.newscientist.com/article/2148503-a-cheap-pollution-sensor-will-keep-you-off-the-dirtiest-roads/.
15. Goldman Sachs, "Drones," 2015, www.goldmansachs.com/our-thinking/technology-driving-innovation/drones/.
16. Zachary Kallenborn, "The Era of the Drone Swarm Is Coming, and We Need to Be Ready for It," US Military Academy Modern War Institute, October 25, 2018, https://mwi.usma.edu/era-drone-swarm-coming-need-ready/.
17. Grant Burningham, "How Satellite Imaging Will Revolutionize Everything from Stock Picking to Farming," *Newsweek*, September 8, 2016, www.newsweek.com/2016/09/16/why-satellite-imaging-next-big-thing-496443.html.
18. Sascha Segan, "What Is 5G?" *PC Magazine*, March 20, 2018, www.pcmag.com/article/345387/what-is-5g.
19. Segan.
20. Matt Artz, "Causality in Time and Space: Agent-Based Modeling for GIS Users" (interview with ESRI's Kevin Johnson), March 5, 2013, https://blogs.ESRI.com/ESRI/ESRI-insider/2013/03/05/causality-in-time-and-space-agent-based-modeling-for-gis-users/.
21. Artz.
22. Sarah Scoles, "It's Sentient," *The Verge*, July 31, 2019, www.theverge.com/2019/7/31/20746926/sentient-national-reconnaissance-office-spy-satellites-artificial-intelligence-ai.
23. Marc Melviez, "The GEOINT Data They Are A-Changin'," *Trajectory*, May 23, 2018, http://trajectorymagazine.com/the-geoint-data-they-are-a-changin/.
24. "The Rise of Machine Learning (ML): How to Use Artificial Intelligence in GIS," *GISGeography*, May 15, 2018, https://gisgeography.com/deep-machine-learning-ml-artificial-intelligence-ai-gis/.
25. Jeremy Hsu, "Wanted: AI That Can Spy," *IEEE Spectrum*, December 2017, 12–13.
26. Harsha Vardhan, "AI and Deep Learning and Geospatial Industry: The Transformation," *Geospatial World*, March 22, 2017, www.geospatialworld.net/article/artificial-intelligence-deep-learning-geospatial-industry-transformation-2016/.
27. Juergen Dold and Jessica Groopman, "The Future of Geospatial Intelligence," *Geo-spatial Information Science* 20, no. 2 (2017): 151–62, www.tandfonline.com/doi/full/10.1080/10095020.2017.1337318.
28. Alliance for Telecommunications Industry Solutions, "Smart Cities Technology Roadmap," 2017, https://access.atis.org/apps/group_public/download.php/34053/ATIS-I-0000058.pdf.
29. Navin Sregantan, "Singapore Tops Global Smart City Performance Ranking in 2017: Study," *Business Times*, March 13, 2018, www.businesstimes.com.sg/government-economy/singapore-tops-global-smart-city-performance-ranking-in-2017-study.

30. Matt O'Brien, "How Much AI Surveillance Is Too Much," *Chicago Sun-Times*, July 3, 2018, https://chicago.suntimes.com/business/how-much-ai-surveillance-is-too-much/.
31. Chris Wilder, "From Smart Cities to Smart Countries: Cisco's CDA Program Fosters Economic Opportunity," *Forbes*, February 16, 2018, www.forbes.com/sites/moorinsights/2018/02/16/from-smart-cities-to-smart-countries-ciscos-cda-program-fosters-economic-opportunity/#3f1ad85e5396.
32. Matthew Zenus, "How Geospatial Analytics Can Give Your Business a Competitive Edge," *CIO*, March 2, 2017, www.cio.com/article/3176116/big-data/how-geospatial-analytics-can-give-your-enterprise-business-a-competitive-edge.html.
33. Shimonti Paul, "BI and Spatial Analytics: A Collaboration We Can't Do Without," *Geospatial World*, September 20, 2017, www.geospatialworld.net/article/business-intelligence-spatial-analytics/.
34. Jenny Rough, "Where You Live Has a Bigger Impact on Happiness and Health than You Might Imagine," *Washington Post*, May 14, 2018, www.washingtonpost.com/national/health-science/where-you-live-has-a-bigger-impact-on-happiness-and-health-than-you-might-imagine/2018/05/11/16be7126-2c55-11e8-8ad6-fbc50284fce8_story.html?noredirect=on&utm_term=.a25999f39194.
35. Susan Scutti, "Five Deaths, 197 Illnesses in Ongoing E. coli Outbreak Tied to Romaine Lettuce," CNN, June 1, 2018, www.cnn.com/2018/06/01/health/e-coli-multistate-outbreak-update-cdc/index.html.
36. Susan Cosier, "The Coming Migration Catastrophe," *MIT Technology Review* 122, no. 3 (May–June 2019): 70.
37. Courtney Humphries, "The 'Mind-Boggling' Problem of Keeping New York Dry," *MIT Technology Review* 122, no. 3 (May–June 2019): 39.
38. James Temple, "Pipe Dreams," *MIT Technology Review* 122, no. 3 (May–June 2019): 79.
39. Temple.
40. David Rotman, "What Would *You* Pay to Save the World?" *MIT Technology Review* 122, no. 3 (May–June 2019): 11.
41. Adam Piore, "The Threat to the World's Breadbasket," *MIT Technology Review* 122, no. 3 (May–June 2019): 34.
42. Rotman, "What Would *You* Pay?"
43. Holly Jean Buck, "How to Cool an Ocean," *MIT Technology Review* 122, no. 3 (May–June 2019): 58–61.
44. Kristin Quinn, "Polar GEOINT," *Trajectory*, no. 2, 2019, 15.
45. Diego Arguedas Ortiz, "Strong Coffee," *MIT Technology Review* 122, no. 3 (May–June 2019): 44.
46. Buck, "How to Cool an Ocean."
47. Tim McDonnell, "GPS for Cows in Western Africa," *MIT Technology Review* 122, no. 3 (May–June 2019): 49.
48. Melissa Dell, "Tracking Networks and the Mexican Drug War," Harvard University, August 2014, https://scholar.harvard.edu/files/dell/files/121113draft_0.pdf.
49. Rukmini Callimachi, "ISIS Leader Paid Rival for Protection but Was Betrayed by His Own," *New York Times*, October 30, 2019, www.nytimes.com/2019/10/30/world/middleeast/isis-leader-al-baghdadi.html.
50. John Walcott, "Here's How US Forces Finally Tracked Down and Killed al-Baghdadi," *Time*, October 27, 2019, https://time.com/5711905/al-baghdadi-capture-isis-intelligence/.
51. Jon Schuppe, "Motorola, Known for Cellphones, Is Fast Becoming a Major Player in Government Surveillance and Artificial Intelligence," NBC News, October 2, 2019, www.nbcnews.com/news/us-news/motorola-company-known-cellphones-fast-becoming-major-player-government-surveillance-n1059551.
52. Gareth Corfield, "Drone 'Swarm' Buzzed Off FBI Surveillance Bods, Says Tech Bloke," *The Register*, May 4, 2018, www.theregister.co.uk/2018/05/04/anti_fbi_drone_swarm_claims/.

53. Sean Broderick, "Drone/Aircraft Encounters Prompt Call for Tighter Regulations," *AINonline*, February 14, 2018, www.ainonline.com/aviation-news/air-transport/2018-02-14/drone/aircraft-encounters-prompt-call-tighter-regulations.
54. Barry Manz, "Dethroning the Drone," *Journal of Electronic Defense* 41, no. 4 (April 2018): 25–33.
55. Emily Dreyfuss, "Google Tracks You Even If Location History's Off; Here's How to Stop It," *Wired*, August 13, 2018, www.wired.com/story/google-location-tracking-turn-off/.
56. Erika Kinetz, "If Your Tesla Knows Where You Are, China May, Too," *Detroit News*, November 29, 2018, www.detroitnews.com/story/business/autos/mobility/2018/11/29/china-spying-cars-tesla/38637687/.
57. Sarah K. White, "What Is Geofencing? Putting Location to Work," *CIO*, November 1, 2017, www.cio.com/article/2383123/mobile/geofencing-explained.html.
58. Borse Manoj V., Bhandure Harshad D., Patil Dhiraj M., and Bhad Pratik B., "Location Based Encryption-Decryption Approach for Data Security," *International Journal of Computer Applications Technology and Research* 3, no. 10 (2014): 610–11, http://ijcat.com/archives/volume3/issue10/ijcatr03101002.pdf.
59. Danica Simic, "Here's How to Fake Location in iOS 11," *Valuewalk*, October 25, 2017, www.valuewalk.com/2017/10/fake-location-in-ios-11-without-jailbreak/.
60. Anthony Heddings, "How to Hide Your IP Address (and Why You Might Want To)," How-to-Geek, October 8, 2018, www.howtogeek.com/363096/how-to-hide-your-ip-address/.
61. Heddings.
62. James Risberg, "Yes, the Blockchain Can Be Hacked," CoinCentral, May 7, 2018, https://coincentral.com/blockchain-hacks/.

Glossary of Terms

activity-based intelligence (ABI). A type of all-source analysis that focuses on activity and transactions associated with an entity, population, or area of interest.
aerostat. A lighter-than-air vehicle that can remain stationary in the air (usually, tethered).
agent-based modeling. A type of simulation that models the behavior of agents (human or animal) in a given area.
all-source analysis. An analytic process that makes use of all available and relevant sources of information.
angle of arrival (AOA). A geolocation technique that determines the direction from which a signal is coming.
augmented reality. A technology that superimposes a computer-generated image on a real-world map or picture.
bathymetry. The study of underwater depth of rivers, lakes, or ocean floors.
big data. Data sets that are both massive in volume and also have complex features, so that normal data-processing applications are inadequate for making valuable use of the information.
cartography. The art and science of graphically representing a geographical area. Historically, cartography once referred to paper maps and charts; modern cartography primarily uses digital tools. The discipline dealing with the conception, production, dissemination, and study of maps.
change detection. Any technique for observing changes in an area over time; typically employs data from remote sensing products such as satellite imagery.
charge-coupled device (CCD). The name applied to an array of solid-state devices that detect incoming photons in an image sensor.
coherent change detection. Using a synthetic aperture radar to, in effect, overlay two radar images, in order to produce a picture of what changed in the time between the two images.
collateral intelligence. Material or information that is extrinsic to a single-source collection organization, usually reporting or intelligence that is produced by another collection organization.
communications intelligence (COMINT). Intelligence information derived from the intercept of voice and data communications by other than the intended recipients.
data analytics. The process of examining data sets in order to derive insights about the information they contain.
digital elevation model. A three-dimensional representation of the Earth's surface.

direction finding (D/F). Determining the direction of arrival of a signal, typically by using an antenna to measure the azimuth at which signal reception is strongest.

dynamic cartography. Applying graphic techniques to inject life and motion into maps.

electronic intelligence (ELINT). Information derived from the intercept of intentional electromagnetic radiation signals, primarily radar, that does not fall into other SIGINT categories.

emissive IR. The optical spectrum band extending from mid-wavelength infrared into the far infrared region.

epidemiology. A branch of medicine concerned with the incidence and geographical spread of disease.

equal area projection. A map projection that represents the size of any given area on Earth accurately, albeit by necessarily creating distortion in the shape of the area.

false color. An image that depicts a subject in colors that differ from those a faithful full-color photograph would show, usually by shifting the colors in each pixel to longer wavelengths.

fleeting target. A moving or transient intelligence target that remains within the observable area for such a short period that little time is available for taking action against it.

full-motion video (FMV). Digital video data that are transmitted or stored for real-time reproduction on a computer (or other multimedia system) at a rate of not less than 25 frames per second, so that objects appear to move smoothly.

fusion. The integration of relevant material from all available sources, both intelligence and nonintelligence.

geodesic. The shortest distance between any two points on the Earth's surface. Also known as a great circle.

geodesy. The science of measuring and understanding the Earth's shape, orientation, and gravitational and magnetic fields.

geodetic datum. A detailed model of the Earth, its size, and its shape, along with a reference point for the coordinate system used.

geofencing. Creating a virtual perimeter around a real-world geographic area, for the purpose of interacting with mobile devices as they enter or exit the area.

geographic information system (GIS). A relational framework that uses technology and digital information to gather, manage, analyze, and present geographic data.

geographic segmentation. In commerce, determining customer preferences that are specific to a location or region.

geography. The study of Earth's physical features, its atmosphere, and its biological and cultural features over time.

geolocation. The process of pinpointing the location of an object on the Earth.

geology. The science and study of Earth's physical properties, including its structure, materials, and processes.

geomatics. The science and technology of collecting, manipulating, presenting, and using spatial and geographic data in digital form.

geomorphology. The study of the physical, chemical, and biological elements of the Earth.

geophone. A type of microphone used to measure seismic disturbances.

geopolitics. The study surrounding the advantages and disadvantages offered by geography in international relations.

georeferencing. Assigning a location to an entity using map coordinates.

geospatial intelligence (GEOINT). Refers to geographically referenced intelligence information about the spatial and temporal aspects of the Earth and human activity on it, which provides an advantage to any decision-maker. In the United States, the term includes both the all-source analytic process and the resulting product.

geosynchronous equatorial orbit (GEO). An orbit above Earth's equator at an altitude of 35,800 kilometers, where the orbital period is 24 hours, equal to that of the Earth's rotation.

geovisualization. A set of technological tools and methods for interactive visualization, used in the analysis of geospatial data.

Global Positioning System (GPS). A satellite-based radio navigation system operated by the US Space Force.

gnomonic projection. A map projection where a great circle (geodesic) is represented by a straight line.

ground-moving-target indicator (GMTI). See **moving-target indicator**.

heat map. A representation of complex data in visual form, wherein data values are represented as colors to simplify understanding.

highly elliptical orbit (HEO). An extremely elongated orbit characterized by a relatively low-altitude perigee and a high-altitude apogee. These orbits can have the advantage of long dwell times during the approach to and descent from apogee.

human geography. The subdiscipline of geography dealing with human cultures and political and economic activities in a spatial and temporal setting.

human intelligence (HUMINT). Intelligence information derived from the use of human beings as both sources and collectors.

hydrography. The scientific discipline concerned with measuring and describing the physical features of bodies of water and the shoreline, primarily for navigational purposes.

hydrophone. A microphone designed to be used underwater for recording or listening to underwater sound.

hyperspectral imagery. Optical imagery that uses hundreds of spectral bands.

imagery intelligence (IMINT). Intelligence information derived from collection by visual photography, infrared sensors, lasers, electro-optics, and radar sensors that results in the creation of images of objects.

inclination. In space systems terminology, the angle of a satellite's orbit measured counterclockwise from the equatorial plane.

incoherent change detection. In imagery, observing changes to a scene between imaging events that are caused by changes in the intensity of energy returned from target pixels. Contrast with **coherent change detection**.

infrared (IR). Electromagnetic radiation having a wavelength longer than that of the red end of the visible light spectrum but shorter than that of microwaves.

low Earth orbit (LEO). Satellite orbits between 200 and 1,500 kilometers above the Earth's surface.

magnetometer. A device that senses changes in the Earth's magnetic field, particularly useful in locating weaknesses or anomalies.

map projection. A technique for representing parts of the surface of the Earth on a plane surface; different types of projections are created to serve specific user needs.

measurement and signature intelligence (MASINT). Intelligence information obtained by quantitative and qualitative analysis of data derived from specific technical sensors for the purpose of identifying any distinctive features associated with the source, emitter, or sender, and to facilitate subsequent identification and/or measurement of the same.

medium Earth orbit (MEO). Satellite orbits typically between 10,000 to 20,000 kilometers in altitude.

Mercator projection. A cylindrical map projection, created by (figuratively) wrapping a sheet of paper around the globe, touching at the equator, and projecting points on the globe's surface on a line from the center of the Earth, through the point, onto the paper.

moving-target indicator (MTI). A feature that allows a radar system to detect target motion. It also is called a ground-moving-target indicator (GMTI).

multispectral imagery (MSI). Imagery collected by a single sensor in multiple regions (bands) of the electromagnetic spectrum. Typically used to refer to the collection of fewer than 100 bands, to distinguish it from hyperspectral imagery.

national technical means (NTM). A euphemism for satellite collection assets, derived from a term used in the Comprehensive Test Ban Treaty of 1963.

open-source (OSINT). An intelligence collection discipline that gathers information from publicly available material in print or electronic media.

orthographic view. A flat-Earth view of a scene.

orthorectification. A process that removes the effects of image perspective, and corrects for terrain elevation, in an image.

overhead collection. A term historically used in the intelligence literature to refer to collection from satellites. Modern usage refers to any aerial- and-space-based collection platforms or systems.

pattern-of-life analysis (POL). Data collection and analysis used to establish past and current patterns of behavior of an individual or a group, and from that to infer potential future behavior.

photogrammetry. The art and science of making measurements from photographs, typically for surveying and mapping.

polar orbit. A 90-degree inclination orbit, which crosses the equator moving directly north or south and directly over the poles.

polarimetry. The measurement and interpretation of the polarization of transverse waves, most notably electromagnetic waves such as radio waves and light.

radio frequency identification (RFID). The use of electromagnetic fields to automatically identify and track tags attached to objects and animals (and humans in voluntary trials).

raster model. Electronic images of paper maps or charts and as such provide no more information than that available on their paper counterparts.

raw intelligence. The product of an intelligence collection that has not yet been exploited or analyzed.

reconnaissance. Periodic observation of a target area; contrast with **surveillance**.

reflective IR. The main infrared component of solar radiation that is reflected from the Earth's surface.

remote sensing. Sensing, primarily of the electromagnetic spectrum, that is done at long distances (on the order of tens to thousands of kilometers).

seismic sensing. Detecting sound that travels through the Earth. (See **seismic waves**.)

seismic waves. Acoustic waves that travel through the Earth (e.g., as a result of an earthquake or explosion).

shutter control. A policy that prohibits a country's commercial imaging satellites from imaging selected areas under certain conditions.

side-looking airborne radar (SLAR). An aircraft- or satellite-mounted imaging radar that is aimed perpendicular to the direction of flight.

signals intelligence (SIGINT). A category of intelligence comprising communications intelligence (COMINT), electronic intelligence (ELINT), and foreign instrumentation signals intelligence (FISINT).

signature. A set of distinctive characteristics of persons, objects, or activities that result from the processing of collected intelligence and that allow the entity to be identified and characterized.

single-source analysis. An analytic process and product of a collection organization that relies primarily on its own collection; communications intelligence, open source, and imagery analysis are examples.

spatial geography. A scientific discipline concerned with identifying the geographic location of features and boundaries on Earth.

spectrometer. An instrument used to measure the intensity and wavelength of light.

stereo image. A three-dimensional image of a scene.

stereograph. A combination of two images of the same scene taken from slightly different angles to provide a three-dimensional view of the scene.

stereoscope. An optical instrument that combines two images to obtain a three-dimensional view of a scene, or stereograph.

stovepipe. A term for maintaining a separate collection or analysis process based on organizational, compartmentation, or technical factors.

sun-synchronous orbit. A satellite orbit designed so that the satellite passes over a given point on the Earth at about the same time every day.

surveillance. The continuous observation of a target or target area by visual, aural, electronic, photographic, or other means. Contrast with **reconnaissance**.

synthetic aperture radar (SAR). A radar system that achieves high-azimuth resolution by obtaining a set of coherently recorded signals such that the radar is able to function as if it had a very large antenna aperture.

teleseismic wave. A seismic wave that is recorded far from its source.

thematic cartography. The art of creating maps that focus on a specific theme or subject area, providing a medium for conveying complex ideas or to send a message, often of political nature.

time difference of arrival (TDOA). A measurement of the relative arrival times of a signal at dispersed geographic points, used to geolocate the signal.

time geography. The study of spatial and time-based processes and events, such as interactions with the environment or interactions with others, that affect human activity.

topography. The study of the shape of the Earth's surface and its physical characteristics (e.g., hills and valleys).

tradecraft. The specialized skills and expertise acquired through long experience in an intelligence discipline, such as in analysis or clandestine operations.

triangulation. The process of geolocating a point by forming triangles to it from known points.

ultraspectral imagery (USI). Optical imagery based on the measurement of thousands of spectral bands.

ultraviolet (UV). That part of the optical spectrum with higher frequencies or shorter wavelengths than the visible light band, and lower frequencies or longer wavelengths than the X-ray band.

universal transverse Mercator projection. A map projection that divides Earth into sixty zones, projecting each zone onto a plane; locations are identified by specifying the zone and the x, y coordinate in that plane.

unmanned aerial vehicle (UAV). An aircraft that either is piloted remotely or operates autonomously, and carries no passengers.

vector model. A digital map model that is created using points, lines, and polygons. Each entry on a vector map has an associated data file that can contain extensive geospatial information.

virtual reality. A computer-generated simulation of a three-dimensional image or environment that immerses the user in the scene.

visual analytics. The science of analytical reasoning supported by interactive visual interfaces.

volunteered geographic information (VGI). Geographic data provided voluntarily by individuals (typically, via social media), harnessed to create, assemble, and disseminate information in service to a cause.

Selected Bibliography

Bolstad, Paul. *GIS Fundamentals: A First Text on Geographic Information Systems*. 5th ed. Ann Arbor, MI: XanEdu, 2016.

Brugioni, Dino. *Eyeball to Eyeball: The Inside Story of the Cuban Missile Crisis*. New York: Random House, 1990.

Card, S., J. Mackinlay, and B. Schneiderman. *Readings in Information Visualization: Using Vision to Think*. San Diego: Academic Press, 1999.

Clapper, James R. *Facts and Fears*. New York: Viking Press, 2018.

Clark, Robert M. *Intelligence Collection*. Thousand Oaks, CA: CQ Press/Sage, 2014.

Clarke, K. C. "Mapping by the US Intelligence Agencies." In *Cartography in the Twentieth Century: The History of Cartography*. Edited by Mark Monmonier. Chicago: University of Chicago Press, 2015.

Day, D. A., J. M. Logsdon, and B. Latell, eds. *Eye in the Sky: The Story of the Corona Spy Satellites*. Washington, DC: Smithsonian Institution Press, 1998.

de Smith, Michael J., Michael F. Goodchild, and Paul A. Longley, *Geospatial Analysis: A Comprehensive Guide*. London: Winchelsea Press, 2018.

DiBiase, D., T. Corbin, T. Fox, J. Francica, K. Green, J. Jackson, J. A. Dykes, A. M. MacEachren, and M. J. Kraak, eds. *Exploring Geovisualization*. Amsterdam: Elsevier, 2005.

Flint, Colin, and Peter J. Taylor. *Political Geography*. 7th ed. New York: Routledge, 2018.

Foresman, Timothy W., ed. *The History of Geographic Information Systems*, Upper Saddle River, NJ: Prentice Hall, 1998.

Gerwehr, Scott, and Russell W. Glenn. *The Art of Darkness: Deception and Urban Operations*. Santa Monica, CA: RAND Corporation, 2000.

Herman, Michael. *Intelligence Services in the Information Age*. London: Routledge, 2001.

Huntington, Samuel P. *The Clash of Civilizations and the Remaking of World Order*. New York: Simon & Schuster, 1996.

Jensen, J. R. *Introductory Digital Image Processing: A Remote Sensing Perspective*, 5th edition. Upper Saddle River, NJ: Prentice Hall, 2015.

———. *Remote Sensing of the Environment: An Earth Resource Perspective*, 2nd edition. Upper Saddle River, NJ: Prentice Hall, 2007.

Lillesand, Thomas, Ralph W. Kiefer, and Jonathan Chipman. *Remote Sensing and Image Interpretation*. 7th ed. New York: John Wiley & Sons, 2015.

Lowenthal, Mark M., and Robert M. Clark, eds. *The Five Disciplines of Intelligence Collection*. Newbury Park, CA: CQ Press/Sage, 2015.

Mikhail, E. M., J. S. Bethel, and J. C. McGlone. *Introduction to Modern Photogrammetry*. New York: John Wiley & Sons, 2001.

Monmonier, Mark. *How to Lie with Maps*. 3rd ed. Chicago: University of Chicago Press, 2018.

National Research Council. *New Research Directions for the National Geospatial: Intelligence Agency: Workshop Report*. Washington, DC: National Academies Press, 2010.

O'Connor, Jack, and Carl W. Ford Jr. *NPIC: Seeing the Secrets and Growing the Leaders: A Cultural History of the National Photographic Interpretation Center*. Alexandria, VA: Acumensa Solutions, 2015.

Rodhan, Nayef al-. *Neo-Statecraft and Meta-Geopolitics*. Geneva: Geneva Centre for Security Policy, 2009.

Riffenburgh, Beau. *Mapping the World: The Story of Cartography*. London: Carlton Books, 2014.

Singleton, Alex David, Seth Spielman, and David Folch. *Urban Analytics (Spatial Analytics and GIS)*. London: Sage Publications Ltd., 2018.

Slocum, T. A., R. B. McMaster, F. C. Kessler, and H. H. Howard, *Thematic Cartography and Geovisualization*. Upper Saddle River, NJ: Pearson/Prentice Hall, 2009.

Torge, W., and J. Müller. *Geodesy*. 4th ed. Berlin: Walter de Gruyter, 2012.

Index

Images are indicated by an *f* or *t* following the page number.

ABI (activity-based intelligence), 113, 132, 224–25
Abraham, 42–43
Abrams P-1 Explorer, 116, 117*f*
Abrams, Talbert, 116
accuracy, 190–91; of angle of arrival, 97; of cyber geolocation, 106; GPS, 59*t*; in mapping, 20–21; of satellite navigation, 57
ACIC (Aeronautical Chart and Information Center), 239
acoustic geolocation, 102–6, 104*f*, 154
activity-based intelligence (ABI), 113, 132, 224–25
advantage, comparative, 133
aerial film cameras, 139–42, 140*f*–42*f*
Aeronautical Chart and Information Center (ACIC), 239
aeronautical charts, 28–29, 86
Aerostatic Corps, 111
aerostats, 113, 114*f*, 147
agent-based modeling, 311
Agincourt, 38–40
AGO (Australian Geospatial-Intelligence Organisation), 261
agriculture, 42–43, 79, 166, 197, 216, 290–91, 295, 317–18
AI (artificial intelligence), 309, 312–13, 316, 321
aircraft, 115–20, 117*f*, 141–43. *See also* lighter-than-air craft
airships, 112–13
AIS (automatic identification system), 101, 195–96
al-Dawadmi, 7–8, 136, 154
al Kibar, Syria, 188, 247
Allen, Charlie, 203, 208, 211–12
all-source analysis, 192, 240, 242, 248, 279
al-Qaeda, 245
Amazon (company), 286–87
Amazon River Basin, 216
AMS (Army Map Service), 238–39
angle of arrival (AOA), 95–97, 96*f*
anticipatory intelligence, 200–201, 213–15, 218, 249–50
AOA (angle of arrival), 95–97, 96*f*

Arab Spring, 212, 301
ArcGIS, 80
Ardennes, 39
ARGUS, 147, 150
Aristotle, 13
arms control, 128, 197
Army Map Service (AMS), 238–39
artificial intelligence (AI), 309, 312–13, 316, 321
Art of War, The (Sun Tzu), 37
Arunachal Pradesh, 32
atomic clocks, 57–58, 98
augmented reality, 82, 86, 89, 207–8, 304, 314
Australia, 29, 30*f*, 261
Australian Geospatial-Intelligence Organisation (AGO), 261
authentication, 324
automatic identification system (AIS), 101, 195–96
automobile insurance, 290
autonomous vehicles, 283, 310, 312

Bacastow, Todd, 249, 272
Baghdadi, Abu Bakr al-, 319–20
Bagley, James W., 140, 150
balloons, 111–12
Baltimore, Maryland, 233
Barrios Altos, 277
Barwell, Clive, 208
bathymetry, 47–49, 48*f*, 247
BeiDou (satellite navigation system), 59, 59*t*
"Belt and Road Initiative," 67, 262–63
BI (business intelligence), 266, 284–88, 295, 297, 306, 309, 315–16
Bible, 42–43
big data, 209–12, 259, 312, 314
bin Laden, Osama, 97, 101, 155, 155*f*, 156, 257
blimps, 112–13
blockchains, 305–6, 325
Bohr, Niels, 302
Bosnian War, 222, 223*f*
Boston Marathon bombing, 227
Bowman, Isaiah, 222
Boyle, Willard S., 145–46
Brazil, 216
Brown, Michael, 233
Buford, John, 110

business intelligence (BI), 266, 284–88, 297, 309, 315–16

cadastral mapping, 41–42, 78
cameras: aerial film, 139–42, 140*f*–42*f*; digital, 145–46, 162–63; satellite film, 142–45, 144*f*; video, 146–48
Canada, 259–61
Canadian Geographic Information System, 79, 260
cartography, 5, 12–15, 15*f*; dynamic, 303–5; establishing claims with, 29–32, 30*f*–31*f*; future of, 303–5; instruments in, 21–23, 22*f*; navigation and, 51; thematic, 67–70, 68*f*–70*f*, 85. *See also* mapping
Caspian Sea Monster, 193–94, 194*f*
Catholic Church, 14
CCD (charge-coupled device), 146–47
celestial navigation, 52–53
Center for Strategic and International Studies (CSIS), 279
Central Imagery Office (CIO), 241
Central Intelligence Agency (CIA), 97, 118, 206, 239–40, 257–58
cesium clock, 57
Chamberlain, Joshua, 110
change detection, 156–57, 178–79, 193, 274, 304
charge-coupled device (CCD), 146–47
charts: aeronautical, 28–29, 86; bathymetric, 48*f*; hydrographic, 48*f*; interactive, 81–83, 82*f*; nautical, 27–28, 86–88, 228, 255
checkbook shutter control, 136
China, 32, 59, 59*t*, 67, 131, 149, 169, 262–63, 323
chronometer, marine, 52–53
Churchill, Winston, 69
CI (competitive intelligence), 284, 295–97
CIA (Central Intelligence Agency), 97, 118, 206, 239–40, 257–58
CIO (Central Imagery Office), 241
cities, smart, 314–15
citizen interaction, 273–74
Civil War, 109–12
claims, with cartography, 29–32, 30*f*–31*f*

Clapper, James, 132, 237, 243–44, 258, 265
Clash of Civilizations and the Remaking of World Order, The (Huntington), 72–73, 73f
climate change, 46, 198, 213, 216, 231, 316–18
Clinton, Bill, 58
clock, 52–53, 57–58
Club of Rome, 217
clutter, 77–78
Cody, Samuel Franklin, 114
Cody War Kite, 114
coherent change detection, 179–180
Cold War, 70, 116, 128–29, 132–34, 139, 141; arms control in the, 197; Cuban missile crisis in the, 155, 185–86; National Photographic Interpretation Center in the, 239–40; seismic geolocation in the, 105; sonar in the, 104; underground facilities in the, 192–93
collateral intelligence, 248
color, false, 163–64, 164f
Columbus, Christopher, 19
COMINT. *See* communications intelligence (COMINT)
COMIREX (Committee on Imagery Requirements and Exploitation), 241
commercial imaging companies, 134–35, 135f
commercial imaging satellites, 133–37, 135f
Committee on Imagery Requirements and Exploitation (COMIREX), 241
communications intelligence (COMINT), 132; activity-based intelligence and, 225; collateral intelligence and, 248; denial and deception in, 189
community planning, 272–73
comparative advantage, 133
compass, 21–22, 22f
competitive intelligence (CI), 284, 295–97
concealing location, 324
conflict resolution, sociocultural factors in, 221–24, 223f
construction planning, 43–44
containment theory, 66
Cook, James, 29, 30f
CORONA, 130–31, 143, 144f, 145
crime, 208–9, 209f. *See also* law enforcement

crime simulations, 214–15
crowdsourcing, 218, 227–31, 234, 256, 286, 297, 304–5
CSIS (Center for Strategic and International Studies), 279
Cuban missile crisis, 155, 185–86, 244–45
Custer, George Armstrong, 112
cyber geolocation, 106–7

Daesh, 166, 319–20. *See also* Islamic State (ISIS)
Dangermond, Jack, 80
dark web, 296–97
DARO (Defense Airborne Reconnaissance Office), 241
DARPA (Defense Advanced Research Projects Agency), 147, 211
data analytics, 211–12, 244, 280, 304, 314
data neutrality, 225
Dayton Peace Accords, 222
D&D (denial and deception), 186–89, 275
dead reckoning, 51
de Blij, Harm, 32
deep web, 296–97
Defence Geographic and Imagery Intelligence Agency, 260
Defence Geospatial Intelligence, 266
Defence Imagery and Geospatial Organisation (DIGO), 261
Defense Advanced Research Projects Agency (DARPA), 147, 211
Defense Airborne Reconnaissance Office (DARO), 241
Defense Intelligence Agency (DIA), 238
Defense Mapping Agency (DMA), 238–39, 241–42
DEM (digital elevation model), 36–37
demographics, 197–98
denial and deception (D&D), 186–89, 275
Depot of Charts and Instruments, 256
Deutch, John, 238, 240
D/F (direction finding), 95–97, 96f
DIA (Defense Intelligence Agency), 238
digital cameras, 145–46, 162–63
digital display technologies, 206
digital elevation model (DEM), 36–37

DigitalGlobe, 134
DIGO (Defence Imagery and Geospatial Organisation), 261
direction finding (D/F), 95–97, 96f
Director of National Intelligence (DNI), 2, 6
disaster relief, 198
disease outbreak, 77, 198–99, 214, 232
disease prediction, 279–80
display technologies, 206
DMA (Defense Mapping Agency), 238–39, 241–42
DNI (Director of National Intelligence), 2, 6
Doctors Without Borders (DWOB), 277–78
Domesday Book, 41, 78–79
"doomsday preppers," 89
Doppler shift, 56–57, 61n9, 176
drinking water, 271–81, 317
Drone Adventures, 276–77
drones. *See* unmanned aerial vehicles (UAVs)
drugs, illicit, 113, 198, 318–19
Duggan, Mark, 232
DWOB (Doctors Without Borders), 277–78
dynamic cartography, 303–5

Earhart, Amelia, 55f, 55, 96
EarthNow, 136–37, 309
Earth Observation Satellite Company (EOSAT), 133–34, 135f
earth resources monitoring satellites, 127–28
EarthViewer, 206–7
Eisenhower, Dwight D., 128
electronic intelligence (ELINT), 2, 132, 225
electro-optical satellites, 131–32
Elektro-L, 146
ELINT (electronic intelligence), 2, 132, 225
emergency management, 229–30
emergency response, 274–75
emissive infrared, 160, 161f
encryption, 324
energy security, 197
enigmas, intelligence, 193–94, 194f
entertainment, geographic information systems and, 88–89
environmental changes, 198
Environmental Systems Research Institute (ESRI), 80–82
EOSAT (Earth Observation Satellite Company), 133–34, 135f

Index 341

epidemiology, 77
Eratosthenes, 13–14
ESRI (Environmental Systems Research Institute), 80–82
Etienne, Jacques, 111
expertise, outside, 191–94, 194f
Explorer 1, 126

facial recognition technology, 233
Fairchild Camera, 141
Fairchild, Sherman Mills, 26, 116, 141
faking location, 324
false color, 163–64, 164f
Ferguson, Missouri, 233
file types, 253
film return satellites, 130–31
financial crime analysis, 85
fires, 229–30
Fisher, Howard, 80
Fitzgerald, Dennis, 130
Five Eyes, 259–62
5G technology, 310–11
flattening of Earth, in imagery, 150–51, 151f
fleeting targets, 189–90
Flightaware, 297
flight simulators, 217–18
flood modeling, 46, 317
FMV (full-motion video), 147
foliage penetration, 177
food security, 197
Foursquare, 230
France, 262
Francica, Joe, 287
French Aerostatic Corps, 111
Frisius, Gemma, 23
full-motion video (FMV), 147
fusion, 186, 189, 200–201, 218, 227, 248

Gaia satellite, 146
Galileo (satellite navigation system), 59, 59t
games, 88–89
Gaofen-5, 169
Gaofen-11, 131, 149
Gates, Bill, 136, 302
Gates, Robert, 188, 237
GEO (geosynchronous equatorial orbit), 125–26, 125f, 132
geocaching, 88
GEOCELL, 244–45
geodesic, 17
geodesign, 78
geodesy, 4, 256
geodetic datum, 191

geoengineering, 318
Geofeedia, 233
geofencing, 293–94, 323
geographic information system(s) (GIS), 205–6; aeronautical charts and, 86; augmented reality in, 82; clutter and, 77–78; defined, 77; entertainment and, 88–89; in financial crime analysis, 85; geospatial intelligence and, 90–91; hard copy layers and, 78–79; interactivity in, 81–83, 82f; layering in, 81–82; nautical charts and, 86–88; power of, 84–85; raster models in, 83–84, 83f, 86, 87f; recreation and, 88–89; rise of, 85–89, 87f; thematic cartography and, 85; Tomlinson and, 79; vector models in, 83–84, 83f, 87f, 88; virtual reality in, 82–83
geographic segmentation, 289
geography, 4
Geography (Ptolemy), 14–15
GEOINT. *See* geospatial intelligence (GEOINT)
geolocation: acoustic, 102–6, 104f, 154; angle of arrival in, 95–97, 96f; basics of, 94; cyber, 106–7; defined, 94; imagery in, 95; mobile phones in, 101; radiofrequency, 95–101, 96f, 98f; radio frequency identification in, 99–101; seismic, 105–6; tagging, 106–7; time difference of arrival in, 97–99, 98f
geomatics, 204–5
geomorphology, 37
geophones, 45
geopolitical strategy, 70–73, 72t, 73t
geopolitics: containment theory in, 66; continuing influence of, 66–67; defined, 63; German, 65–66; heartland theory and, 64–65, 65f, 66, 71; meta-, 71–72, 72t; rimland theory in, 66–67; sea power theory and, 64; thematic cartography and, 67–70, 68f–70f
Geopolitik, 65–66
georeferencing: business and, 315; defined, 94; geovisualization and, 206; mapping and, 84
George III of England, 53
geosocial networks, 230–31
geospatial intelligence (GEOINT): accuracy in, 190–91; analysis boundary in, 248–49; boundaries of, 245–51; challenge of ubiquitous, 321–25; commercial, 283–98; competitive, 295–97; defined, 2, 6–8; definition boundary in, 249–51; drivers of, 185–201, 194f–95f; explosion, 253–67, 258f; geographic information systems and, 90–91; nonnational, 271–81; operational, 292–95; organizations, 254–59, 257f; outside expertise in, 191–94, 194f; precision in, 190–91; reification of, 5–6; sociocultural, 221–34, 223f; source boundary in, 246–48; strategic, 288–92; tools of, 203–18; transnational organizations and, 265–66
geospatial simulation modeling, 213–18
geospatial terminology, 4–5
Geospatial World Forum, 266
geosynchronous equatorial orbit (GEO), 125–26, 125f, 132
geovisualization, 206–9, 207f, 209f
Gettysburg, 109–10
GI Jane (film), 7
GIS. *See* geographic information system(s) (GIS)
Global Cultural Knowledge Network, 224
global navigation satellites, 59, 59t
Global Positioning System (GPS), 56, 58–60, 88, 98–101, 106, 230, 256, 275, 322
GLONASS (satellite navigation system), 59, 59t
GMTI (ground-moving-target indicator), 178
gnomonic projection, 16–17, 17f
Google Earth, 207
GPS (Global Positioning System), 56, 58–60, 88, 98–101, 106, 230, 256, 275, 322
Gray, Freddie, 233
Great Grain Robbery, 197, 290–91
ground-moving-target indicator (GMTI), 178
Gulf War, 39–40, 93–94, 203–4, 237

hacking, of satellite navigation systems, 60
Hägerstrand, Torsten, 213–14
hard copy layers, 78–79
Harris, Jeffrey, 237
Harrison, John, 52–53

Harrison, Richard Edes, 69–70, 69f
Harvard University, 79–80
Hassler, Ferdinand, 27–28, 255
Haushofer, Karl, 65–66
Haushofer, Max, 65
health, 198–99, 232, 316
heartland theory, 64–65, 65f, 66, 71
heat map, 230, 275, 290
HEO (highly elliptical orbit), 125, 125f
Herman, Michael, 197
Herodotus, 11–12
Herschel, William, 159
hierarchy of needs, 302–3, 303f
high ground: aerostats and, 113–14, 114f; aircraft and, 115–20, 117f; blimps and, 112–13; exotic approaches to, 113–15, 114f; at Gettysburg, 109–10; hot-air balloons and, 111–12; kites and, 114–15; lighter-than-air craft, 111–13; observation towers in, 110–11; pigeons and, 115; reconnaissance and, 109; surveillance and, 109; survivability and, 119; unmanned aerial vehicles and, 119–20
highly elliptical orbit (HEO), 125, 125f
Hood, John Bell, 110
Hootenanny, 229
hot-air balloons, 111–12
HSI (hyperspectral imagery), 167, 167f, 170
human geography, 221
human intelligence (HUMINT): in Ancient Greece, 38; and boundaries of spatial intelligence, 245–46; commercial entities and, 283; defined, 2; "Five Eyes" and, 259; imagery deception and, 189; ISIS and, 200, 319–20
humanitarian operations, 198
human rights, 198, 277
Human Terrain System, 223–24
HUMINT. See human intelligence (HUMINT)
Huntington, Samuel, 72–73, 73f
Hurricane Florence, 275
Hussein, Saddam, 93–94, 203
hydrography, 47
hydrophone, 103–4, 104f
hyperspectral imagery (HSI), 167, 167f, 170
hyperspectral sensing, 309

IAPF (International Anti-Poaching Foundation), 278
Idrisi, Muhammad al-, 15, 21, 227
image analysis, 153–57, 153f, 155f
image quality, 148–50, 150t
imagery, in geolocation, 95
imagery intelligence (IMINT): defined, 2; denial and deception with, 186–87; film return satellites and, 130. See also visible imaging
imagery standards, 253
IMINT. See imagery intelligence (IMINT)
inclination, 125–26
incoherent change detection, 178
India, 24, 25f, 32, 67, 187–88, 263–64
infectious disease, 198–99
Influence of Sea Power upon History, The: 1660-1783 (Mahan), 64
infrared (IR) light: bands, 160–62, 161f; discovery of, 159; emissive, 160, 161f; long-wave, 161–62, 161f, 169; mid-wave, 161, 161f; near, 160–61, 161f, 163–65; photography, 162–63; reflective, 160; short-wave, 161, 161f
In-Q-Tel, 206–7
insurance, 289–90, 294–95
intelligence: analysis, 3–4; anticipatory, 4, 200–201, 213–15, 249–50; boundaries of, 1–4; business, 266, 284–88, 297, 309, 315–16; competitive, 284, 295–97; defined, 1; descriptive, 4; electronic, 2, 132, 225; fusion, 186, 189, 200–201, 218, 224, 227, 248; geospatial, 2; human, 2; imagery, 2; measurement and signal, 2; open-source, 2; prescriptive, 4; raw, 2–3, 224, 248, 259; signals, 2; sources, 2–3. See also geospatial intelligence (GEOINT); human intelligence (HUMINT); imagery intelligence (IMINT); measurement and signature intelligence (MASINT); open-source intelligence (OSINT); signals intelligence (SIGINT)
intelligence enigmas, 193–94, 194f
interactive maps and charts, 81–83, 82f
Intergraph, 80–81
International Anti-Poaching Foundation (IAPF), 278
International Cartographic Association, 14

internet of things (IoT), 306–8, 323
involuntary geographic information (iVGI), 231–34
IR. See infrared (IR) light
Iran, 8, 32, 60, 129, 155, 193
Iran-Iraq War, 203
Iraq, 31–32, 39–40, 93–94, 203–4, 237
Iraq War, 113, 147, 222–24, 223f
Islamic State (ISIS), 166, 200, 319–20
Islamist 70
iVGI (involuntary geographic information), 231–34

Japan, 32, 64, 264
Jefferson, Thomas, 27
Jesusita Fire, 229–30
Jobs, Steve, 302
Joint Air Reconnaissance Intelligence Centre, 260
Joint STARS (JSTARS), 119, 178, 190
Jones, R. V., 192

Kaye, Thom, 90
Kettering Aerial Torpedo, 119
Keyhole (company), 206–7
KH-9 HEXAGON, 131, 144–45
KH-11, 131
Khrushchev, Nikita, 185
King, Samuel Archer, 111
kites, 114–15
Kompass Sender, 54
Korean Air Lines Flight 007, 58
Kratt, Jake, 118
Kuwait, 31–32, 93–94, 203–4

Land Data Assimilation System (LDAS), 280
Land Remote Sensing Commercialization Act, 133
Land Remote Sensing Policy Act, 134
Landsat, 45–46, 127–28, 134
laser radar (LIDAR), 180–81, 181f
latitude, 52
law enforcement, 232–33, 275–76. See also crime
layering, 78–79, 81–82, 82f
LDAS (Land Data Assimilation System), 280
learning, simulation-based, 217–18
Lee, Robert E., 109–10
"left hook" maneuver, 40
legislation, and commercial satellite industry, 135–36

LEO (low Earth orbit), 124, 125f, 126, 131–32
Levy, Walter, 258
Lewis, Jeffrey, 278–79
Liberia, 271
Libyan Civil War, 197–98
LIDAR (laser radar), 180–81, 181f
lighter-than-air craft, 111–13
light table, 153–54, 153f
Lima, Peru, 277
Limits to Growth, The, 217
Lincoln, Abraham, 112
line of position (LOP), 54, 59
local government, 272–76
location analytics, 284–87
location history, 322–23
longitude, 52
Longitude Act (Britain), 52
long-wave infrared (LWIR) band, 161–62, 161f, 169
LOP (line of position), 54, 59
Loran-C, 55–56
Louisiana Purchase, 42
Louis IX of France, 27
low Earth orbit (LEO), 124, 125f, 126, 131–32
Lowe, Thaddeus, 112–113
Lundahl, Arthur C., 149, 240
LWIR (long-wave infrared band) 161–62, 161f, 169

M-1 Abrams tank, 40
Mackinder, Halford John, 64–65, 65f, 66, 71
Magellan, Ferdinand, 16
magnetometers, 46
Mahan, Alfred Thayer, 63–64, 71
Manning, Warren H., 78, 153
MapD, 212
mapping, 11; cadastral, 41–42, 78; interactive, 81–83, 82f; as volunteered geographic information, 228–29. *See also* cartography
map projections, 16–20, 21f; defined, 16; equal area, 18–19, 19f; gnomonic, 16–17, 17f; Mercator, 18, 18f, 70; prime meridian and, 19–20; Universal Transverse Mercator, 20
Mariel Boatlift, 197
Marine Asset Tag Tracking System (MATTS), 100–101
marine chronometer, 52–53
Masback, Keith J., 283
MASINT. *See* measurement and signature intelligence (MASINT)
Maslow, Abraham, 302–3, 303f

MATTS (Marine Asset Tag Tracking System), 99–100
Maury, Matthew Fontaine, 256
Maxar Technologies Ltd., 134
measurement and signature intelligence (MASINT): big data and, 210; and boundaries of geospatial intelligence, 247; defined, 2; "Five Eyes" and, 259; National Geospatial-Intelligence Agency and, 244; signatures and, 247–48; underground facilities and, 193
medium Earth orbit (MEO), 124–25, 125f
megacities, 216
Memex, 211
MEO (medium Earth orbit), 124–25, 125f
Mercator projection, 18, 18f, 70
meta-geopolitics, 71–72, 72t
Mexican-American War, 42
mid-wave infrared (MWIR) band, 161, 161f
migration, 197–98, 213–14, 317
military geospatial simulations, 215
Military Survey Defence Agency, 260
Millennium Tower, 44
Minard, Charles Joseph, 68, 68f
mineral exploration, 45–46. *See also* natural resources
money laundering, 85
Montgolfier, Michel, 111
Morehouse, Scott, 80
Mostak, Todd, 212
mountainous terrain, 199–200
moving-target indicator (MTI), 178
multispectral imagery (MSI), 167, 167f, 170
MWIR (mid-wave infrared band), 161, 161f

Napoleon, 68, 68f
National Geodetic Survey (NGS), 255–56
National Geospatial-Intelligence Agency (NGA), 6, 237–51
National Imagery and Mapping Agency (NIMA), 240–43
National Imagery Intelligence Rating Scale (NIIRS), 149, 150t
National Oceanic and Atmospheric Administration (NOAA), 127, 133, 255
National Photographic Interpretation Center (NPIC), 238–42, 257

National Reconnaissance Office (NRO), 129–30, 132, 254
National Spatial Reference System, 256
National System for Geospatial Intelligence, 258–59
national technical means (NTM), 129
natural gas, 32, 44–45, 165, 214, 263
natural resources, 67, 78, 198, 216, 272
nautical charts, 27–28, 86–88, 228, 234, 255
Naval Oceanographic Office, 256–57
navigation: cartography and, 51; celestial, 52–53; dead reckoning in, 51; latitude in, 52; longitude in, 52; marine chronometer in, 52–53; on open ocean, 51; radio, 53–56, 55f; satellite, 56–60, 59t
near infrared (NIR), 160–61, 161f, 163–65
needs, hierarchy of, 302–3, 303f
Nett Warrior, 307
New Zealand, 259, 261
NGA (National Geospatial-Intelligence Agency), 237–51
NGOs (nongovernmental organizations), 276–80
NGS (National Geodetic Survey), 255–56
NIIRS (National Imagery Intelligence Rating Scale), 149, 150t
Nikumaroro (island), 55
NIMA (National Imagery and Mapping Agency), 240–43
1984 (Orwell), 301
NIR (near infrared), 160–61, 161f, 163–65
NOAA (National Oceanic and Atmospheric Administration), 127, 133, 255
nongovernmental organizations (NGOs), 276–80
Noonan, Fred, 55
North Korea, 171–72, 188, 278–79
Northwest Territory, 30–31, 31f
NPIC (National Photographic Interpretation Center), 238–42, 257
NRO (National Reconnaissance Office), 129–30, 132, 254
NTM (national technical means), 129
nuclear testing, 171

observation towers, 110–11
oceanographic terrain, 46–49, 48f
ocean traffic, 194–96, 195f
Odyssey project, 80
Office of Strategic Services (OSS), 257–58, 258f
Office of the Director of National Intelligence. *See* Director of National Intelligence (DNI)
Open Geospatial Consortium (OGC), 264
OpenSeaMap, 228
open-source intelligence (OSINT): big data and, 210; and boundaries of geospatial intelligence, 245; in the Cold War, 185; commercial entities and, 283; defined, 2; "Five Eyes" and, 259
OpenStreetMap, 228
operational geospatial intelligence, 292–95
Operation Desert Storm, 39–40, 93
Orbital Imaging Corporation, 134, 135f
orbits, satellite, 124–26, 125f
orthographic view, 151, 151f
orthorectification, 151
Orwell, George, 301
OSINT. *See* open-source intelligence (OSINT)
OSS (Office of Strategic Services), 257–58, 258f
outside expertise, 191–94, 194f
overhead collection assets, 129

Pakistan, 67, 97, 121, 187–88, 215, 263
paper maps, 89–90
Paradis, Michel, 204–5
Parkinson, Brad, 58
pattern-of-life (POL) analysis, 225–27
Patton, George, 90
Perry, William, 238, 240
Persian Royal Road, 11–12
personal sensors, 308
Phoenicians, 27
phones, mobile, 51, 81, 101, 171, 285–86, 288, 292, 323–24
photogrammetry, 25–27, 204–5, 239
Picard, Jean, 24
Pickett, George, 110
pigeons, 115
piracy, 198
Planet Labs, 136, 309

PMESII (political, military, economic, social, infrastructure, and information), 71, 72t, 285–87
poaching, 278
Pokémon Go, 88–89
Pokhran, 187–88
POL (pattern-of-life analysis), 225–27
polarimetry, 152–53, 177, 179
polar orbit, 126
population movements, 197–98
Powell, Colin, 237–38
Pratt, Lee, 79
precision, 190–91
preppers, 89
prime meridian, 13, 19–20, 253
prisoners of war, 197, 199
property insurance, 289–90
provincial government, 272–76
Ptolemy, 14–15
public health, 198–99, 232

radar: artifacts, 177–78; change detection and, 178–79; civil applications of, 179–80; conventional, 172–73; foliage penetration with, 177; laser, 180–81, 181f; polarimetry and, 177; side-looking airborne, 173–74, 173f–74f; synthetic aperture, 174–80
radar fingerprinting, 196
radio direction finders (RDFs), 54–55, 55f, 95–97, 96f
radiofrequency geolocation, 95–101, 96f, 98f
radio frequency identification (RFID), 99–101
radio navigation, 53–56, 55f
raster models, 83–84, 83f, 86, 87f
Ratzel, Friedrich, 65–66
raw intelligence, 2–3, 224, 248, 259
RDFs (radio direction finders), 54–55, 55f, 95–97, 96f
real estate assessment, 274
reconnaissance: balloons in, 111–12; in the Cold War, 129; fleeting targets and, 189–90; high ground and, 109; kites in, 114; radar in, 173–74; satellites in, 124–25, 128–29, 131–33; U-2 program and, 118, 129, 141–42; unmanned aerial vehicles in, 119–20; in World War I, 115–16; in World War II, 117, 192
recreation, geographic information systems and, 88–89

reflective infrared, 160
relief operations, 198
remote sensing, 5, 308–10
remote-sensing satellites, 123–26, 124f–25f
resources, natural, 67, 78, 198, 216, 272
retail, 288–89, 293–94
Revolutionary War, 42
RFID (radio frequency identification), 99–101
ride-hailing, 291–92, 295
rimland theory, 66–67
Ritter, Johann, 159–60
Rodhan, Nayef Al-, 71–72
Roger II of Sicily, 15
Roman Empire, 41
Rossmo, Kim, 208, 209f, 276
Roy, William, 24
Rubens, Paul, 226
rubidium clock, 57–58
Russia, 59–60, 73, 85, 90, 97, 120, 193, 258f, 263

SAR (synthetic aperture radar), 174–80
Sasson, Steven, 146
satellite film cameras, 142–45, 144f
satellite navigation, 56–60, 59t
satellite navigation systems, 57–58
satellites: commercial imaging, 133–37, 135f; earth resources monitoring, 127–28; electro-optical, 131–32; film return, 130–31; government nonmilitary applications of, 126–28, 127f; military applications of, 128–32; orbits of, 124–26, 125f; reconnaissance, 129; remote-sensing, 123–26, 124f–25f; signals intelligence, 132; small, 136–37; in spectral imaging, 169; weather observation, 126–27, 127f
Satellite Sentinel Project, 277
Saudi Arabia, 7–8, 73, 93, 154
Scales, Robert, 223
Schwarzkopf, Norman, 40
Scilly incident, 52
Sea of Japan, 32
sea power theory, 64
Seasat, 179–80
Second Life, 304
seismic geolocation, 105–6
seismic sensing, 105
seismic survey, 45f
seismic waves, 43

self-driving car, 310
sensors, personal, 308
Shalikashvili, John, 238, 240
Shea, Stu, 265–66
short-wave infrared (SWIR) band, 161, 161f
shutter control, 135–36
side-looking airborne radar (SLAR), 173–74, 173f–74f
SIGINT. *See* signals intelligence (SIGINT)
signals intelligence (SIGINT): aerostats and, 113; automated identification system and, 195–96; big data and, 210; in the Cuban missile crisis, 185–86; defined, 2; "Five Eyes" and, 259; fleeting targets and, 190; imagery analysis and, 185–86; imagery deception and, 189; ocean traffic and, 195–96; radiofrequency tagging and, 100; satellites, 132; single-source analysis and, 248; underground facilities and, 193; unmanned aerial vehicles and, 119–20
signatures, 5, 102, 147, 149, 154–56, 155f, 167–68, 169f, 172, 174, 188, 247–48
Silk Road, 11–12, 12f, 67
silver chloride, 159–60
simulation-based learning, 217–18
simulation on demand, 311–12
Singapore, 171, 198, 314–15, 324
single-source analysis, 246, 248
SLAR (side-looking airborne radar), 173–74, 173f–74f
smallsats, 136–37
smart cities, 314–15
smartphones, 88–89, 230–31, 253, 283, 285
Smith, George E., 145–46
Snow, John, 77, 279
social media, 106–7, 227, 230–34, 298, 301, 315, 320
sociocultural factors, in conflict resolution, 221–24, 223f
sonar, 103–5
soundings, 47–49
sound surveillance system (SOSUS), 104–5, 104f
South China Sea, 32
Soviet Union. *See* Cold War; Great Grain Robbery
spatial geography, 213
spectral imaging: applications, 164–66, 165f; false color in, 163–64, 164f; hyperspectral, 167, 167f, 170; imagers in, 163–69, 164f–65f, 167f, 169f; infrared bands and, 160–62, 161f; mechanisms of, 166–67, 167f; multispectral, 167, 167f, 170; platforms, 168–69; satellites, 169; "seeing," 163–64, 164f; signatures in, 167–68, 169f; ultraspectral, 167, 167f, 170; ultraviolet spectrum and, 161f, 162
spectral sensing, 309–10
spectrometer, 166–67
SPOT 1, 133–34
Sputnik 1, 56, 123, 126, 129
Spykman, Nicholas, 66
state government, 272–76
stereograph, 152
stereo imaging, 151–52
stereoscope, 152
Stopher, John, 265
strategic geospatial intelligence, 288–92
Strava, 230
sun angle, 152
sun shadow, 152
sun-synchronous orbit, 126
Sun Tzu, 37–38, 215
surveillance: activity-based intelligence and, 225; aerostats in, 113; countering, 323; disease outbreak, 232; high ground and, 109; ocean, 179; remote sensing and, 190; satellites in, 125–26, 131–32; sonar in, 104; unmanned aerial vehicles in, 120, 147, 309; video, 321
surveys, 41–42
survivability, high ground and, 119
survivalists, 89
SYMAP, 80
synthetic aperture radar (SAR), 174–80
Syria, 60, 147, 188, 197–98, 200–201, 233–34, 277, 317, 319–20

Tabula Rogeriana (al-Idrisi), 15, 15f, 227
Tardivo, Cesare, 25
TARS (Tethered Aerostat Radar System), 113, 114f
Tasman, Abel, 29
Tasmania, 29
TDOA (time difference of arrival), 97–99, 98f, 102–3, 106
Tea Fire, 229
Telefunken Kompass Sender, 54
teleseismic wave, 105
Television Infrared Observation Satellite (TIROS-1), 126–27, 127f, 145
TELs (transporter-erector-launchers), 93–94
temporal dimension, 213–14
terminology, geospatial, 4–5
terrain: agriculture and, 42–43; in bathymetry, 47–49, 48f; civil use of, 41–46, 45f; in construction planning, 43–44; digital elevation models in, 36–37; in flood modeling, 46; hydrography and, 47; measurement of, 35–36; military use of, 37–40; mineral exploration and, 45–46; natural gas and, 44–45; oceanographic, 46–49, 48f; oil and, 44–45; representation of, 35–36. *See also* topography
terrorism, 60, 85, 199–200, 225, 245, 264
Tethered Aerostat Radar System (TARS), 113, 114f
Thales, 16–17
thematic cartography, 67–70, 68f–70f, 85
theodolite, 22f, 22–23
Thermopylae, 38
Thurston County, Washington, 273
time, 51–53, 57–58
time difference of arrival (TDOA), 97–99, 98f, 102–3, 106
time geography, 213
timeliness, of visible imagery, 156
TIROS-1 (Television Infrared Observation Satellite), 126–27, 127f, 145
Tomlinson, Roger, 79
topography: in construction planning, 43; defined, 35; digital elevation models in, 36–37; for targeting, 40. *See also* terrain
Torrens, Paul, 214
Tournachon, Félix, 111
towers, observation, 110–11
trade, 11–12, 65, 67, 90, 198, 207, 207f
transit, 22f, 23
Transit (navigation system), 56–57
transportation, 287–88, 292–93
transporter-erector-launchers (TELs), 93–94
treaty monitoring, 197

triad, 5–6
triangulation, 15, 23–24, 24f–25f, 28, 35, 54, 95, 102, 111
Tsarnaev, Dzhokar, 227
Tsarnaev, Tamerlan, 227

U-2 program, 118, 129, 131, 141–43
UAVs. *See* unmanned aerial vehicles (UAVs)
Uber, 291–92
UGFs (underground facilities), 192–93
ultraspectral imagery (USI), 167, 167f, 170
ultraspectral sensing, 310
ultraviolet (UV) radiation: discovery of, 160; in laser radar, 181; spectrum, 161f, 162
underground facilities (UGFs), 192–93
underwater sound, 103–5
United Kingdom, 228, 259–60
Universal Transverse Mercator projection (UTM), 20, 21f
unmanned aerial vehicles (UAVs), 46, 147, 190, 276–77; image quality and, 150; proliferation of, 321–22; remote-sensing technology and, 308–10; spectral imaging and, 168–69; in World War I, 119
urban terrain, 200
US Geological Survey (USGS), 254–55
US Geospatial Intelligence Foundation (USGIF), 265–66
USI (ultraspectral imagery), 167, 167f, 170
US Naval Observatory and Hydrographical Office, 256–57
UTM (Universal Transverse Mercator projection), 20, 21f

Vajpayee, Atal Bihari, 187
value, of data, 210
variety, of data, 210
vector models, 83–84, 83f, 87f, 88
velocity, of data, 210
veracity, of data, 210
Versailles Treaty, 221–22
VGI (volunteered geographic information), 227–31
video cameras, 113, 120, 146–48
video surveillance, 137, 307, 309, 321
Vietnam War, 239, 257
virtual reality, 82–83, 303–4
visible imaging: aerial film cameras in, 139–42, 140f–42f; change detection in, 156–57; digital cameras in, 145–46; flattening of Earth in, 150–51, 151f; image analysis in, 153–57, 153f, 155f; image quality in, 148–50, 150t; polarimetry in, 152–53; satellite film cameras in, 142–45, 144f; stereo imaging in, 151–52; sun angle in, 152; sun shadow in, 152; timeliness of, 156; video cameras in, 146–48
visible signatures, 154–56, 155f
visual analytics, 205, 212
volume, of data, 210
volunteered geographic information (VGI), 227–31

war crimes, 198
water: drinking, 271, 317; sound in, 103–5
Watson-Watt, Robert, 172–73
weapons of mass destruction, 171–72, 198
weather observation satellites, 126–27, 127f
WikiMapia, 228–29

wildfires, 229–30
Wiley, Carl, 175–76
William the Conqueror, 41–42
Wilson, Woodrow, 222
wireless communication, 310–11
Wood, Robert Williams, 162
WorldView Imaging Corporation, 134
World War I, 102, 150–52, 157; acoustic geolocation in, 102; aerial film cameras in, 139–41, 140f–41f; aircraft in, 115–16; airships in, 112; angle of arrival systems in, 96; Ardennes in, 39; kites in, 114; photogrammetry in, 26; sonar in, 103; Versailles and, 221–22
World War II, 42, 152, 160; aerial film cameras in, 141; aircraft in, 117–18; airships in, 112–13; angle of arrival systems in, 97; Ardennes in, 39; Army Map Service in, 238–39; bathymetry in, 49; denial & deception in, 186; map projection needs in, 20; Office of Strategic Services in, 257–58; paper maps in, 90; photogrammetry in, 26; radio frequency identification in, 99; thematic cartography in, 69–70, 69f; unmanned aerial vehicles in, 119
Wullenweber, 97
Wyler, Greg, 137

Xi Jinping, 67

Young, William J., 23

Zarqawi, Abu Musab al-, 245
Zeppelin, Ferdinand von, 112

About the Author

Robert M. Clark currently teaches intelligence courses as adjunct faculty for Johns Hopkins University. His US intelligence career spans six decades: he was a CIA senior analyst and group chief, the president of Scientific and Technical Analysis Corp., and an independent consultant performing threat analyses. He received an SB from the Massachusetts Institute of Technology, a PhD from the University of Illinois, and a JD from George Washington University. He retired from the US Air Force at the rank of lieutenant colonel. He is the author of *Intelligence Analysis: A Target-Centric Approach* (6th ed., 2019); *The Technical Collection of Intelligence* (2010); and *Intelligence Collection* (2014). He is coauthor, with William Mitchell, of *Target-Centric Network Modeling* (2015) and *Deception: Counterintelligence and Counterdeception* (2018); and he is coeditor, with Mark Lowenthal, of *Intelligence Collection: The Five Disciplines* (2015).